Practical Protein Chromatography

Methods in Molecular Biology

John M. Walker, Series Editor

Methods in Molecular Biology • 11

Practical Protein
Chromatography

Edited by

Andrew Kenney

Drew Scientific Limited, London

Susan Fowell

Leicester Polytechnic, Scraptoft, Leicester, UK

Humana Press ✳ Totowa, New Jersey

© 1992 The Humana Press Inc.
999 Riverview Drive
Totowa, New Jersey 07512

Printed in the United States of America

Library of Congress Cataloging-in-Publication Data
Main entry under title:

Methods in molecular biology.

Practical protein chromatography / edited by Andrew Kenney, Susan Fowell.
        p. cm. — (Methods in molecular biology : 11)
    Includes index.
    ISBN 0-89603-213-2
    1. Proteins—Analysis. 2. Affinity chromatography. I. Kenney, Andrew. II. Fowell, Susan.
  III. Series : Methods in molecular biology (Totowa, N.J.) : 11.
  QP550.P73 1992
  574.19′245—dc20   92-3609
                                                                                                                CIP

Preface

One of the many impacts of recombinant DNA technology over the last 15 years has been a strongly refreshed interest in methods for the separation and purification of proteins. This interest has encompassed not only analytical separations, but also small- and large-scale preparative methods directed to both pure and applied research throughout biology and biomedicine.

Many of the new or substantially modified techniques developed have been reported in the literature, but a sufficiency of *detailed* practical help in establishing these methods for the first time in a new laboratory has often been difficult to find. With these problems in mind, we expect that *Practical Protein Chromatography*, designed as a key volume in the *Methods in Molecular Biology* series, will provide concise practical help to those carrying out new techniques for the first time.

Each chapter has been written by expert authors known to have direct and regular practical experience with their chosen techniques. The structure of each chapter is designed to make it easy for a worker new to the method to follow it to an effective conclusion. An Introduction treats the theory behind the method being described. The Materials and Methods sections allow the reader to prepare for, and then perform techniques in a rational stepwise manner. The Notes sections provide the sort of background 'hints' and 'tricks' that are so often essential for success, but are rarely reported in the literature. They also contain information about modifications to the basic methods that may help the reader to apply the technique in novel ways to new problems.

It is anticipated that *Practical Protein Chromatography* will be of use to research workers at all levels wishing to use a method for the first time. It is also hoped that the wide variety and range of methods and procedures covered here will encourage the reader to innovate with them in their own research, thereby extending and developing still more fully the inherent capabilities of the techniques described.

Andrew Kenney
Susan Fowell

v

Contents

Contributors

NICHOLAS H. BATTEY • *School of Plant Sciences, University of Reading, Whiteknights, Reading, Berkshire, UK*

STEVEN J. BURTON • *Affinity Chromatography Ltd., Girton, Cambridge, UK*

YANNIS D. CLONIS • *Department of Chemistry, University of Patras, Patras, Greece*

STEVEN M. CRAMER • *Department of Chemical Engineering, Rensselaer Polytechnic Institute, Troy, NY*

KEVIN FORD • *Krebs Institute, University of Sheffield, UK*

OWEN GOLDRING • *Department of Biological Sciences, North East Surrey College, Ewell, Surrey, UK*

JAMES H. HAGEMAN • *Department of Chemistry, New Mexico State University, Las Cruces, NM*

DAVID A. HARRIS • *Department of Biochemistry, Oxford University, South Parks Road, Oxford, UK*

DAVID HORNBY • *Krebs Institute, University of Sheffield, UK*

T. WILLIAM HUTCHENS • *Department of Pediatrics, Baylor College of Medicine, Houston, TX*

GEORGE W. JACK • *Centre for Applied Microbiology and Research, Porton Down, Salisbury, Wiltshire, UK*

ANDREW C. KENNEY • *Drew Scientific Limited, Chiswick, London, UK*

GLENN D. KUEHN • *Department of Chemistry, New Mexico State University, Las Cruces, NM*

CHEE MING LI • *Department of Pediatrics, Baylor College of Medicine, Houston, TX*

SUDESH B. MOHAN • *Department of Chemical Engineering, Birmingham University, Birmingham, West Midlands, UK*

STEN OHLSON • *HyClone Laboratories, Inc., Logan, UT*

DUNCAN S. PEPPER • *Scottish National Blood Transfusion Service, Edinburgh, UK*

ROBERT KERRY SCOPES • *Department of Biochemistry, La Trobe University, Bundoora, Victoria, Australia*

ROGER F. SHERWOOD • *Centre for Applied Microbiology and Research, Porton Down, Salisbury, Wiltshire, UK*

PAUL SHORE • *Krebs Institute, University of Sheffield, UK*

MICHAEL A. VENIS • *Department of Plant Physiology, Institute for Horticultural Research, East Malling, Maidstone, Kent, UK*

MOOKAMBESWARAN A. VIJAYALAKSHMI • *Laboratoire de Technologie des Separations, Universite de Technologie de Compiegne*

IRIS WEST • *Department of Biological Sciences, North East Surrey College, Ewell, Surrey, UK*

TAI-TUNG YIP • *Department of Pediatrics, Baylor College of Medicine, Houston, TX*

CHAPTER 1

Thiophilic Adsorption Chromatography

T. William Hutchens

1. Introduction

Thiophilic adsorption is useful for the purification of immuno-globulins under mild conditions (e.g., *see* ref. *1*). Although there are several established procedures for the purification of immunoglobu-lins *(2–5)*, thiophilic adsorption appears thus far to be unique in its capacity to adsorb three major classes of immunoglobulins (and their subclasses) *(6–8)*. Furthermore, in contrast to other affinity purifica-tion methods (e.g., *see* refs. *3,4*), recovery of the adsorbed (purified) immunoglobulins from the thiophilic adsorption matrix is accom-plished efficiently at neutral pH, without the need for perturbation of protein structure *(1)*. The most important utility of thiophilic adsorp-tion is perhaps its use for the selective depletion of immunoglobulins from complex biological fluids (e.g., calf serum and hybridoma cul-ture media, colostrum and milk) *(6,7,9)*. This latter development has been particularly useful with hybridoma cell culture applications *(9,10)*, and in the investigation of milk-immunoglobulin function during early periods of human infant nutrition *(6,7)*.

Thiophilic adsorption actually describes the affinity of proteins for a specific set of sulfur-containing (i.e., thioether-sulfone) immobilized ligands, which is observed in the presence of certain salts. Since the first demonstration of this adsorption phenomenon *(11)*, it has been further developed and used primarily for the selective adsorption of

From: *Methods in Molecular Biology, Vol 11: Practical Protein Chromatography*
Edited by: A. Kenney and S. Fowell Copyright © 1992 The Humana Press Inc., Totowa, NJ

immunoglobulins (e.g., *see* refs. *1,6–8*). Even though thiophilic adsorption chromatography, like hydrophobic interaction chromatography, is a salt-promoted adsorption process, hydrophobic proteins, such as serum albumin, are not thiophilic *(10)*. The mechanism of thiophilic adsorption is not understood thoroughly at the present time *(10,12)*. In practice, we do know that the selectivity of thiophilic adsorption is dependent upon the type and density of immobilized thiophilic ligand, as well as the concentration and type of water-structure promoting (i.e., antichaotropic) salt used to promote adsorption. The F_c, F_{ab}, and $F_{(ab')2}$ immunoglobulin fragments are each thiophilic, although different concentrations of antichaotropic salt are required to promote adsorption. Only those procedures for the thiophilic adsorption of intact immunoglobulins are presented here.

Stationary phases for thiophilic adsorption chromatography are not presently commercially available. Preparation and use of the thiophilic adsorbent is, however, relatively simple even for the nonorganic chemist. This chapter presents a detailed description of the synthesis of both polymeric- and silica-based thiophilic adsorbents and of the use of these thiophilic adsorbents for the selective and reversible adsorption of multiple immunoglobulin classes.

2. Materials

2.1. Materials for the Synthesis of Thiophilic (T-Gel) Adsorbents

2.1.1. Agarose-Based T-Gel

1. Agarose (6%) or Pharmacia Sepharose 6B.
2. Divinyl sulfone.
3. 2-Mercaptoethanol.
4. Ethanolamine.

2.1.2. Silica-Based T-Gel

1. Porous silica (e.g., LiChrosorb 60).
2. γ-glycidoxypropyltrimethoxysilane.
3. Sodium hydrosulfide.
4. Divinyl sulfone.
5. 2-Mercaptoethanol.
6. Dithiothreitol.
7. 5,5'-Dithiobis (2-nitrobenzoic acid) (DTNB: Ellman's reagent).

8. Methanol.
9. Acetone.
10. Toluene.
11. Diethyl ether.

2.1.3. Buffers and Eluents

1. 2, 4-Hydroxyethyl-1-piperazine ethanesulfonic acid (HEPES), sodium phosphate, tris [hydroxymethyl] aminomethane (Tris) (*see* Section 2.3.1.).
2. Ethylene glycol (*see* Section 2.3.2.).
3. Isopropanol (*see* Section 2.3.3.).

2.2. Chromatographic Columns and Equipment

The thiophilic adsorbents described here, if used under the conditions specified, are quite selective and have a high capacity for immunoglobulins *(1)*. Conventional open-column procedures with an agarose-based T-gel have been used in my laboratory with column bed volumes ranging from 1 to 2000 mL. High flow rates can be used *(1)*: the stability of the agarose beads is increased substantially because of the crosslinking with divinyl sulfone. Since this is an adsorption—desorption procedure (no resolution required), specific column dimensions are not critical to the success of the operation. Note, however, that the adsorption of other proteins can be induced with higher concentrations of ammonium sulfate (or other water-structure-forming salts) *(12)*; isocratic elution (e.g., milk lactoferrin) and gradient elution of proteins are also possible. Under these conditions column dimensions (length vs diameter) may be much more important.

Monitor protein elution with any type of flow-through UV (280 nm) detector. Different types of peristaltic pumps may be required to deliver the appropriate buffer flow rate for the various column diameters used. The step-wise elution protocol described here does not require a fraction collector, although the use of a fraction collector may help eliminate dilution during sample recovery.

Dialysis (or diafiltration) of the immunoglobulin-depleted sample (e.g., bovine calf serum or milk) may be required if subsequent utilization is desired. This requires dialysis membranes or some type of diafiltration apparatus (e.g., Amicon, Beverly, MA). Alternatively, size-exclusion chromatography may be used. Isolated immunoglobulins can be recovered in various buffers at low salt concentrations (even water) so that dialysis is not required.

2.3. Thiophilic Adsorption and Elution Buffer

The thiophilic adsorption of proteins to the T-gel is a salt-promoted process at neutral pH. Protein desorption simply requires removal of the adsorption-promoting salt, also at neutral pH. The salts best suited for the promotion of thiophilic adsorption include combinations of anions and cations of the Hoffmeister series that are counter the chaotropic ions. Ammonium sulfate and potassium sulfate are excellent choices for work with immunoglobulins. The inclusion of sodium chloride (e.g., 0.5M) does not have a negative influence on either adsorption or desorption. We normally include sodium chloride to improve the solubility of the purified and concentrated immunoglobulins.

2.3.1. Sample Preparation and T-Gel Column Equilibration Buffer

Sample: Add solid ammonium sulfate to the sample to a final concentration of between 5% (e.g., milk) and 10% (e.g., serum or cell culture media) (w/v) and adjust the pH to 7.0 if neccssary (pH 7–8 is optimal). Filter (e.g., 0.45-μm Millipore HAWP filters) or centrifuge (e.g., 10,000g for 10 min) the sample to remove any insoluble material. Column equilibration buffer: 20 mM HEPES (pH 8.0), 0.5M NaCl, 10% (w/v) ammonium sulfate. Sodium phosphate (20–50 mM) or Tris-HCl (50 mM) may be substituted for the HEPES buffer.

2.3.2. Immunoglobulin Elution Buffer

Immunoglobulin elution buffer is the same as column equilibration buffer except that the ammonium sulfate is eliminated. A secondary elution buffer of 50% (v/v) ethylene glycol in 20 mM HEPES (pH 7–8) is sometimes useful to elute immunoglobulins of higher affinity.

2.3.3. T-Gel Column-Regeneration Buffers

The agarose- or silica-based T-gel columns may be regenerated and used (100–300 times) over extended periods. Wash the column with 30% isopropanol and water before reequilibration. We have washed columns with 6M guanidine HCl, 8M urea, and even 0.1% sodium dodecylsulfate (SDS) without noticeable changes in performance (capacity or selectivity). Wash the silica-based columns with 0.5N HCl (e.g., 1 h at room temperature). There should be no loss of performance.

2.4. Reagents and Material Required for Analysis of Immunoglobulin Content, Purity, and Class

Estimate the total protein in your starting samples and isolated fractions using the protein–dye binding assay of Bradford *(13)* or the bicinchoninic acid method as described by Smith et al. *(14)* and modified by Redinbaugh and Turley *(15)* for use with microtiter plates. The reagents for these assays are commercially available from BioRad (Richmond, CA) and Pierce (Rockford, IL), respectively (*see* Chapter 20).

Immunoglobulins are a heterogenous population of proteins. Even monoclonal antibodies can exhibit microheterogeneity. Thus, the methods and criteria used to verify immunoglobulin purity are more subjective than usual.

SDS polyacrylamide gel electrophoresis (*see* Chapter 20) is used frequently for this purpose *(16)*. The silver-staining method of Morrissey (17) works well to reveal impurities. An alternate method, Coomassie blue staining, is easier to use, but is much less sensitive. Because of the presence of disulfide–linked heavy and light chains immunoglobulin sample preparation (denaturation) in the presence and absence of reducing agents, such as 2-mercaptoethanol, will affect the electrophoretic profile significantly.

The quantitive determination of isolated immunoglobulin class and subtypes may be accomplished by the use of enzyme-linked immunosorbent assay (ELISA) methods. A variety of species- and class-specific antisera are available commercially. It is beyond the scope of this presentation to review these methods.

3. Methods

3.1. Synthesis of the Agarose-Based Thiophilic Adsorbent or T-Gel

Synthesize the agarose-based sulfone-thioether stationary phase or T-gel ligand,

$$agarose–OCH_2 \, CH(SO_2)CHCH_2 \, SCH_2CH_2OH$$

as follows:

1. Suspend suction-dried Sepharose 6B (1 kg/100 mL) in 0.5 M sodium carbonate (pH 11.0) and incubate with divinyl sulfone (DVS) for 20 h at room temperature (21–24°C) on a rotary shaker.

2. Wash the divinylsulfone-crosslinked and activated gel exhaustively with water. After suction-drying on a sintered-glass funnel, suspend it in 1 L of $0.1M$ sodium bicarbonate (pH 9.0) with 100 mL of 2-mercaptoethanol and incubate overnight at room temperature on a rotary shaker.

3. Wash away excess 2-mercaptoethanol with distilled water and store the gel as a suction-dried or semidry material at 4°C (stable for several months).

4. Elemental (C, H, N, and S) analyses should be performed after each step in the T-gel synthesis (we use Galbraith Laboratories in Knoxville, TN) to estimate immobilized ligand density. The T-gel adsorbents used in most of the investigations cited here contained from 7.6 to 9.4% S and were calculated to have between 320 and 890 μmol of the sulfone-thioether ligand per g of dried gel. As an independent verification of terminal ligand density, test aliquots of the DVS-crosslinked agarose were reacted with ethanolamine instead of 2-meraptoethanol to produce "N-gel." The "N-gel" was evaluated for %N and calculated to contain the same terminal ligand density (e.g., 320 mol N vs 319 mol S/g dried gel). The agarose T-gel ligand density for the procedures described here was 750 μmol/g dried gel.

3.2. Synthesis of the Silica-Based Thiophilic Adsorbent or T-Gel

The silica-based thiophilic adsorbent or T-gel may be synthesized by the procedure described above for the agarose-based adsorbent. Alternatively, a modified synthetic route has recently been described by Nopper et al. *(8)*. This procedure, summarized below, produces the following thiophilic adsorbent:

$$silica-OSiCH_2CH_2CH_2OCH_2CHOHCH_2SCH_2CH_2SO_2CH_2CH_2SCH_2CH_2OH$$

1. Silanize the silica gel under anhydrous conditions according to the description of Larsson et al. *(18)*. Wash 20 g of LiChrospher™ silica gel with 20% HNO_3, water, $0.5M$ NaCl, water, acetone, and diethyl ether. Dry in a 500-mL 3-neck reaction flask for 4 h at 150°C under vacuum. Cool the flask and suck 300 mL of sodium-dried toluene into the flask. Add 5 ml of γ–glycidoxypropyltrimethoxysilane (Dow Corning Z 6040) and 0.1 mL triethanolamine to the reaction mixture, stir (overhead), and reflux for 16 h under a slow stream of dry (H_2SO_4) nitrogen gas to maintain anhydrous conditions.

2. Wash the silanized epoxy-silica gel sequentially with toluene, acetone, and diethyl ether. Dry the washed silica gel under vacuum.

3. The number of epoxide groups introduced may be assayed by titration according to the method of Axen et al. (19). This step is optional.
4. Cleave the epoxide groups with sodium hydrosulfide (tenfold molar excess) in 0.2 M Tris-HCl buffer (pH 8.5) for 1 h at room temperature.
5. Wash the hydrosulfide modified silica gel with water, methanol, acetone, and diethyl ether.
6. Assay the number of thiol groups introduced into the gel by the Ellman DTNB method *(20)*. This step is optional.
7. Stir the gel with divinyl sulfone (1.5 mL/g gel) in 0.2 M Tris-HCl buffer, pH 8.8, for 1 h at room temperature.
8. Wash the DVS-activated gel an a sintered-glass funnel with water, acetone, and diethyl ether. Dry under vacuum.
9. The quantity of double bonds available for reaction with 2-mercaptoethanol (next step) can be estimated after reaction of a portion of the activated gel with dithiothreitol followed by the assay of thiol groups by the method of Ellman *(20)*.
10. Incubate the DVS-activated silica gel with excess 2-mercaptoethanol in 0.2 M Tris-HCl, pH 8.5.
11. Finally, wash the thiophilic silica gel product with water, acetone, and diethyl ether, and dry under vacuum.

The ligand density of this product should approach 800 mol/g dry silica.

3.3. Column Packing and Equilibration

Suspend T-gel in column-equilibration buffer (*see* Section 2.3.1.), pour into a column of desired dimensions, adjust the flow rate to maximum (<60 cm/h), and equilibrate with column-equilibration buffer (monitor elution pH and conductivity). High-performance (i.e., HPLC) stationary phases (i.e., silica-based) for thiophilic adsorption should be packed into columns specifically designed for HPLC applications. Use packing pressures recommended by the manufacturer or supplier of the silica particles.

3.4. Preparation of Sample for Thiophilic Adsorption

Add solid ammonium sulfate to the sample (up to 10–12% w/v) and adjust the pH to 7.0 if necessary (pH 7–8 is optimal). The final concentration of ammonium sulfate in the sample varies with (1) sample type (e.g., 5–8% ammonium sulfate for colostrum or milk; 10–12% ammonium sulfate for bovine or human serum, ascites fluid or hybridoma cell-culture media), (2) immobilized T-gel ligand density, and (3) desired selectivity and immunoglobulin extraction efficiency (*see* Notes). There should be little or no protein precipitation with the concentrations of ammonium sulfate. Nevertheless, filter (e.g., 0.45-μm Millipore HAWP filters) or centrifuge (e.g., 10,000g for 10 min) the sample to remove any particulate material.

3.5. Thiophilic Adsorption of Immunoglobulin

As outlined below, procedures for the preparation of immuno-globulin-depleted fluids (e.g., serum or milk) vary slightly from the procedure to obtain the highly purified antibodies. Three specific examples are presented.

3.5.1. Purification of Monoclonal Antibodies from Serum, Ascites Fluid, or Hybridoma Cell-Culture Media

1. Add solid potassium sulfate to the sample (serum, ascites fluid, or culture media); a final concentration of $0.5M$ is suitable (10% ammonium sulfate may also be used).
2. Load the sample (e.g., 20–25 mL) onto a T-gel column (1.0 cm id; 5 mL) at 20 cm/h.
3. Wash away unbound protein with column elution buffer until no absorbance is detected at 280 nm (>20–50 column vol).
4. Elute the adsorbed antibodies with 20 mM HEPES buffer (pH 7.4), with $0.5M$ NaCl.
5. Introduce 50% ethylene glycol in 20 mM HEPES, pH 7.4 to initiate column regeneration and elute remaining immunoglobulins.

3.5.2. Selective Depletion of Immunoglobulins from Fetal or Newborn Calf Serum or Hybridoma Cell Culture Media

1. Prepare columns of T-gel (1-3 cm id) with settled bed volumes of 5–40 mL, depending upon sample type (e.g., serum or cell-culture fluid) and volume.
2. Equilibrate the T-gel column with (for example) $0.5M$ potassium sulfate (or 10% ammonium sulfate), $0.5M$ NaCl, and 20 mM HEPES at pH 7.4–7.6. Use the same salt that was added to the sample.
3. Add solid potassium sulfate (to a final concentration of $0.5M$) or solid ammonium sulfate (10% w/v) to the sample. Load the sample onto a T-gel column with a bed volume equal to 1-2 vol of applied sample

(the ratio of sample volume to T-gel volume may vary depending upon the immunoglobulin content to be removed). Apply the sample to the T-gel column at a linear flow rate between 6 and 20 cm/h.

4. Collect the peak column flow-through fractions.
5. Remove the added salt from the immunoglobulin-depleted sample by Sephadex G-25 (Pharmacia) chromatography or dialysis using phosphate-buffered saline at pH 7.4. Immunoglobulin titers should be monitored before and after thiophilic adsorption chromatography using ELISA techniques referred to below.
6. If you wish to recover the removed (adsorbed) immunoglobulins, wash away unbound and loosely bound proteins with column equilibration buffer (until column eluate has little or no absorbance at 280 nm). This usually requires several hours (e.g., overnight), since 20–50 column vol are necessary.
7. Elute bound immunoglobulins with column-elution buffer (i.e., column-equilibration buffer without the ammonium sulfate).

3.5.3. Preparation
of Immunoglobulin-Depleted Colostrum or Milk

1. Preparation of colostral whey: Obtain colostrum (porcine colostrum described here) on the first postpartum day. Two methods are used to remove casein from the colostrum: (1) decreasing its pH to precipitate casein and (2) coagulation of casein by adding 5 mg of rennin (Sigma Chemical Company, St. Louis, MO) per two liters of colostrum (7). In the pH method, frozen porcine colostrum (–80°C) is thawed at 37°C and the pH lowered to 4.3 for incubation at 4°C (1 h). The preparation is centrifuged at 45,000 rpm for 1 h at 4°C in a Beckman Ty45Ti rotor. The fatty layer is removed and the whey decanted. In the rennin method, frozen porcine colostrum (–80°C) is thawed at 37°C, and defatted by centrifugation at 10,000 rpm (Beckman JA-10 rotor) at 4°C for 20 min. Rennin is added to the defatted colostrum and the mixture is stirred gently (at 34°C) until coagulation of casein is complete. The coagulated colostrum is centrifuged at 10,000 rpm at 4°C for 20 min before the whey is decanted.
2. Selective depletion of immunoglobulins: Add solid ammonium sulfate (to 5 or 8% w/v) and sodium chloride (to 0.5M) to the whey sample and adjust to pH 7.0. Equilibrate the thiophilic adsorption gel column (7.5 cm id × 24.5 cm) with 20 mM HEPES buffer (pH 8.0) containing 10% ammonium sulfate and 0.5M NaCl. Load samples onto the column and wash with column equilibration buffer over-

night. Elute the adsorbed proteins by removal of ammonium sulfate, that is, elute with 20 mM HEPES buffer containing 0.5M NaCl. Introduce 50% ethylene glycol in 20 mM HEPES, pH 8.0 to initiate column regeneration. Used columns should be washed with 30% isopropanol and water before reequilibration with column-equilibration buffer. Collect eluted fractions. Immunoglobulin-depleted colostral whey proteins (T-gel column flow-through fractions) and isolated immunoglobulins can be concentrated and dialyzed using an Amicon Model CH2 PRS spiral-cartridge concentrator with a 3000 or 10,000 MW cutoff. An example is provided in Figs. 1 and 2.

3.6. Protein Analyses of Isolated Fractions and Quantitation of Eluted Immunoglobulins

Monitor the protein elution profile during thiophilic adsorption chromatography by detection of UV absorbance at 280 nm. Determine the elution properties of the various immunoglobulins using ELISA techniques, immuno "dot" blotting procedures (21), or Western (electrophoretic) transfer of the eluted proteins from SDS gels to nitrocellulose and immune blotting (22). The protein or immunoglobulin concentrations of pooled fractions should be evaluated as described by Bradford (13) (see vol. 3) or Smith et al. (14) to determine overall protein and immunoglobulin recoveries; these values should routinely exceed 90%.

3.6.1. ELISA Methods for the Detection of Specific Immunoglobulins

Coat 96-well microtiter plates (50 µL/well) with affinity-purified antibodies (example presented here: rabbit antibodies to bovine immunoglobulins obtained from Jackson Immunoresearch Laboratories, Inc., West Grove, PA) diluted to the appropriate titer with phosphate-buffered saline (PBS) pH 7.4. The plates can be stored for several days (at 4°C) before use. Wash the plates 3X in distilled water then add a 100 µL aliquot of diluted sample to each well. Human plasma and unfractionated fetal calf serum (originating from Australia) should be used at various dilutions as negative and positive controls, respectively. Dilute purified bovine IgG (Jackson Immunoresearch Laboratories) with 3% PBS to a final concentration of 0.5 ng/mL. Use this solution for calibration. Incubate the loaded plates for 1 h at 37°C before washing (5X) with distilled water containing 0.05% Tween 80. Dilute horse-

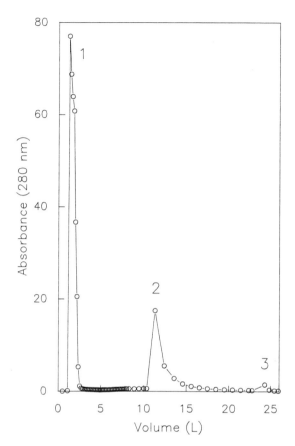

Fig. 1. Thiophilic adsorption of immunoglobulins from porcine colostral whey. Chromatography was performed at 4°C using columns (7.5 cm id × 24 cm) packed with 1 L of T-gel at a loading flow rate of 8.0 cm/h and an elution flow rate of 16 cm/h. Equilibration buffers consisted of 20 mM HEPES containing 12% ammonium sulfate (w/v) and 0.5M NaCl at pH 8.0. Porcine colostral whey (1 L) containing 8% ammonium sulfate (w/v) and 0.5M NaCl was loaded to the T-gel columns. Immunoglobulins were eluted by removal of ammonium sulfate (Peak 2) with 20 mM HEPES buffer containing 0.5M NaCl at pH 8.0. Following removal of NaCl, tightly bound immunoglobulins were eluted using 50% ethylene glycol (Peak 3). Reproduced with permission from ref. 7.

radish peroxidase conjugated rabbit antibodies against bovine IgG with 2% PBS. Add 50 µL to each well. Incubate the plates for 1 h at 22°C before washing 5X each with 0.05% Tween-distilled water, and then with distilled water only. Finally, add 20/1 mixture of 20 mM

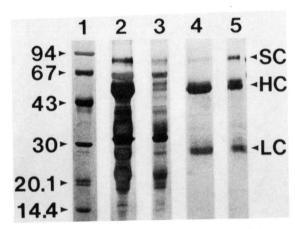

Fig. 2. Electrophoretic pattern of whey proteins and immunoglobulins on SDS-PAGE. Electrophoresis was performed on a 10–20% T 3% C gel under reducing and denaturing conditions at a constant current of 40 mA/ gel slab. Samples were dialyzed against 20 m*M* HEPES buffer, pH 8.0, in dialysis tubing with a mol wt cutoff of 3500. The gel was silver-stained following a modification of the method described by Morrissey *(17)*. Lane 1 represents low-mol-wt standards (in kDa) (Pharmacia), lane 2 represents porcine colostral whey, lane 3 represents the proteins unretained by the T-gel (Peak 1, Fig. 1), lane 4 represents the proteins eluted from the T-gel with the removal of ammonium sulfate (Peak 2, Fig. 1), and lane 5 represents the proteins eluted by the addition of ethylene glycol (Peak 3, Fig. 1). Secretory component, heavy chain, and light chain are indicated by the arrows. Reproduced with permission from ref. 7.

tetramethylbenzidine/0.1*M* calcium citrate (pH 4.5) (100 µL/well). Incubate for 10-20 min at room temperature. Stop the reaction by adding 50 µL H_2SO_4 (1*M*) to each well. Measure the developed color at 450 nm using a Titertek® Multiscan photometer.

3.6.2. Verification of Protein Purity

Polyacrylamide gradient gel electrophoresis (e.g., Pharmacia PAA 4/30 gels) should be performed using conditions specified by the manufacturer or as described by Laemmli *(16)*. (*See* Chapter 3 and vol. 1.) Standard reference proteins may include albumin (68,000 Da), lactate dehydrogenase (140,000 Da), catalase (230,000 Da), ferritin (450,000 Da), and thyroglobulin (668,000 Da). Protein bands can be detected by staining overnight with a 0.1% solution of Coomassie brilliant blue R-250 or by silver-staining *(17)*. (*See* vols. 1 and 3 and Chapter 20.)

3.7. T-Gel Column Regeneration

Wash the agarose-based T-gel column with 30% isopropanol and water. Occasional washing with 6M guanidine HCl may be required. Wash the silica-based T-gel column with 0.5M HCl and water. Reequilibrate the T-gel column with column equilibration buffer.

4. Notes

1. Maximizing immunoglobulin adsorption efficiency and selectivity: The specific and efficient removal of all classes of immunoglobulins, but not other proteins, from a complex biological fluid, such as serum or milk, is difficult. To maximize efficiency compromises absolute selectivity. Conversely, highly selective adsorption is possible, but there will be some decrease in efficiency. There is no substitute for trial and error in this process. The specific procedures detailed here are meant to serve as guidelines for other specific applications.

    It is essential that comparisons of thiophilic adsorption efficiency (i.e., immunoglobulin binding capacities) between various stationary phases take into consideration stationary-phase ligand density. Any decrease in stationary phase ligand density will increase the concentration of a given salt necessary to promote the adsorption of protein. Stated another way, with a given set of buffer conditions (i.e., salt concentration), immunoglobulin adsorption efficiency will decrease with decreasing ligand density on the stationary phase.

2. Depletion of immunoglobulins from the serum used to culture hybridoma cells facilitates subsequent purification of monoclonal antibodies. Fetal and newborn calf serum has been shown to contain immunoglobulins. These immunoglobulins can be removed (selective depletion) prior to the use of that serum for hybridoma cell culture. The production of monoclonal antibodies is not diminished. The subsequent purification of specific monoclonal antibodies from the hybridoma cell culture media is greatly simplified by prior removal of the bovine antibodies.

Acknowledgment

This work was supported, in part, by the US Department of Agriculture, Agricultural Research Service Agreement No. 58-6250-1-003. The contents of this publication do not necessarily reflect the views or policies of the US Department of Agriculture nor does mention of trade names, commercial products, or organizations imply endorsement by the US Government.

References

1. Hutchens, T. W. and Porath, J. (1986) Thiophilic adsorption of immunoglobulins—Analysis of conditions optimal for selective immobilization and purification. *Anal. Biochem.* **159,** 217–226.
2. Cohn, E. J., Strong , L. E., Hughes, W. L. Jr., Mulford, D. J., Ashworth, J. N., Melin, M., and Taylor, H. L. (1946) Preparation and properties of serum and plasma proteins. IV. A system for the separation into fractions of the protein and lipoprotein components of biological tissues and fluids. *J. Am. Chem. Soc.* **68,** 459–473.
3. Hjelm, H., Hjelm, K., and Sjoquist, J. (1972) Protein A from staphylococcus aureus. Its isolation by affinity chromatography and its use as an affinity adsorbent for isolation of immunoglobulins *FEBS Lett.* **28,** 73–76.
4. Roque-Barreira, M. C., and Campos-Neto, A. (1985) Jacalin: An IgA-binding lectin. *J. Immunol.* **134,** 1740–1743.
5. Regnier, F. E. (1988) Liquid chromatography of immunoglobulins. *LC-GC* **5,** 962–968.
6. Hutchens, T. W., Magnuson, J. S., and Yip, T.-T. (1989) Selective removal, recovery and characterization of immunoglobulins from human colostrum. *Pediatr. Res.* **26,** 623–628.7.
7. Hutchens, T. W., Magnuson, J. S., and Yip, T.-T. (1990) Secretory IgA, IgG, and IgM immunoglobulins isolated simultaneously from colostral whey by selective thiophilic adsorption. *J. Immunol. Methods* **128,** 89–99.
8. Nopper, B., Kohen, F., and Wilchek, M. (1989) A thiophilic adsorbent for the one-step high-performance liquid chromatography purification of monoclonal antibodies. *Anal. Biochem.* **180,** 66–71.
9. Hutchens, T. W. and Porath, J. (1987) Protein recognition in immobilized ligands: Promotion of selective adsorption. *Clin. Chem.* **33,** 1502–1508.
10. Porath, J. and Hutchens, T. W. (1987) Thiophilic adsorption: A new kind of molecular interaction. *Int. J. Quant. Chem.: Quant. Biol. Symp.* **14,** 297–315.
11. Porath, J., Maisano, F., and Belew, M. (1985) Thiophilic adsorption—a new method for protein fractionation. *FEBS Lett.* **185,** 306–310.
12. Hutchens, T. W. and Porath, J. (1987) Thiophilic adsorption. A comparison of model protein behavior. *Biochemistry* **26,** 7199–7204.
13. Bradford, M. M. (1976) A rapid and sensitive method for the quantitation of micrgram quantities of protein utilizing the principle of protein-dye binding. *Anal. Biochem.* **72,** 248–254.
14. Smith, P. K., Krohn, R. I., Hermanson, G. T., Mallia, A. K., Gartner, F. H., Provenzano, M. D., Fujimoto, E. K., Goeke, N. M., Ohlson, B. J., and Klenk, D. C. (1985) Measurement of protein using bicinchoninic acid. *Anal. Biochem.* **150,** 76–85.
15. Redinbaugh, M. G. and Turley, R. B. (1986) Adaptation of the bicinchoninic acid protein assay for use with microtiter plates and sucrose gradient fractions. *Anal. Biochem.* **153,** 267–271.
16. Laemmli, U. K. (1970) Cleavage of structural proteins during the assembly of the head of bacteriophage T4. *Nature* **227,** 680–685.

17. Morrissey, J. M. (1981) Silver stain for proteins in polyacrylamide gels: A modified procedure with enhanced uniform sensitivity. *Anal. Biochem.* **117,** 307–310.
18. Larsson, P.-O., Glad, M., Hansson, L., Mansson, M.-O., Ohlson, S., and Mosbach, K. (1983) High-performance liquid chromatography. *Adv. Chromatogr.* **21,** 41–85.
19. Axen, R., Drevin, H., and Carlsson, J. (1976) Preparation of modified agarose gels containing thiol groups. *Acta Chem. Scand.* **82,** 471–474.
20. Ellman, G.L. (1959) Tissue sulfhydryl groups. *Arch. Biochem. Biophys.* **82,** 70–77.
21. Hawkes, R., Niday, E., and Gordon, J. (1982) A dot-immunobinding assay for monoclonal and other antibodies. *Anal. Biochem.* **119,** 142–147.
22. Towbin, H., Staehelin, T., and Gorden, J. (1979) Electrophoretic transfer of proteins from polyacrylamide gels to nitrocellulose sheets: Procedure and some applications. *Proc. Natl. Acad. Sci.* USA **76,** 4350–4354.

CHAPTER 2

Immobilized Metal Ion Affinity Chromatography

Tai-Tung Yip and T. William Hutchens

1. Introduction

Immobilized metal ion affinity chromatography (IMAC) *(1,2)* is also referred to as metal chelate chromatography, metal ion interaction chromatography, and ligand-exchange chromatography. We view this affinity separation technique as an intermediate between highly specific, high-affinity bioaffinity separation methods, and wider spectrum, low-specificity adsorption methods, such as ion exchange. The IMAC stationary phases are designed to chelate certain metal ions that have selectivity for specific groups (e.g., His residues) in peptides (e.g., *3–7*) and on protein surfaces *(8–13)*. The number of stationary phases that can be synthesized for efficient chelation of metal ions is unlimited, but the critical consideration is that there must be enough exposure of the metal ion to interact with the proteins, preferably in a biospecific manner. Several examples are presented in Fig. 1. The challenge to produce new immobilized chelating groups, including protein surface metal-binding domains *(14,15)* is being explored continuously. Table 1 presents a list of published procedures for the synthesis and use of stationary phases with immobilized chelating groups. This is by no means exhaustive, and is intended only to give an idea of the scope and versatility of IMAC.

The number and spectrum of different proteins (Fig. 2) and peptides (Fig. 3) characterized or purified by use of immobilized metal ions are increasing at an incredible rate. The three reviews listed

From: *Methods in Molecular Biology, Vol. 11: Practical Protein Chromatography*
Edited by: A. Kenney and S. Fowell Copyright © 1992 The Humana Press Inc., Totowa, NJ

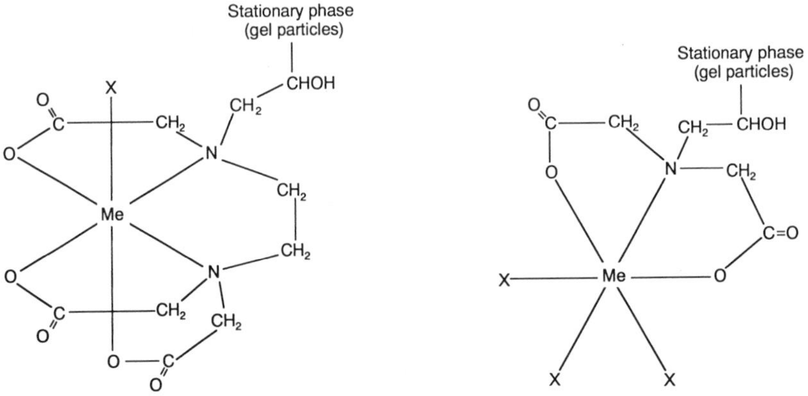

A) Tris(carboxymethyl)ethylenediamine (TED)

B) Iminodiacetate

Agarose-GHHPHG
Agarose-GHHPHGHHPHG
Agarose-GHHPHGHHPHGHHPHG
Agarose-GHHPHGHHPHGHHPHGHHPHGHHPHG

C) Immobilized metal-binding peptides

Fig. 1. Schematic illustration of several types of immobilized metal-chelating groups, including, iminodiacetate (IDA), tris(carboxymethyl) ethylenediamine (TED), and the metal-binding peptides $(GHHPH)_n G$ (where n = 1,2,3, and 5) *(14,15)*.

(8,18,19) barely present the full scope of activities in this field. Beyond the use of immobilized metal ions for protein purification are several analytical applications, including mapping proteolytic digestion products *(5)*, analyses of peptide amino acid composition (e.g., *5,6*), evaluation of protein surface structure (e.g., *8–11*), monitoring ligand-dependent alterations in protein surface structure (Fig. 4) *(12,20)*, and metal ion exchange or transfer (e.g., *14,21*).

The versatility of IMAC is one of its greatest assets. However, this feature is also confusing to the uninitiated. The choice of stationary phases and the metal ion to be immobilized is actually not complicated. If there is no information on the behavior of the particular protein or peptide on IMAC in the literature, use a commercially available stationary phase (immobilized iminodiacetate), and pick the relatively stronger affinity transitional metal ion, Cu(II), to immobilize. If the interaction with the sample is found to be too strong, try other metal ions in the series, such as Ni(II) or Zn(II), or try an immobilized

Table 1
Immobilized Chelating Groups and Metal Ions Used
for Immobilized Metal Ion Affinity Chromatography

Chelating group	Suitable metal ions	Reference	Commercial source
Iminodiacetate	Transitional	1,2	Pharmacia LKB Pierce Sigma Boehringer Mannheim TosoHaas
2-Hydroxy-3[N-(2-pyridylmethyl) glycine]propyl	Transitional	3	Not available
α-Alkyl nitrilo-triacetic acid	Transitional	4	Not available
Carboxymethylated aspartic acid	Ca(II)	13	Not available
Tris (carboxy-methyl) ethylene diamine	Transitional	2	Not available
(GHHPH)$_n$G*	Transitional	14,15	Not available

*Letters represent standard 1-Letter amino acid codes (G = glycine; H = histidine; P = proline). The number of internal repeat units is given by n (n = 1, 2, 3, and 5)

metal chelating group with a lower affinity for proteins (2,22). An important contribution to the correct use of IMAC for protein purification is a simplified presentation of the various sample elution procedures. This is especially important to the first-time user. There are many ways to decrease the interaction between an immobilized metal ion and the adsorbed protein. Two of these methods are efficient and easily controlled; they will be presented in detail in this chapter. Interpretation of IMAC results for purposes other than separation (i.e., analysis of surface topography and metal ion transfer) has been discussed elsewhere and is beyond the scope of this contribution.

2. Materials

The following list of materials and reagents is only representative. Other stationary phases, metal ions, affinity reagents, and mobile phase modifiers are used routinely.

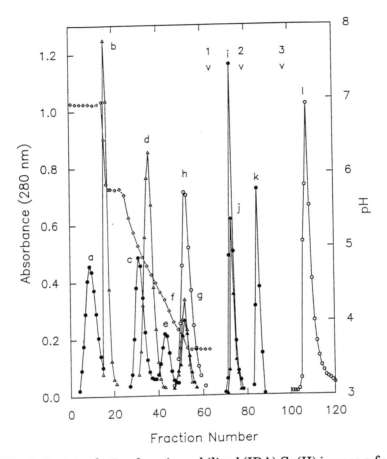

Fig. 2. Protein elution from immobilized (IDA) Cu(II) ions as a function of decreasing pH and increasing imidazole concentration. Proteins were eluted in the following order: chymotrypsinogen (a), chymotrypsin (b), cytochrome c (c), lysozyme (d), ribonuclease A (e), ovalbumin (f), soybean trypsin inhibitor (g), human lactoferrin (h), bovine serum albumin (i), porcine serum albumin (j), myoglobin (k), and transformed (DNA-binding) estrogen receptor (l). Open triangles represents pH values of collected fractions. Arrows 1–3 mark the introduction of 20 mM, 50 mM, and 100 mM imidazole, respectively, to elute high-affinity proteins resistant to elution by decreasing pH. Protein elution was evaluated by absorbance at 280 nm. In the case of the [^{3}H]estradiol-receptor complex, receptor protein elution was determined by liquid scintillation counting. Except for the estrogen receptor (l), protein recovery exceeded 90%. Only 50–60% of the DNA-binding estrogen receptor protein applied at pH 7.0 was eluted with 100–200 mM imidazole. Reproduced with permission from ref. *16*.

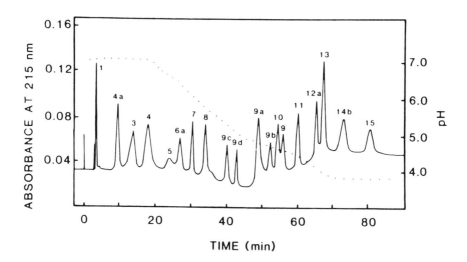

Fig. 3. Separation of bioactive peptide hormones by immobilized metal ion affinity chromatography (IMAC). The pH-dependent separation of a synthetic peptide hormone mixture (19 peptides) was accomplished using a TSK chelate-5PW column (8 × 75 mm, 10-µm particle diameter) loaded with Cu(II). A 20-µL sample (1–4 µg of each peptide) was applied to the column equilibrated in 20 mM sodium phosphate containing 0.5M NaCl, pH 7.0. After 10 min of isocratic elution, pH-dependent elution was initiated with a 50-min gradient to pH 3.8 using 0.1M sodium phosphate containing 0.5M NaCl at a flow rate of 1 mL/min. Peptide detection during elution was by UV absorance at 215 nm (0.32 AUFS). The pH profile of effluent is indicated by the dotted line. Sample elution peaks were identified as: 1, neurotensin; 4a, sulfated [leu[5]] enkephalin; 3, oxytocin; 4, [leu[5]] enkephalin; 5, mastoparan; 6a, tyr-bradykinin; 7, substance P; 8, somatostatin; 9c, [Asu[1.7]] eel calcitonin; 9d, eel calcitonin (11–32); 9a, [Asu[1.7]] human calcitonin; 9b, human calcitonin (17-32); 10, bombesin; 9, human calcitonin; 11, angiotensin II; 12a, [Trp (for) [25,26]] human GIP (21–42); 13, LH-RH; 14b, human PTH (13–34); 15, angiotensin I. Reproduced with permission from reference *17*.

2.1. Stationary Phase for IMAC

2.1.1. Conventional Open Column Stationary Phases (Agarose)

One example is Chelating Sepharose Fast Flow (Pharmacia), which uses the IDA (iminodiacetate) chelating group. Another example is

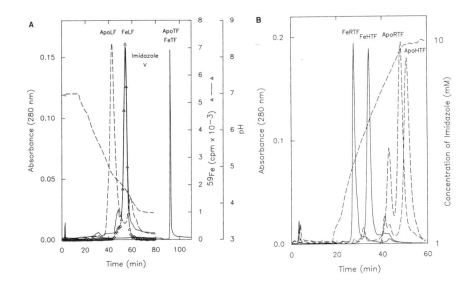

Fig. 4. Panel **A**: Separation of apolactoferrin (dashed line) and iron-saturated (solid line) human lactoferrins on high-performance IDA-Cu(II) columns with a phosphate-buffered gradient of decreasing pH in the presence
of $3M$ urea. Both apotransferrin (ApoTF) and iron-saturated human serum
transferrins (FeTF) were eluted only upon introduction of 20 mM imidazole
(imidazole-labeled arrow). UV absorbance was monitored at 280 nm. The
elution of iron-saturated human lactoferrin was determined by measuring
protein-bound ^{59}Fe radioactivity (open triangles). Panel **B**: Separation of
iron-free (dashed line) and iron-saturated (solid line) transferrins on a high-
performance IDA-Cu(II) affinity column using an imidazole elution gradient protocol. The apo and holo forms of human serum transferrin (hTF)
and rabbit serum transferrin (rTF) are shown. The Fe and Apo prefixes to
these abbreviations designate iron-saturated and metal-free transferrins,
respectively. Reproduced with permission from reference *12*.

Tris (carboxymethyl)ethylenediamine (TED) (*see* Table 1 and Fig. 1).
This staionary phase is used for proteins whose affinity for IDA-metal
groups is too high *(2,22)*.

2.1.2. High-Performance Stationary
Phases (Rigid Polymer)

One example is TSK Chelate 5PW (TosoHaas) (iminodiacetate
chelating group).

2.1.3. Immobilized Synthetic Peptides
as Biospecific Stationary Phases

Stationary phases of this type are designed based upon the known sequence of protein surface metal-binding domains *(14,15)*. The metal-binding sequence of amino acids is first identified (e.g., *14, 23*). The synthetic protein surface metal-binding domain is then prepared by solid-phase methods of peptide synthesis *(14)* and verified to have metal-binding properties in solution *(14,24)*. Finally, the peptides are immobilized (e.g., to agarose) using chemical coupling procedures consistent with the retention of metal-binding properties *(15)* (Table 1 and Fig. 1).

The procedures outlined in this chapter emphasize the specific use of agarose-immobilized iminodiacetate metal chelating groups. In general, however, these procedures are all acceptable for use with a wide variety of different immobilized metal chelate affinity adsorbents.

2.2. Metal Ion Solutions in Water

1. 50 mM CuSO$_4$.
2. 50 mM ZnSO$_4$.
3. 50 mM NiSO$_4$.

2.3. Buffers

1. 20 mM Sodium phosphate (or HEPES), 0.5 M sodium chloride, pH 7.0.
2. 0.1 M Sodium acetate, 0.5 M sodium chloride, pH 5.8.
3. 0.1 M Sodium acetate, 0.5 M sodium chloride, pH 3.8.
4. 50 mM Sodium dihydrogen phosphate; 0.5 M sodium chloride. Add concentrated HCl until pH is 4.0.
5. 20 mM Imidazole (use the purest grade or pretreat with charcoal), 20 mM sodium dihydrogen phosphate (or HEPES); 0.5 M NaCl. Adjust pH to 7 with HCl.
6. 50 mM EDTA; 20 mM sodium phosphate, pH 7.
7. 200 mM Imidazole, 20 mM sodium phosphate; 0.5 M sodium chloride, pH 7.
8. Milli-Q (Millipore) water or glass distilled, deionized water.

Urea (1–3M) and ethylene glycol (up to 50%) have been found useful as additives to the abovementioned buffers. (*See* Notes 7,8.)

2.4. Columns and Equipment

1. 1 cm inner diameter columns, 5–10 cm long for analytical and micropreparative scale procedures.

2. 5–10 cm inner diameter columns, 10–50 cm long for preparative scale procedures.
3. Peristaltic pump.
4. Simple gradient forming device to hold a vol 10–20X column bed vol (if stepwise elution is unsuitable).
5. Flow-through UV detector (280 nm) and pH monitor.
6. Fraction collector.

3. Methods

3.1. Loading the Immobilized Metal Ion Affinity Gel with Metal Ions and Column Packing Procedures (Agarose-Based Iminodiacetate Chelating Gel)

1. Suspend the iminodiacetate (IDA) gel slurry well in the bottle supplied. Pour an adequate portion into a sintered glass funnel. Wash with 10 bed vol of water to remove the alcohol preservative.
2. Add 2–3 bed vol of 50 mM metal ion solution in water. Mix well.
3. Wash with 3 bed vol of water (use 0.1M sodium acetate, 0.5M NaCl, pH 3.8 for IDA-Cu^{2+}) to remove excess metal ions.
4. Equilibrate gel with 5 bed vol of starting buffer.
5. Suspend the gel and transfer to a suction flask. Degas the gel slurry.
6. Add the gel slurry to column with the column outlet closed. Allow the gel to settle for several minutes; then open the outlet to begin flow.
7. When the desired volume of gel has been packed, insert the column adaptor. Pump buffer through the column at twice the desired end flow rate for several minutes. Readjust the column adapter until it just touches the settled gel bed. Reequilibrate the column at a linear flow rate (volumetric flow rate/cross-sectional area of column) of approx 30 cm/h.

3.2. Elution of Adsorbed Proteins

3.2.1. pH Gradient Elution (Discontinuous Buffer System)

1. After sample application, elute with 5 bed vol of 20 mM phosphate, 0.5M NaCl, pH 7.
2. Change buffer to 0.1M sodium acetate, 0.5M NaCl, pH 5.8, and elute with 5 bed vol (*see* Note 1).
3. Elute with a linear gradient of 0.1M sodium acetate, 0.5M NaCl, pH 5.8 to 0.1M sodium acetate, 0.5M NaCl, pH 3.8. Total gradient vol should be equal to 10–20 bed vol.
4. Finally, elute with additional pH 3.8 buffer until column effluent pH is stable and all protein has eluted (recovery should exceed 90%).

3.2.2. pH Gradient Elution
(Continuous Buffer System)

1. After sample application, elute with 2.5 bed vol of 20 m*M* sodium phosphate, 0.5*M* NaCl, pH 7.
2. Start a linear pH gradient of 20 mM sodium phosphate, 0.5*M* NaCl, pH 7 to 50 m*M* sodium phosphate, 0.5*M* NaCl, pH 4. Total gradient vol should be equal to approx 15 bed vol (*see* Note 2).
3. Elute with additional pH 4 buffer until the column effluent pH is stable.
4. If the total quantity of added protein is not completely recovered, elute with a small vol (<5 bed vol) of the 50 m*M* phosphate buffer adjusted to pH 3.5.

3.2.3. Affinity Gradient Elution with Imidazole

1. Equilibrate the column first with 5 bed vol of 20 m*M* sodium phosphate (or HEPES), 0.5*M* NaCl, pH 7, containing 20 m*M* imidazole. Now, equilibrate the column with 5–10 bed vol of 2 m*M* imidazole in the same buffer. (*See* Note 3.)
2. After sample application, elute with 2.5 bed vol of 2 m*M* imidazole in 20 m*M* sodium phosphate (or HEPES), 0.5*M* NaCl, pH 7.
3. Now elute with a linear gradient to 20 m*M* imidazole in 20 m*M* sodium phosphate (or HEPES), 0.5*M* NaCl, pH 7. Total imidazole gradient vol should equal 15 bed vols. (*See* Notes 4,5.)

3.3. Evaluation of Metal Ion Exchange
or Transfer from the Stationary Phase
to the Eluted Peptide/Protein

1. Use trace quantities of radioactive metal ions (e.g., ^{65}Zn) to label the stationary phase (i.e., immobilized) metal ion pool (*see* Section 3.1.). After elution of adsorbed proteins (*see* Section 3.2.), determine the total quantity of radioactive metal ions transferred to eluted proteins from the stationary phase (by use of a gamma counter).
2. To avoid the use of radioactive metal ions, the transfer of metal ions from the stationary phase to apo (metal-free) peptides present initially in the starting sample may be determined by either of two methods of soft ionization mass spectrometry. Both electrospray ionization mass spectrometry *(25)* and matrix-assisted UV laser desorption time-of-flight mass spectrometry *(14, 23–25)* have been used to detect peptide-metal ion complexes (Fig. 5). Both techniques are rapid (<10 min), sensitive (pmoles), and are able to address metal-binding stoichiometry.

3.4. Column Regeneration

1. Wash with 5 bed vol of 50 mM EDTA (EDTA should be dissolved in 20 mM sodium phosphate, 0.5M NaCl, pH 7).
2. Wash with 10 bed vol of water. The column is now ready for reloading with metal ions.

3.5. High-Performance IMAC

For example, use a TSK chelate 5PW column (7.5 mm id × 750 mm) 10 µm particle size.

1. High-performance liquid chromatography (HPLC) pump system status.
   a. Flow rate: 1 mL/min.
   b. Upper pressure limit: 250 psi.
2. Metal ion loading.
   a. Wash the column with 5 bed vol of water.
   b. Inject 1 mL of 0.2M metal sulfate in water.
   c. Wash away excess metal ion with 3 bed vol of water; for Cu(II), wash with 3 bed vol of 0.1M sodium acetate, 0.5M NaCl, pH 3.8.
3. pH gradient elution: phosphate buffers pH 7.0 (A) and pH 4.0 (B).
   a. 100% A, 5–10 min.
   b. 0–80% B, duration 25 min.
   c. 80–100% B, duration 20 min.
   d. 100% B until column effluent pH is constant or until all proteins have been eluted.
4. pH gradient elution in 3M urea: phosphate buffers pH 7.5 (A) to pH 3.8 (B).
   a. 100% A, 5 min.
   b. 0–10% B, duration 10 min.
   c. 10–80% B, duration 18 min.
   d. 80–100% B, duration 25 min.
   e. 100% B until eluent pH is constant or all proteins have been eluted.
5. Imidazole gradient elution: 1 mM imidazole (5% B in A) to 20 mM imidazole (100% B). (*See* Notes 5,6.)
   a. 5–10% B, immediately after sample injection, duration 10 min.
   b. 10–100% B, duration 30 min.
   c. 100% B until all samples have been eluted.

4. Notes

1. The discontinuous buffer pH gradient is ideal for the pH 6 to 3.5 range. We have observed that acetate is also a stronger eluent than phosphate.

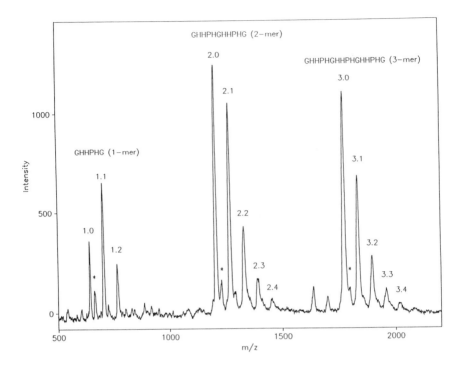

Fig. 5. Matrix-assisted UV laser desorption time-of-flight mass spectrometry (LDTOF) of a mixture of three synthetic metal-binding peptides (1-mer, 2-mer, and 3-mer) *after* elution from a column of immobilized GHHPHGHHPHG (2-mer) loaded with Cu(II) ions *(14)*. The synthetic peptide-metal ion affinity column used for metal ion transfer was prepared by coupling GHHPHGHHPHG (2-mer) to Affi-10 (Bio-Rad). CU(II) ions were loaded as described in Section 3.1. The column was equilibrated with 20 mM sodium phosphate buffer (pH 7.0) with 0.5M NaCl. An equimolar mixture of the three different synthetic peptides (free of bound metal ions) was passed through the column unretained. Flow-through peptide fractions were anaylzed directly by LDTOF *(23–25)*. The metal ion-free peptides GHHPHG (1-mer peak 1.0), GHHPHGHHPHG (2-mer peak 2.0), and GHHPHGHHPHGHHPHG (3-mer peak 3.0) are observed along with peptides with 1, 2, 3, or 4 bound Cu(II) ions (e.g., 3.1, 3.2., 3.3., 3.4). The small peaks marked by an asterisk indicate the presence of a peptide-sodium adduct ion. A detailed description of these results is provided in ref. *14* (reproduced with permission).

2. The phosphate buffer pH gradient is good for the pH 7–4.5 range. It has the advantage of UV transparency and is particularly suitable for peptide analysis *(17)*.

3. HEPES can be used instead of phosphate for both discontinuous buffer pH gradients and in the imidazole gradient elution mode. HEPES is a weaker metal ion "stripping" buffer than is phosphate. HEPES is also good for preserving the metal binding capacity of some carrier proteins such as transferrin *(12)*.

4. For well-characterized proteins, a stepwise gradient of either pH or imidazole can be used to eliminate the need for a gradient-forming device.

5. For the affinity elution method with imidazole, the imidazole gradient actually formed must be monitored by, e.g., absorbance at 230 nm or by chemical assay, if reproducible results are desired. Even when the column is presaturated with concentrated imidazole, and then equilibrated with buffers containing a substantial amount (2 mM) of imidazole, the immobilized metal ions can still bind additional imidazole when the affinity elution gradient is introduced. As a result, when simple (nonprogrammable) gradient-forming devices (typical for open-column chromatography) are used, a linear imidazole gradient is not produced; a small imidazole elution front (peak) at the beginning of the gradient will cause some proteins to elute "prematurely." The multistep gradient described for use with the high-performance chromatography systems is designed to overcome this problem. However, we emphasize that this particular program is custom designed only for the high-performance immobilized metal ion column of given dimensions and capacity. The program must be adjusted for other column types.

6. The imidazole gradient of up to 20 mM is only an example. Quite often, much higher concentrations (up to 100 mM) are required to elute higher affinity proteins *(9,10,16)* *(see* Fig. 2).

7. To facilitate the elution of some proteins, mobile phase modifiers (additives), such as urea, ethylene glycol, detergents, and alcohols, can be included in the column equilibration and elution buffers *(10,16,17,26)*.

8. To ensure reproducible column performance for several runs without a complete column regeneration in between, low concentrations of free metal ions can be included in the buffers to maintain a fully metal-charged stationary phase. This will not affect the resolution and elution position of the proteins *(7,16,27)*. On the other hand, if free metal ions are not desired in the protein eluent, a separate metal ion scavenger column (e.g., a blank or metal-free IDA-gel or TED-gel column) may be connected in series.

9. Batch-type (i.e., nonchromatographic) equilibrium binding assays

have been described *(16,28)*. This is useful in screening different types of immobilized metal ion–protein interaction variables. These variables include the selection of appropriate immobilized metal ion type *(16)*, the effects of temperature and mobile phase conditions *(16,26)*, and free metal ions *(7)* on protein-immobilized metal ion interaction capacity and affinity.

10. Immobilized metal ions may be useful for the reversible site- or domain-specific immobilization of functional receptor proteins or enzymes. Data collected with the estrogen receptor protein suggest that receptor immobilization on IDA-Zn(II) *(29)* and IDA-Cu(II) (unpublished) does not impair receptor function (ligand-binding activity).

Acknowledgment

This work was supported, in part, by the US Department of Agriculture, Agricultural Research Service Agreement No. 58-6250-1-003. The contents of this publication do not necessarily reflect the views or policies of the US Department of Agriculture, nor does mention of trade names, commercial products, or organizations imply endorsement by the US Government.

References

1. Porath, J., Carlsson, J., Olsson, I., and Belfrage, G. (1975) Metal chelate affinity chromatography, a new approach to protein fractionation. *Nature* **258,** 598,599.
2. Porath, J. and Olin, B. (1983) Immobilized metal ion affinity adsorption and immobilized metal ion affinity chromatography of biomaterials. Serum protein affinities for gel-immobilized iron and nickel ions. *Biochemistry* **22,** 1621–1630.
3. Monjon, B. and Solms, J. (1987) Group separation of peptides by ligand-exchange chromatography with a Sephadex containing *N*-(2-pyridylmethyl)glycine. *Anal. Biochem.* **160,** 88–97.
4. Hochuli, E., Dobeli, H., and Schacher, A. (1987) New metal chelate adsorbent selective for proteins and peptides containing neighbouring histidine residues. *J. Chromatogr.* **411,** 177–184.
5. Yip, T.-T. and Hutchens T. W. (1989) Development of high-performance immobilized metal affinity chromatography for the separation of synthetic peptides and proteolytic digestion products, in *Protein Recognition of Immobilized Ligands.* UCLA Symposia on Molecular and Cellular Biology, vol. 80 (Hutchens, T. W., ed.), Alan R. Liss, New York, pp. 45–56.
6. Yip, T. T., Nakagawa, Y., and Porath, J. (1989) Evaluation of the interaction of peptides with Cu(II), Ni(II), and Zn(II) by high-performance immobilized metal ion affinity chromatography. *Anal. Biochem.* **183,** 159–171.

7. Hutchens, T. W. and Yip, T. T. (1990) Differential interaction of peptides and protein surface structures with free metal ions and surface-immobilized metal ions. *J. Chromatogr.* **500**, 531–542.

8. Sulkowski, E. (1985) Purification of proteins by IMAC. *Trends Biotechnol* **3**, 1–7.

9. Hutchens, T. W. and Li, C. M. (1988) Estrogen receptor interaction with immobilized metals: Differential molecular recognition of Zn^{2+}, Cu^{2+}, and Ni^{2+} and separation of receptor isoforms. *J. Mol. Recog.* **1**, 80–92.

10. Hutchens, T. W., Li, C. M., Sato, Y., and Yip, T.-T. (1989) Multiple DNA-binding estrogen receptor forms resolved by interaction with immobilized metal ions. Identification of a metal-binding domain. *J. Biol . Chem.* **264**, 17,206–17,212 .

11. Hemdan, E. S., Zhao, Y.-J., Sulkowski, E., and Porath, J. (1989) Surface topography of histidine residues: A facile probe by immobilized metal ion affinity chromatography. *Proc. Natl. Acad. Sci. USA* **86**, 1811–1815.

12. Hutchens T. W. and Yip, T.-T. (1991) Metal ligand-induced alterations in the surface structures of lactoferrin and transferrin probed by interaction with immobilized Cu(II) ions. *J. Chromatogr.* **536**, 1–15.

13. Mantovaara-Jonsson, T., Pertoft, H., and Porath, J. (1989) Purification of human serum amyloid ccmponent (SAP) by calcium affinity chromatography. *Biotechnol. Appl. Biochem.* **11**, 564–571.

14. Hutchens, T. W., Nelson, R. W., Li, C. M., and Yip, T.-T. (1992) Synthetic metal binding protein surface domains for metal ion-dependent interaction chromatography. I. Analysis of bound metal ions by matrix-assisted UV laser desorption time-of-flight mass spectrometry. *J. Chromatogr.* (in press).

15. Hutchens, T. W. and Yip, T.-T. (1992) Immobilization of synthetic metal-binding peptides derived from metal ion transport proteins. II. Building models of bioactive protein surface domain structures. *J. Chromatogr.* (in press).

16. Hutchens, T. W. and Yip, T.-T. (1990) Protein interactions with immobilized transition metal ions: Quantitative evaluations of variations in affinity and binding capacity. *Anal. Biochem.* **191**, 160–168.

17. Nakagawa, Y., Yip, T. T., Belew, M., and Porath, J. (1988) High performance immobilized metal ion affinity chromatography of peptides: Analytical separation of biologically active synthetic peptides. *Anal . Biochem.* **168**, 75–81.

18. Fatiadi A.J. (1987) Affinity chromatography and metal chelate affinity chromatography. *CRC Critical Rev. Anal. Chem.* **18**, 1–44.

19. Kagedal, L. (1989) Immobilized metal ion affinity chromatography, in *High Resolution Protein Purification* (Ryden, L. and Jansson, J.-C., eds.), Verlag Chemie Inst., Deerfield Beach, FL, pp. 227–251.

20. Hutchens, T. W., Yip, C., Li, C. M., Ito, K., and Komiya, Y. (1990) Ligand effects on estrogen receptor interaction with immobilized Zn(II) ions. A comparison of unliganded receptor with estrogen-receptor complexes, and antiestrogen-receptor complexes. submitted.

21. Muszynska, G., Zheo., Y.-J., and Porath, J. (1986) Carboxypeptidase A: A model for studying the interaction of proteins with immobilized metal ions. *J. Inorg. Biochem.* **26**, 127–135.
22. Yip, T.-T. and Hutchens, T. W. (1991) Metal ion affinity adsorption of a ZN(II)-Transport protein present in maternal plasma during lactation: Structural characterization and identification as histidine-rich glycoprotein. *Protein Expression and Purification* **2**, 355–362.
23. Hutchens, T. W., Nelson, R. W., and Yip, T.-T. (1992) Recognition of transition metal ions by peptides: Identification of specific metal-binding peptides in proteolytic digest maps by UV laster desorption time-of-flight spectrometry. *FEBS Lett.* (in press).
24. Hutchens, T. W., Nelson, R. W., and Yip, T.-T. (1991) Evaluation of peptide-metal ion interactions by UV laser desorption time-of-flight mass spectrometry. *J. Mol. Recog.* **4**, (in press).
25. Hutchens, T. W., Nelson, R. W., Allen, M. H., Li, C. M., and Yip, T.-T. (1992) Peptide metal ion interactions in solution: Detection by laser desorption time-of-flight mass spectrometry and electrospray ionization mass spectrometry. *Biol. Mass Spectrom.* (in press).
26. Hutchens, T. W. and Yip, T.-T. (1991) Protein interactions with surface-immobilized metal ions: Structure-dependent variations in affinity and binding capacity constant with temperature and urea concentration. *J. Inorg. Biochem.* **42**, 105–118.
27. Figueoroa, A., Corradini, C., Feibush, B., and Karger, B. L. (1986) High-performance immobilized metal ion affinity chromatography of proteins on iminodiacetic acid silica-based bonded phases. *J. Chromatogr.* **371**, 335–352.
28. Hutchens, T. W., Yip, T.-T., and Porath, J. (1988) Protein interaction with immobilized ligands. Quantitative analysis of equilibrium partition data and comparison with analytical affinity chromatographic data using immobilized metal ion adsorbents. *Anal. Biochem.* **170**, 168–182.
29. Hutchens, T. W., and Li, C. M. (1990) Ligand-binding properties of estrogen receptor proteins after interaction with surface-immobilized Zn(II) ions: Evidence for localized surface interactions and minimal conformational changes. *J. Mol. Recog.* **3**, 174–179.

CHAPTER 3

Histidine Ligand Affinity Chromatography

Mookambeswaran A. Vijayalakshmi

1. Introduction

In biospecific ligand affinity chromatography, the interaction between the immobilized ligand and the solute molecule is based on their complementarity of charge, hydrophobicity, shape, and so on. The same forces that govern these interactions play a role in pseudo-biospecific systems, but perhaps differ in their relative magnitude (1).

Histidine residues in proteins are involved in the interactions in Immobilized Metal Affinity Chromatography systems, and sometimes in the dye-ligand system but, histidine can also be immobilized and act as a ligand to adsorb proteins (1).

Histidine has many properties that make it unique among the amino acids; these include its mild hydrophobicity, weak charge transfer possibilities owing to its imidazole ring, the wide range of its pKa values, and its asymmetric carbon atom. Histidine residues also play a charge relay role in acid–base catalysis (2). These properties mean that it can interact in many ways with proteins depending on conditions, such as pH, temperature, and ionic strength. Moreover, when immobilized through the appropriate groups to a polyhydroxy matrix, like Sepharose or silica, specific dipole-induced interactions with proteins can occur.

Several proteins and peptides have been purified using histidine-ligand affinity chromatography, both in analytical HPLC systems and on preparative scales. In all cases, the protein molecules were retained

From: *Methods in Molecular Biology, Vol. 11: Practical Protein Chromatography*
Edited by: A. Kenney and S. Fowell Copyright © 1992 The Humana Press Inc., Totowa, NJ

on histidine-ligand affinity chromatography at or around the isoelectric point of the protein/peptide *(3)*.

2. Materials

2.1. Chemicals

1. Sepharose 4B or 6B (Pharmacia, Uppsala, Sweden).
2. Silica-Spherosil XOB30 (Rhone-Poulenc, France).
3. Lichrosorb 60 (Merck, France).
4. L. Histidine, epichlorhydrin, 1,4 butanediol diglycidyl ether, sodium hydroxide (NaOH), sodium chloride (NaCl), sodium borohydride, and all other reagents are obtained either from Sigma or from Merck and are of Analar purity grade.

2.2. Apparatus

1. Stirring water bath or a shaft stirrer (OSI, France).
2. Normal laboratory vacuum filtration equipment.
3. Column, pump, detector, recorder, and fraction collector (LKB, Sweden).
4. Radial flow column: Sepragen was a kind gift from Touzart et Matignon, France.

3. Method

A typical gel is prepared by first introducing oxirane active groups onto a polysaccharide based (e.g., Sepharose 4B®) or silica based, OH containing insoluble matrices at basic pH (*see* Note 1). Then, the active oxirane ring is opened and coupled to the primary amine group of the amino acid histidine. The proposed structure of such an adsorbent is represented in Fig. 1.

1. Sepharose 4B is supplied as an aqueous suspension containing 20% ethanol as a preservative. So, to start with, wash the gel as supplied by the manufacturer with water to remove the ethanol and suction-dry on a sintered glass. The surface cracking of the gel cake is taken as the indication of the end of suction drying. To 10 g of suction-dried gel contained in a reactor or an Erlenmeyer flask, add 5 mL of $2M$ sodium hydroxide along with 0.5 mL of epichlorhydrin and 100 mg sodium borohydride to avoid any oxidation of the primary alcohol groups and keep under stirring. Avoid magnetic stirring, which will disrupt the soft gel beads; use lateral or shaft stirring. Then, add progressively another 5 mL of $2M$ sodium hydroxide and 2.5 mL of

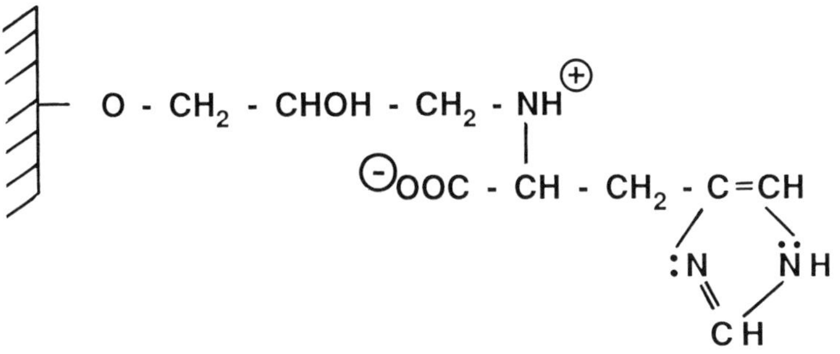

Fig. 1. Chemical structure of histidine ligand affinity adsorbent.

epichlorhydrin. The progressive addition ensures a more uniform activation. The reaction medium should be maintained at alkaline pH (9–11) by the addition of $2M$ sodium hydroxide. Otherwise, the decrease in pH caused by the HCl produced from the reaction is not favorable for the epoxy activation. After about 8 h of reaction under stirring, wash the contents abundantly with water. This gel is called epoxy Sepharose 4B, and is now ready for coupling with histidine.

2. To 10 g of suction-dried epoxy Sepharose 4B, add 15 mL of a 20% solution of L-histidine in $2M$ sodium carbonate (Na_2CO_3) containing 100 mg sodium borohydride and raise the temperature to $65 \pm 5°C$ and keep under stirring for about 24 h at $65 \pm 5°C$ (*see* Note 2). Avoid magnetic stirring. Then, wash the contents abundantly with water to remove the excess histidine and sodium carbonate until the washings are neutral.

   This gel can then be stored suspended either in water or in the equilibration buffer chosen for the chromatographic step. The amount of ligand coupled can be determined by the nitrogen estimation of an aliquot of the gel using a micro Kjeldahl method (4).

3. Suspend the histidyl Sepharose 4B (10 g) prepared as above, in the starting buffer (20–25 mL). Degas the slurry using a water trap and pour, if possible at one stretch, into a 1X 15-cm column fitted with a lower piston. Then, fill the upper empty space of the column with the same buffer degassed prior to use, and insert the piston, taking care to avoid the introduction of any air bubbles.

4. Connect the column as prepared above to a peristaltic pump in an upward flow mode. Connect the top outlet of the column to the inlet of a UV detector for monitoring the UV absorbance at 280 nm. The

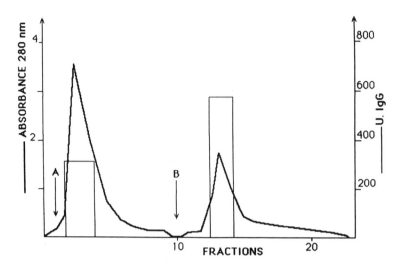

Fig. 2. A typical chromatogram of IgG$_1$ purification using histidine ligand affinity adsorbent.

    outlet from the UV detector is connected to a fraction collector and the fraction collector is set to collect fraction of ~1 mL each.

5. Adjust the flow rate to a linear flow of 20 cm/h (~25 mL/h). Usually, the whole setup is maintained at ~+4°C either in a cold room or in a special chromatographic chamber (*see* Note 8 for the temperature variables and the choice of temperatures). Pump about 2–3 column vol of the starting buffer through the column to equilibrate the gel bed at the chosen pH and ionic concentration.

6. Ideally, pump through the column about 10 mg of the extract containing the molecule to be separated (for example, IgG$_1$ subclass) in a minimum vol of not more than 0.5 mL, equilibrated as described above, at the same flow rate. Take care not to introduce any bubbles. Then, continue to pump the equilibrating buffer (50 m*M* Tris-HCl, pH 7.4, in the case of IgG$_1$ purification), until no UV absorbing material comes out of the column. Then, start the elution with buffers containing increasing amounts of NaCl (0.1–0.5*M*). Monitor the elution for UV absorption and collect fractions for further analysis and characterization. A typical chromatograph of IgG$_1$ purification from a placental extract is represented in Fig. 2.

7. Verify the total recovery of proteins injected by the cumulative absorbance units of the fraction eluted with the different buffers. Any strongly retained protein(s) should be removed before reusing the same column. In the case of agarose based gels, this is usually done

by washing the column with 1–2 column vol of 0.1 M sodium hydroxide solution. Wash the column with water to neutral pH immediately afterward.

4. Notes

1. Two types of –OH group containing matrices (silica-based and agarose-based) can be successfully used for coupling the histidine ligand. However, the chemistry used is different in each case. In the case of silica, the oxirane group is introduced by reacting with γ glycidoxy trimethoxy silane according to Chang et al. *(5)*. Then, the histidine is coupled through this oxirane by reacting the silanated matrix with a solution of 20% L-Histidine in 50% dimethylformamide at room temperature for about 48 h.

    The amide group containing matrices, e.g., acrylamide, were, however, found to be less interesting, perhaps, because of the amide groups interfering (proton extraction) with the histidine-protein recognition.

    TSK Fractogel is also found to be suitable, showing identical behavior to that of Sepharose 4B. The chemistry used for coupling histidine is similar to that used with agarose-based matrices.

2. The L-Histidine is invariably coupled to the OH-containing matrix via epoxy (oxirane) groups. The structure of the adsorbent is represented in Fig. 1. The histidine is coupled through its α amino group, leaving the imidazole ring totally available for the interaction. When histidine was coupled to trisacryl matrix using glutaraldehyde as the bifunctional reagent, the resulting adsorbent did not show any selectivity for the proteins, although the amount of coupled histidine was comparable to the case of coupling via oxirane groups. Coupling via CNBr groups has not been attempted so far.

3. With the activation chemistry chosen being epoxy activation with coupling through the oxirane group; the simplest means to introduce a long carbon chain between the matrix and the ligand is to use a bisoxirane. This is done by using 1–4 butanediol diglycidyl ether instead of epichlorhydrin for the activation as described by Sundberg and Porath *(6)*.

    In the case of silica based matrices, introduction of a spacer arm by coupling aminocaproic acid, prior to the coupling of histidine did not improve the performance of the adsorbent *(7)*.

4. In an exploratory study, Sepharose-based adsorbents were compared at different ligand concentrations, for their capacity to retain IgG$_1$ from the placental serum. The results are given in Table 1. In this

Table 1
Comparison of Sepharose-Based Adsorbents
at Different Concentrations[a]

Ligand conc., $\mu M/g$ dry gel	% Purity of the final product	Yield, %
10	98.9	9.2
25	95.5	4.5

[a]Effect of ligand concentration expressed as μM histidine/g gel dry wt on the purification and yield of IgG_1.

case, the lower ligand concentration (10.5 μM his/g dry gel) gives a higher recovery (9.2% IgG_1) and a higher purity (98.9% pure) compared to a gel with a higher ligand concentration (24.3 μM his/g dry gel). However,. this factor may vary with each protein studied (unpublished data).

5. The upstream extraction methods used to prepare the crude extract of the protein to be purified play an important role in its retention behavior. The following examples illustrate this.

Chymosin from calf or kid abdomen (sodium chloride extract) was selectively retained on a histidyl-Sepharose or histidyl silica column (8) but, when the extraction was done with sodium chloride in the presence of sodium benzoate, no specific retention could be obtained (unpublished data).

In the case of purification of IgG_1 from human placental serum, a comparative study of the ammonium sulfate precipitation and ethyl alcohol precipitation gave the results, shown in Fig. 3.

In the case of an ammonium sulfate extract, all the proteins including the IgG subclasses other than IgG_1, were found in the breakthrough fractions (peak 1, Fig. 2). Whereas, in the case of ethanol-precipitated extracts, the albumin was not retained (peak 1, Fig. 3) and the IgG fractions other than IgG_1 were slightly retarded (peak 2, Fig. 3) while IgG_1 was strongly retained and eluted with 200 mM sodium chloride (peak 3, Fig. 3).

The purity of the extract used for the chromatographic step is another important factor. Table 2 shows the comparison of the purification of IgG_1 from different placental extracts, namely A, S_2, and S_4, which had the initial purities of 58 and 90%, respectively.

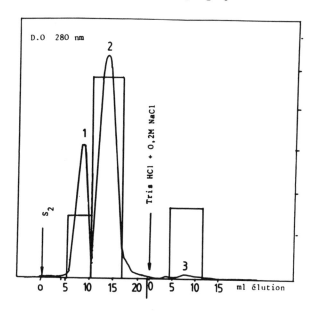

Fig. 3. Elution pattern of ethanol precipitated placental serum extract on histidyl Sepharose.

The weak (binding strength) affinities with this kind of pseudobiospecific ligand in the order of 10^{-3}–$10^{-4}M$ *(9)* can explain this relation between the degree of purity of the initial extract and the efficiency of the chromatographic step. Different compounds may compete for the same sites, which in turn decrease the binding capacity for the protein of interest. This is similar to the observation with metal-chelate adsorbents *(10)*, where the adsorption capacity of a purified enzyme was in the range of 50 mg protein/mL gel, whereas in the case of a crude extract, it was only about 30 mg/mL, albeit with the specificity conserved.

6. The specific retention capacities of the histidyl liganded adsorbents vary from one protein to another. Table 3 summarizes the capacities obtained for different proteins with a histidyl-Sepharose adsorbent having 10 µmol his/mL gel. It is to be noted that this adsorbent, though very specific to certain groups of IgG$_1$ subclass from the human placental sera, shows rather a low capacity. However, the capacity can be improved by increasing the OH groups on the support matrix prior to histidine coupling, as described by Vijayalakshmi and Thomas *(11)* and by Wilchek *(12)*.

Table 2
The IgG content and Purity of Different Starting Preparations

Precipitation method	Yield, %	Purity, %	Purif.[a] fold	Purity % After HLAC
$(NH_4)_2 SO_4$ (0–50%)	88	26	1.88	75
Ethanol Fra. S2	84	58	4.16	95.6
Ethanol Fra. S4	58	90	6.45	98.9

[a]Purification fold is calculated with reference to a chloroform extract. Purity of chloroform extract, 14%, NaCl extract 1.3%.

Table 3
The Capacity of the Histidyl Sepharose 4B
for Different Proteins Tested[a]

Protein	Capacity mg/mL
IgG1	0.5
Myxaline	0.5–1.0
Chymosin	1.0

[a]Capacity is calculated from the amount of pure product recovered in real experiments where crude fractions containing other proteins/peptides and molecules were injected.

Typically, 7 mL of extract containing 40 mg total protein is injected on a 35-mL column at an average linear flow of 20 cm/h.

7. The specific retention of proteins onto a histidine-coupled adsorbent (e.g., histidyl-Sepharose or histidyl-silica) is rather closely related to their isoelectric pH. Thus, only the IgG_1, which has a pI of 8.0 + 0.05, is selectively retained on a histidyl-Sepharose column at pH 7.4. Below pH 7.4, no IgG fractions are retained (2). Similarly, other proteins/peptides that are successfully purified on histidyl-Sepharose columns are retained only at pH values at or around their isoelectric pH (Table 4). So, ideally, the equilibrating adsorption pH for a given protein is chosen, based on the knowledge of their pI values.

8. The theoretical consideration and understanding of the mechanism(s) of interaction between the proteins/peptides and the matrix coupled histidine ligands have shown the mechanism to be water-mediated, involving the changes in the dielectric constants at the adsorption interface owing to the combined electrostatic, hydrophobic, and to some extent, charge transfer interactions between his-

Table 4
Proteins and Peptides Purified by HPLAC

Substance purified	p I	Adsorption pH
Chymosin	6.0	5.5
Acid protease A. *Niger*	3.6	3.0–4.0
Yeast carboxy peptidase	3.6	3.0–4.0
Human placental IgG$_1$	6.8–7.2	7.4
Myxaline (bacterial human blood anticoagulant)	4.0	5.5
Phycocyanin chromopeptides	5.8	5.0–6.0

tidine and the specific amino and residues available on the protein surface (e.g., tyr, his, and so on). The temperature parameter does modify these interactions and hence, contributes to the selectivity of the adsorption of the selected protein molecule. As an example, a detailed study carried out in the case of the milk-clotting enzyme, chymosin, showed that the selective adsorption of chymosin in the presence of other proteases, such as pepsin, was favored at +4°C, whereas at a higher temperature, +20°C, both pepsin and chymosin are equally retained on the gel. This temperature dependency is different for different molecules and different support matrices. As an example, in the case of IgG$_1$ specific adsorption onto a histidyl-silica was favored at +20°C rather than at +4°C *(7)*.

9. As a general rule, low ionic strength of the chosen buffer (in the range of 20–50 mM) is favorable for the protein retention on the histidine-coupled matrices. However, in the case of highly aromatic molecules, high salt concentrations, particularly by the addition of 1.0M sodium chloride into the adsorption buffer, favors their retention because of charge transfer interaction. This was well demonstrated in the case of a selective retention (retardation) of serotonin (most aromatic) in the presence of dopamine and tryptamine (unpublished data).

10. Desorption of proteins is invariably achieved by the addition of sodium chloride to the adsorption buffer. However, the optimal concentration of sodium chloride varies from 0.2M to 0.5M with different proteins. IgG$_1$, and some proteases were successfully eluted by adding 0.2M sodium chloride to the initial buffer and chymosin can be desorbed with 0.5M sodium chloride. In the case of "Myxaline" (glycopeptide showing blood anticoagulant activity) from the culture medium of *Myxococcus xanthus,* "Myxaline" was eluted with 0.5M and

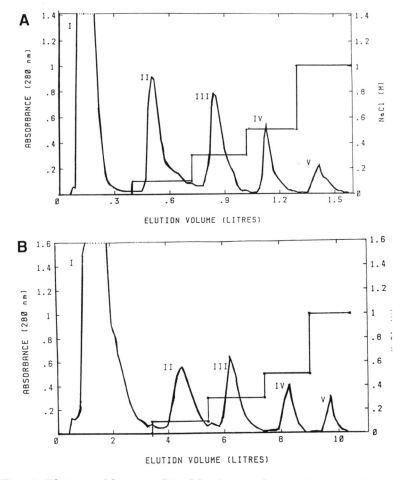

Fig. 4. Elution of heparin-like blood coagulation factor with histidyl Sepharose using: **A,** a linear flow column; **B,** a radial flow column.

with $1.0M$ sodium chloride, using a stepwise gradient from 0 to $1.0M$ sodium chloride while the contaminants were eluted at a lower sodium chloride concentration *(13)*.

11. An easy scale- up from 30 mL to 500 mL of adsorbent was achieved in the case of "Myxaline" recovery from the *Myxococcus xanthus* culture broth. The laboratory scale (30 mL bed vol) allowed the recovery of 15 mg "Myxaline" from 1500 mL microfiltered, heat-treated culture broth, whereas at a pilot scale we used a "Sepragen" radial flow column to recover 200 mg of "Myxaline" from 20 L of the microfiltered, heat-treated culture broth (Fig. 4). The efficiency of the operation in terms of both yield and purity of the product were absolutely

Table 5
Purification of Myxaline by Two-Step Chromatography
Using Linear and Radial Flow Columns

	Activity, IU		Purification factor		Activity yield, %	
	R.F.	L.F.	R.F.	L.F.	R.F.	L.F.
Crude extract	52	10.4	1	1	100	100
Chromatography on Hist. Seph.						
Fraction IV	31	4.2	13	9.2	59.6	40.4
Fraction V	41.8	5.6	33.5	21.5	80.4	53.8
Gel filtration (Sephadex G25)						
*Fraction IV'	32.8	4.4	18.5	12.8	63	42.3
*Fraction V'	44.6	6	45.2	27.8	85.8	57.7

R.F. Radial flow. L.F. Linear flow. *The fractions IV' and V' correspond to fraction IV and V, respectively, after desalting on a Sephadex G.25 gel filtration column.

unimpaired or even improved during the scaled up operation (Table 5). So, this technique developed at a lab scale, can be directly scaled up without difficulty *(14)*.

12. The scaling down of this pseudoaffinity technique can be acheived by coupling histidine to an appropriate particle size silica matrix after introducing the oxirane active groups to the silica (*see* Note 1). As an example, we used a histidine coupled to the Lichrosorb Si 60 matrix (0.8 × 10 cm) in a stainless steel column at 100 bar back pressure and connected to a Waters HPLC system. The detection of residual chymosin activity in milk whey from cheese manufacture could be realized by using the same chromatographic conditions (pH, ionic strength, and temperature) with a flow rate of 1.0 mL/min.

13. Note 7 mentioned that the retention mechanism is based on multiple interactions, such as ionic, hydrophobic, and charge transfer. The potential chemical functions in coupled histidine (Fig. 1) for ionic interactions are the COOH (negative charge) and the residual imino group of $-NH_2$ (positive charge). The free electron pairs on the imidazole NH groups are potentially mild charge transfer groups that can interact with the aromatic residues of proteins, while the relative hydrophobicity of the imidazole will contribute to the hydrophobic recognition of the proteins. In order to understand and appreciate their individual contribution, histamine and imidazole were coupled directly to Sepharose 4B and tested for the retention of pla-

cental IgG. The histamine-coupled gel retained a much higher amount of IgG, but the specificity to IgG_1 was enhanced with the imidazole *(15)*.

References

1. Vijayalakshmi, M. A. (1989) Pseudo-biospecific ligand affinity chromatography. *Tibtech* **7(3)**, 71–76.
2. Lehninger, A. L. (1976) *Biochemistry.* Worth, New York, 230,231.
3. Kanoun, S., Amourache, L., Krishnan, S., and Vijayalakshmi, M. A. (1986) A new support for the large scale purification of proteins. *J. Chromatogr.* **376**, 259–267.
4. Hjalmarsson, S. and Akesson, R. (1983), Modern Kjeldahl procedure. *Int. Lab.* 70–76.
5. Chang, S. H., Gooding, K. M., and Regnier, F. E. (1976) Use of oxiranes in the preparation of bonded phase supports. *J. Chromatogr.* **120**, 321–333.
6. Sundberg, L. and Porath, J. (1974) Preparation of adsorbents for biospecific affinity chromatography. 1. Attachment of amino group containing ligands to insoluble polymers by means of bifunctional oxiranes. *J. Chromatogr.* **90**, 87–98.
7. Kanoun S. (1988) Contribution à la chromatographie de pseudo-affinité. Modele de protéines: Les immunoglobulines G placentaires. PhD Thesis, University of Compiegne, 95–99.
8. Amourache, L. and Vijayalakshmi, M. A. (1984) Affinity chromatography of kid chymosin on histidyl-sepharose. *J. Chromatogr.* **303**, 285–290.
9. Melin, S. and Vijayalakshmi, M. A. (1987) Zonal elution analysis of IgG retention on histidyl-sepharose. *Biol. Chem. Hoppe-Seyler's* **368 (7)**, 762.
10. Krishnan, S., Gaehel-Vignais, I., and Vijayalakshmi, M. A. (1987) A semi-preparative isolation of carboxypeptidase. Isoenzymes from *Aspergillus Niger* by a single immobilized metal ion affinity chromatography. *J. Chromatogr.* **397**, 339–346.
11. Vijayalakshmi, M. A. and Thomas, D. (1985) Affinity chromatography as a production technique. *The World Biotech. Report Online* **1 (Europe)**, 131–140.
12. Cabrera, K. E. and Wilchek, M. (1987) Coupling of ligands to primary hydroxyl containing silica for high performance affinity chromatography. *J. Chromatogr.* **397**, 187–196.
13. Akoum, A., Vijayalakshmi, M. A., Cardon, P., Fournet, B., Sigot, M., and Guespin-Michel, J. F. (1987) *Myxococcus xanthus* produces an extracellular glycopeptide that displays blood anticoagulant properties. *Enz. Microb. Technol.* **9**, 426–429.
14. Akoum, A., Vijayalakshmi, M. A., and Sigot, M. (1989) Scale-up of "Myxaline" purification by a pseudo-affinity method using a radial flow column. *Chromatographia* **28**, 157–160.
15. Elkak, A. and Vijayalakshmi, M. A. (1991) Study of the separation of mouse monoclonal antibodies by pseudobioaffinity chromatography using matrix-linked histidine and histamine. *J. Chromatogr.* **570**, 29–41.

CHAPTER 4

Boronic Acid Matrices
for the Affinity Purification
of Glycoproteins and Enzymes

James H. Hageman and Glenn D. Kuehn

1. Introduction

Boronic acids immobilized on insoluble matrices have been used for over a decade to purify proteins and enzymes (1,2). Boronic acid matrices have been used to purify:

1. Glycoproteins,
2. Enzymes for which a boronic acid moiety is an inhibitor, and
3. Enzymes that bind to specific diol-containing cofactors (such as UTP or NADP[+]).

In all instances, the mechanisms by which the purifications are achieved are presumed to involve the unique capacity of boronic acids to bind reversibly to biomolecules bearing diols. The great recent success of boronic acid matrices in clinical screening for elevated levels of glycosylated hemoglobin, which occur in diabetes melitus, may give some impetus for extending the use of these materials in the purification of glycoproteins generally. Because of the relative recency of boronate-matrix chromatography and our belief that the methods described will be of wide applicability, we give below a larger introduction than is usual for this series.

In aqueous solutions, the only stable complex formed between diols, such as neutral sugars and boronic acids, is that of the boronate

From: *Methods in Molecular Biology, Vol. 11: Practical Protein Chromatography*
Edited by: A. Kenney and S. Fowell Copyright © 1992 The Humana Press Inc., Totowa, NJ

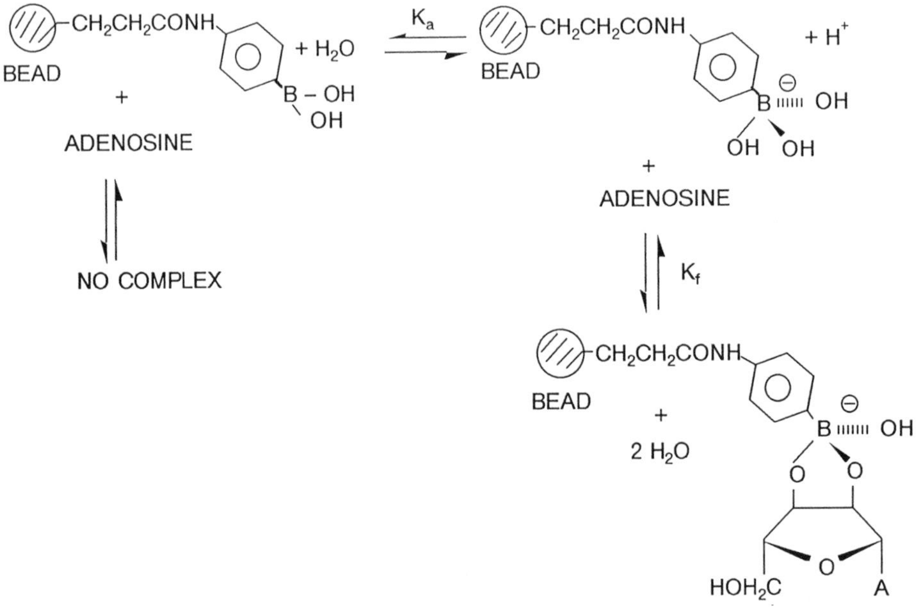

Fig. 1. Ionization and diester formation, with adenosine, of an immobilized boronic acid. The "BEAD" represents any insoluble matrix to which the aryl boronic acid is attached. In aqueous solution, only the ionized, tetrahedral boronate forms appreciable amounts of stable complex.

anion, and thus, the process of diesterification that occurs is necessarily pH-dependent. Since most work has been conducted with immobilized 3-aminobenzeneboronic acid, the chemistry of ionization and complexation for this boronic acid is illustrated below. The pK_a of the boronic acid in Fig. 1 is about 9 *(3)*, unless shifted by the presence of competing diols on the matrix or by residual positive charges from pendant amino groups *(4)* attached to the matrix. Thus, the working pH for the commercially available boronic acid matrices cannot be much lower than 8.0–8.5 in order to have 10–25%, respectively, of the boronic acid in the boronate anion form.

1.1. Variety of Matrices

Quite a variety of matrices, boronic acid derivatives, and coupling methods have been employed to prepare immobilized boronic acid chromatography media (Table 1), and recently, several boronic acid derivatives have been evaluated with respect to their suitability for use

Table 1
Types of Solid Matrices That Have Been Derivatized with Boronate Ligands (*See* Note 4)

Matrix	Boronate ligand coupled to matrix	Coupling method	Reference
3-Phenylenediamine and formaldehyde	3-Aminobenzeneboronic acid	Copolymerization	5
Styrene	4-Vinylbenzeneboronic acid	Copolymerization	6,7
Carboxymethyl cellulose	3-Aminobenzeneboronic acid	Succinic anhydride hydrazine activation	8
Aminoethyl cellulose	3-Aminobenzeneboronic acid	Succinic anhydride, carbodiimide	8
O-(2-Diethylaminoethyl) Sephadex[a]	4-(Bromomethylene) benzeneboronic acid	Direct alkylation	9
Hydroxypropylated Sephadex[b]	4-(Bromomethylene) benzeneboronic acid	Direct alkylation	9
O-[2-(Ethylamino) ethyl] cellulose	4-(Bromomethylene) benzeneboronic acid	Direct alkylation	9
Aminoethylpolyacrylamide	3-Aminophenylboronic acid	Succinic anhydride, carbodiimide	10
Carboxymethyl agarose[c]	4-(ω-Aminomethyl) benzeneboronic acid	Carbodiimide	11
Carboxymethyl agarose	3-Aminobenzeneboronic acid	Carbodiimide	12
Porous glass beads with alkylamine	N-(3-Dihydroxyborylbenzene) succinamic acid	Carbodiimide	13,14
Epoxide-activated silica	3-Aminobenzeneboronic acid	Epoxide condensation with amine	15
Polychlorotrifluoro-ethylene beads	3-(Decanoylamino)-benzeneboronic acid	Reversed-phase adsorption	14
Aminoethylpolyacrylamide	2-Nitro-3-succinamyl-benzeneboronic acid 4 nitro-3-succinamyl benzeneboronic acid	Carbodiimide	16
Carboxymethylstyrene[d]	3-Aminobenzeneboronic acid	Carbodiimide	17
Phenylboronic polyacrylamide	Benzeneboronic acid acrylamide	Polymerization	18

[a]DEAE-Sephadex A-25, Pharmacia Fine Chemicals, Piscataway, NJ. [b]DEAE-Sephadex LH-20, Pharmacia Fine Chemicals, Piscataway, NJ. [c]CH-Sepharose, Pharmacia Fine Chemcials, Piscataway, NJ. [d]Bio-Rex 70, Bio-Rad Laboratories, Richmond, CA.

Table 2
Commercial Sources of Benzeneboronate Matrices

Matrix	Trade name	Commercial sources
Cellulose	Indion PhenylBoronate (PB 0.8)	Chemical Dynamics Corp., Plainfield, NJ, USA
Agarose	Matrex Gel (PBA-10, PGA-30, PBA-60)	Amicon Corp., Danvers, MA, USA
Agarose	Glyco Gel B	Pierce Chemical, Rockford, IL, USA
Agarose	Glyc-Affin	Isolab Inc., Akron, OH, USA
Polyacrylamide	Affi-Gel 601 (P-6 acrylamide)	Bio-Rad Laboratories, Richmond, CA, USA
Methacrylate	Boric Acid Gel (0.1–0.4 mm)	Aldrich Chemical, Milwaukee, WI, USA
Methacrylate	Boric Acid Gel (0.1–0.4 mm)	Sigma Chemical, Inc., St. Louis, MO, USA
Silica	SelectiSpher-10™ Boronate	Pierce Chemical, Rockford, IL, USA
Silica	Progel-TSK Boronate-5PW (prepacked column)	Supelco, Inc., Bellefonte, PA, USA

in chromatography (19). Workers wishing more detailed comparisons of advantages and disadvantages of various boronate matrices may consult Note 9., and workers wishing to prepare specific types of boronate matrices are directed to the references given in Table 1. In contrast to what has appeared in the literature, commercial sources of boronates universally use 3-aminobenzeneboronic acid to link to solid supports; nevertheless, five types of matrices can be purchased, and among these, different chemistry has been used to effect coupling of the 3-aminobenzeneboronic add to the matrices. Table 2 lists commercial sources for benzeneboronate matrices. The advantages and disadvantages of various boronate matrices that have been made have not been formally reviewed. Therefore, a brief comment concerning each type of matrix will be helpful to investigators contemplating their use for the first time.

Virtually every support matrix available for general chromatography has been derivatized with boronate ligands by a variety of coupling procedures. Table 1 summarizes most of the types of matrices that have been exploited. These include cellulose, styrene, polyacrylamide, agarose, glass, silica, and dextrans. An evaluation of the development of different boronated supports indicates that the most successful matrices are those in which the solid matrix contains functional side-chain amine or carboxyl groups that can serve as sites for subsequent coupling to boronated ligands. Thus, carboxymethylated derivatives of cellulose *(8)*, agarose *(11,12)*, and styrene *(17)*, or aminoethylated derivatives of cellulose *(8)* and polyacrylamide *(10)* have produced the best boronated matrices. Carboxymethylated side chains on specific matrices are readily coupled directly onto a boronate ligand, such as 3-aminobenzeneboronic acid, using a carbodiimide condensing reagent. Aminoethylated side chains can be coupled with equal facility onto boronated ligands, such as 4-carboxybenzeneboronic acid, also by carbodiimide-mediated condensation. In some cases, it has been advantageous to insert spacer molecules between the matrix and the boronate ligand.

The first boronated chromatographic matrices reported in the literature with extensive characterization were carboxymethyl and aminoethyl cellulose *(8)*. However, these boronated celluloses retained high residual charge after derivatization with the boronate ligand and resisted attempts to reduce the charges by acetylation *(20)*. Sephadex-based matrices *(9)*, which employed Sephadex A-25 or LH-20, demonstrated unpredictable behavior toward certain *cis*-diol compounds and have never achieved extensive utility. An acrylic-based, boronated matrix prepared by coupling Bio-Rex 70 to 3-aminobenzeneboronic acid has a large capacity for binding high-mol-wt solutes *(17)*. This boronated medium does not denature proteins as can the styrene-based resins. Considerable potential exists for further development of this medium. Polyacrylamide matrices have been prepared by coupling boronic acid ligands to aminoethylpolyacrylamide marketed as aminoethyl Bio-Gel P-2, aminoethyl Bio-Gel P-150 *(3,10)*, or others *(21)*. Quantitative coupling can be achieved with these matrices, thus obviating the problems of residual charge. Agarose matrices such as CH-Sepharose 4B *(11)*, Sepharose 4B *(14)*, CM-Sepharose 4B *(12)*, and CM-Sepharose CL-6B *(12)*, have all been exploited for specialized applications. Recent successes in the preparation of boronated silica matrices *(15,22)* have extended applications of boronate-mediated chromatography into high-pressure and high-flow-rate-protocols.

As a guide to workers unfamiliar with the Lewis acid behavior and aqueous complexation chemistry of boronates with diols, each of the three cases listed in the Introduction will be briefly described. It should be stressed that the capacity of di- or triols to complex with a boronate is strongly influenced by configurations of the alcohols.

1.2. Glycoproteins

The capacity of a boronate to complex with a neutral carbohydrate depends strongly on the stereochemistry of the diols and, consequently, on whether a sugar is in a pyranose or furanose form. Because the boronate can be expected to alter the equilibrium between these two forms, a knowledge of the binding strength of boronates to sugars with free anomeric carbons will not be an infallible guide to the capacity of proteins having these bound sugars to be retained on a boronate matrix. For the most part, however, binding data for free sugars is all that is available. Table 3 gives most of the currently available binding data for sugars and related compounds to the boronate group or borate group.

By far the best-studied glycoprotein that is retained on a boronate matrix is glucosylated hemoglobin *(see below)*. This is an atypical glycoprotein because a free glucose reacts with an amino terminus of the hemoglobin and undergoes an Amadori rearrangement, presenting an arabitol-like structure and an adjacent positive charge to the boronate anion. Note 5 cites other more typical glycoproteins that can bind to a boronate matrix.

1.3. Boronate-Sensitive Enzymes

A number of classes of enzymes have been reported to be inhibited by aryl and alkylboronic acids. In practice, serine proteinases have been purified by their affinity for matrices bearing benzeneboronate. In principle, quite a number of other enzymes might be purified by use of appropriate boronate matrices. Table 4 summarizes the relatively few examples of immobilized boronate inhibitors that have been used for affinity chromatography and suggests other examples where specific enzymes might be purified with matrices containing appropriate boronic acids. Quite a number of hydrolases can be expected to be retarded on a matrix containing a benzeneboronate, even though more potent boronate inhibitors can usually be synthesized. The reader may refer to the references given in Table 4 to prepare specific boronic acids not commercially available.

Table 3
Binding Strengths and Probable Sites of Binding of Benzeneboronate
to Sugars and Related Compounds in Aqueous Solution

Compound	K_f^a	Probable diols in ester bonds [b]	References
D-Glucose	110	1,2-pyranoid and 1,2-furanoid	23,24
D-Galactose	276	3,4-and 5,6-pyranoid	23,24
D-Mannose	172	2,3-mannopyranoid and furanoid	23,25
D-Fructose	5200	α-1,3,6- or β-2,3,6-furanoid	25,26,27
Methyl β-D-ribofuranoside	430	2,3-furanoid	28
Sucrose	v. weak or no binding		25
L-Arabinose	391	1,2- and 2,3-pyranoid	24,25
Glycerol	19.7	?	24
myo-Inositol	25[c]	1,2- and 2,3-pyranoid (?)	29
cis-Inositol	1,100,000	1,3,5-pyranoid	29
Lactate	18	1,2-hydroxyls	30
Adenosine	1047	2,3-furanoid	28

[a]K_f = (benzeneboronate-sugar⁻)/(sugar benzeneboronate⁻) for 1:1 complex.
[b]Inferred in some cases from results reported for borate anion.
[c]1:1 Complexes with borate anion.

1.4. Exchangeable-Ligand Chromatography (ELC)

The third general use of boronate matrices for enzyme-affinity chromatography is based on the fact that a large array of enzyme substrates and cofactors are capable of esterifying to immobilized boronates *(3)*. Enzymes that bind to these esterified substrates or cofactors can be retained on the matrix, and can be discharged by exchanging the liganded cofactor or substrate with another ligand, such as mannitol or glucose, or by shifting the pH to values below the pK_a of the boronic acid *(3,4)*. A number of substrates and cofactors that have been shown to bind to immobilized boronates are listed in Table 5, and others that would be predicted to bind are listed in Table 6.

Although very few of these cofactors have been used in exchangeable-ligand chromatography, the ease of preparing a boronate matrix

Table 4
Boronic Acid Enzyme Inhibitors Potentially Useful
for Affinity Chromatography

Enzyme	Inhibitory boronic acid	K_I	Matrix	Reference
α-Chymotrypsin	Benzeneboronic acid	0.2 mM	CH-Sepharose	11,31
Subtilisin	Benzeneboronic acid	0.4–0.8 mM	CH-Sepharose	11
Trypsin	Benzeneboronic acid	1 mM	CH-Sepharose	11,32
α-Lytic protease	N-t-Boc-A-P-V-boronic acid	0.35 nM	None	33
Urease	Benzeneboronic acid	1.26 mM	None	34
	4-bromobenzene boronic acid	0.12 mM	None	34
β-Lactamase I (Bacillus cereus)	CF$_3$-CH$_2$-CONH-CH$_2$-B(OH)$_2$	33 μM	None	35
α-Chymotrypsin	Bz-phe-B(OH)$_2$	0.65 μM	None	36
Alanine racemase	1-Aminoethyl boronic acid	20 mM	None	37
D-Ala-D-ala ligase	1-Aminoethyl boronic acid	35 μM	None	37
Pancreatic lipase	p-Bromobenzene boronic acid	0.24 μM	None	38
Serum cholinesterase	p-Bromobenzene boronic acid	6.9 μM	None	39
Aminopeptidase (Zn^{2+})	1-butane boronic acid	9.6 μM	None	40
Chymase (mast cells)	MeO-Suc-Ala-Ala Pro-(L) boroPhe OH	57.9 nM	None	41

Table 5
Enzyme Substrates and Cofactors Demonstrated
to Bind to a Boronate Matrix

Compound or class of compounds	Reference
NAD⁺	*3,4*
NADH	*3,4*
FAD	*3,4*
NADP⁺	*4*
UTP	*3*
ATP	*4, 10*
Ribonucleotides	*18*
Ribonucleotides	*18*
2,4-Dihydroxycinnamic acid	*18*
DOPA, dopamine	*18*
Pyridoxal	*3*
Epineprine	*3*
Citric acid	*3*
Lactic acid (2-hydroxy acids)	*3*
Gluconic acid	*18*

See Note 3.

containing any of these substrates or cofactors makes boronate exchangeable-ligand chromatography a very attractive method to examine in developing an enzyme purification scheme. For each of the separations (of glycoproteins, of boronate-sensitive enzymes, and of enzymes bound to exchangeable ligands), specific conditions for the enzyme in question will have to be developed to suit the enzyme at hand. In Section 3, which follows, method details are given for the isolation of glycosylated from nonglycosylated hemoglobin *(42–44)* and for purification of uridine diphosphoglucose (UDP glucose) phosphorylase on a UTP-boronate matrix *(3)*.

2. Materials

2.1. Isolation of Glycosylated Hemoglobin

1. Polypropylene or glass columns (0.5 cm × 7 cm) containing 1 mL of agarose-boronate gel (Pierce Glyco-Gel) equilibrated in loading buffer (*see* Note 1).
2. About 100 μL of wet-packed human erythrocytes (washed 3X with a solution of NaCl, 8.5 g/L).

Table 6
Enzyme Substrates, Cofactors, and Other Compounds
Predicted to Bind to a Boronate Matrix

Compound or class of compounds	Examples
Nucleotide sugars or alcohols	CDP-ethanolamine, GDP-glactose and dephospho coenzyme A
Aromatic 1,2 diols	Tannins, gossypol
Aromatic o-hydroxy acids and amides	Salicylic acid 3-methoxy-4-hyrdoxyphenylethylene-glycol
Diketo and triketo compounds	Alloxan, benzil, dehydroascorbic acid
Steroids	Aldosterone, ecdysones
Drugs and toxins	Chloramphenicol tetrodotoxin
Other coenzymes	Adenosyl cobalamin, flavin mononucleotide
Inositols	Phosphatidyl inositol and various inositides

3. Stock solutions
   a. Loading/washing buffer (250 mL): 4.817 g ammonium acetate, 1.19 g magnesium chloride (anhydrous), 0.2 g sodium azide. These are dissolved in 200 mL of deionized water, adjusted to pH 8.5 with NaOH, and adjusted to 250 mL final vol. This results in a buffer that is $0.25M$ $NH_4(OOCCH_3)$, 50 mM $MgCl_2$ and 12 mM NaN_3.
   b. Eluting buffer (250 mL): 3.03 g Tris (Tris[hydroxymethyl]amino-methane, free base), 9.11 g sorbitol, 0.2 g sodium azide. These are dissolved in 200 mL of deionized water, adjusted to pH 8.5 with HCl and made to a final vol of 250 mL with deionized water, yielding a solution with final concentrations of $0.1M$ Tris, $0.2M$ sorbitol, and 12 mM sodium azide.

2.2. Purification of UDP-Glucose Pyrophosphorylase

1. A glass column (1 cm × 20 cm).
2. Cell extraction buffer (this will vary depending on properties of enzyme in the organism studied).
3. Dialysis/equilibration buffer (1.0 L): 11.92 g of HEPES (*N*-2-

hydroxyethyl piperazine-*N*-ethanesulfonic acid), 7.10 g of magnesium acetate (anhydrous). These are dissolved in 950 mL of deionized water, adjusted to pH 8.5 with KOH, and brought to a final vol of 1.0 L with deionized water, yielding a solution with final concentrations of 50 m*M* HEPES and 50 m*M* magnesium acetate.

4. Elution buffer (500 mL): 6.80 g of KH_2PO_4, 9.00 g of D-glucose. Dissolve these in 450 mL of deionized water, adjust pH to 7.0 with KOH, and bring vol to 0.50 L with deionized water, yielding a solution with final concentrations of 0.1*M* phosphate and 0.1*M* D-glucose.

Both procedures require the use of a UV-visible spectrophotometer, reagents for carrying out a standard protein assay, and a fraction collector.

3. Methods

3.1. Isolation of Glycosylated Hemoglobin (44)

1. Place the column packed with agarose-boronate gel, which has been equilibrated with the loading/washing buffer, in a vertical position over a fraction collector loaded with tubes of 4–5 mL vol. Wash the loaded column with 5 mL of loading/washing buffer immediately before using.
2. Prepare the hemolysate by suspending the 100 µL of washed erythrocytes in 1.9 mL of deionized water; after hemolysis has occurred, pellet the debris by centrifugation (800*g*, 5 min), and use the supernatant fraction at once or store frozen (–20°C for up to a week or –70°C for up to 6 mo).
3. Apply 100 µL of supernate to the column, and allow it to enter column; apply 0.5 mL of loading/washing buffer; elute column with an additional 20.0 mL of loading/washing buffer. Fractions of 1–2 mL may be collected and should contain all of the nonglycosylated hemoglobin. Hemoglobin concentrations can be measured spectrophotometrically by following the absorbance in the fractions at a wavelength of 414 nm. After the absorbance at 414 nm returns to the baseline value in the fractions, elute the column with 5–10 mL of eluting buffer; collect and measure the absorbance of the eluted fractions. These fractions will contain the glycosylated hemoglobin.
4. The columns can be regenerated by washing with 8 mL of 0.1*M* HCl (reagent grade), followed by 8 mL of loading/washing buffer (store at 4°C). (*See* Notes 1 and 2.)
5. Because the nature and local environment of the carbohydrate on any particular glycoprotein (*see* Note 5) will be unique in each case,

no universally applicable protocol can be given. Nevertheless, the guiding principles will remain the same if a suitable carbohydrate is present and has reactive diols available (*see* Table 3). Appropriate modifications (pH, ionic strength, and so on) of what is given here should result in a satisfactory protocol. (*See* Notes 6–8.)

3.2. Purification of UDP-Glucose Pyrophosphorylase (3)

1. Pack the column with 14 mL of boronate gel suspended in dialysis/equilibration buffer. Commercial agarose gels have appropriate pore sizes, as does polyacrylamide P-150-boronate; the latter can be made in 1 d by methods described elsewhere *(10)*. All work is conducted in a cold room maintained at 4–6°C.
2. Load the column with UTP by passing 10 mL (or more) of the dialysis/equilibration buffer, which is made up of 20 mM UTP (0.117 g of the trisodium, dihydrate salt of uridine-5'-triphosphate dissolved in 10 mL of dialysis/equilibration buffer, readjusted to pH 8.5), followed by 10–20 mL of the dialysis/equilibration buffer. Check loading of UTP by monitoring the UV-absorbance of the eluants ($\varepsilon_M^{260} = 9.9 \times 10^3 M^{-1} \, cm^{-1}$ at pH 7.0). A loading of 2–5 µmol of UTP/mL of gel is sufficient for good binding of the enzyme.
3. Break cells in 0.1M Tris-HCl, pH 7.8 buffer by passage through a French pressure cell. Centrifuge at 40,000 g for 1 h, and decant the supernatant fraction. Dialyze the supernatant fraction against dialysis/equilibration buffer. (Further purification by ammonium sulfate fractionation, and so on, may be carried out, if desired, before applying to column.)
4. Apply 5–25 mg of protein to the column (less may be used if highly purified), and elute with dialysis/equilibration buffer. Collect 3-mL fractions until the protein concentration in the tubes falls to background levels.
5. Elute column with elution buffer, collecting 3-mL fractions. Measure the protein concentration and enzyme activity *(3)* in all fractions. The enzyme activity should be retained on the column until the glucose in the elution buffer displaces the covalently bound UTP.

This method will be generally applicable to any enzyme with currently available commercial boronate gels, provided the enzyme of interest is stable for a reasonable period of time at above pH 8.0.

4. Notes

1. Boronate-containing matrices can be quite stable if properly handled and stored; we have kept some fully active materials for 4 yr in neutral

solutions at 4°C, with a few drops of toluene on the surface to retard microbial growth. However, the C—B bond is extremely sensitive to oxidation by hydrogen peroxide. The C—B bond can also be cleaved by certain transition metals, such as Ag^+, Hg^{2+}, Cu^{2+}, and Zn^{2+} *(45)*, especially when heated. The presence of these ions should be excluded from buffers used for chromatography.

2. The binding capacity of boronate columns for diol-containing compounds can be routinely monitored by addition of a standard vol of a stock solution of pyrocatechol violet (Aldrich Chemical, Co., Milwaukee, WI) through 1 mL of the gel and determining the amount of dye discharged upon washing with $0.05M$ acetic acid.

3. Although a large number of trigonal benzeneboronate-sugar esters have been characterized in the solid state *(46)*, they are often hydrolytically unstable and often are not the same as the tetrahedral esters, which form in aqueous solution. However, Table 5 lists compounds that have been demonstrated to bind to boronates or boronate matrices in aqueous solution. Thus, columns with a boronate matrix, when loaded with one of these compounds, might be used for affinity chromatography of any enzyme recognizing one of these.

   In addition, Table 6 provides a catalog of compounds that are expected from their known structural chemistry to bind to boronate matrices. It should be stressed that these have never been tested in this regard and, thus, should be checked for binding before proceeding with affinity chromatography.

4. A potential problem associated with carboxymethylated or aminoethylated boronate matrices can arise from field effects *(44)* owing to residual, charged, underivatized carboxyl or amino groups that remain after coupling the boronate ligand to a given matrix. Thus, residual aminoethyl groups may retain significant amounts of anionic solutes regardless of *cis*-diol content or may shift apparent pK_a's to lower values by stabilizing the boronate. Frequently *N*-acetylation of the aminoethylated matrix has been performed after the coupling reaction with the unreacted amino groups to eliminate these positive charges *(20)*. Although such treatment does reduce the amount of nonspecific adsorption resulting from residual charge groups, in many cases it has proven to lack the capacity to prevent it completely *(12)*. On the other hand, residual negative charges can arise from carboxyl groups that are not fully derivatized on carboxymethyl matrices. Since the phenylboronate groups are themselves negatively charged, field effects generated by the proximity of neighboring negatively charged groups would tend to raise the pK_a's

of the ionization of the matrix-bound dihydroxylboryl groups above that to be expected for the ionization of corresponding unbound dihydroxylboryl ligand. An increase in the ionic strength of the solvent used (addition of 0.1–0.2M KCl, for example) will diminish these field effects resulting in a decrease in the pK_a of the ionization of the dihydroxylboryl group. This should increase the extent of complex formation with a 1,2 *cis*-diol, which will be reflected in an increased retention volume from column separations *(8)*. Because of this behavior, separations on boronated chromatographic supports are usually conducted in solvents with elevated ionic strength, which is usually achieved by addition of KCl or NaCl to elution buffers *(18)*. Alternatively, virtually quantitative coupling has been achieved between aminoethylated polyacrylamide and 3-aminobenzeneboronic acid by exhaustive treatment of the matrix with carbodiimide condensing agent *(10)*.

5. In addition to hemoglobin, stressed above, other glycated serum proteins, such as albumin *(47)*, can be recovered with benzeneboronic acid supports. Glycosylated membrane proteins *(48)* and peptide hormones, such as human chorionic gonadotropin *(49)* can be isolated with boronate matrices. In addition, ADP-ribosylated proteins *(17, 50)* have also been found to bind to boronate columns.

6. For both glycosylated proteins and for enzymes bound to coenzymes esterified to boronates a variety of buffers can be successfully employed. It is important, however, to avoid the use of Tris or related buffers with polyols, since these greatly reduce the binding capacity of the boronated matrix by direct competition *(10)*.

7. Two alternatives to the one shown here for eluting enzymes from ligands bound to boronate columns may in some instances improve resolution: (a) a pH gradient elution from 8.5 to 6.0 (provided the enzyme of interest is stable in an acidic range) or (b) a gradient elution with increasing concentrations of the bound ligand (UTP in the example given above). In addition to glucose, such compounds as sorbitol, glycerol, Tris, or mannitol may be employed to elute the ligand and, hence, the enzyme from the column.

8. In addition to the exchangeable ligand approach, direct enzyme binding to boronates can be used. Table 4 lists enzymes that do or might bind directly to the boronate group. In these limited instances, one may be able to purify enzymes directly through their affinity for boronic acids. In some instances, the boronic acids are very powerful inhibitors and should strongly bind to the boronate matrices. Competing diols, acid pH, or competing boronates should serve to elute these enzymes.

9. The following are factors to consider in choosing a boronate matrix
   The experiences of many laboratories that developed the various types
   of boronated matrices listed in Table 1 permit a discussion of some
   of the advantages and disadvantages associated with each matrix.
   a. *Cellulose matrices.* The first boronated chromatographic matrices
      reported in the literature with extensive characterization were
      carboxylmethyl and aminoethyl celluloses *(51).* The amino-
      ethylcellulose matrix was prepared with a spacer succinamyl group
      between the primary amine of the cellulose and the primary amine
      of aminobenzeneboronic acid. This boronated matrix product con-
      tained 0.6 mmol of the dihydroxyboryl group/g of dry cellulose.
      This yield corresponded to a 60% substitution of the amino groups
      originally present in the cellulose. Suspensions of this cellulose
      derivative exhibited properties expected of a polycation owing
      to high residual charge from underivatized groups. The
      carboxymethyl cellulose was coupled directly to amino-benzene-
      boronic acid by reaction of the acylazide cellulose derivative. This
      product contained only about 0.2 mmol of dihydroxyboryl
      group/g of dry cellulose. This cellulose derivative was anionic in
      character in aqueous suspensions of pH values above 5. Thus, an
      appreciable fraction of the original carboxyl groups remained
      unsubstituted.
         Although separations of various ribonucleosides and polyol sug-
      ars were achieved on these matrices in a manner consistent with
      complexation of the covalently bound boronate group with *cis-*
      glycol systems in the sugar moieties, several limitations were
      apparent. First, flow rates for columns packed with boronate-cel-
      lulose were very low. For example, columns of relatively large
      dimensions, such as 1 cm diameter × 55 cm height, gave flow rates
      of only 10 mL/h. Second, the derivatized celluloses demonstrated
      low binding capacities owing to the small amount of covalently
      linked dihydroxylboryl group. Third, all of the celluloses retained
      high residual charge after derivatization with the boronate ligand.
         Attempts have been made to eliminate or minimize the effect
      of residual charge resulting from underivatized amino groups on
      aminoethylcellulose through *N*-acetylation *(52).* However, even this
      procedure failed to eliminate the capacity of the cellulose deriva-
      tives to retain significant amounts of RNA, regardless of *cis*-diol
      content *(53).* All of these limitations soon led workers to develop
      other boronated matrices. (*See* Note 4.)
   b. *Sephadex-based matrices.* Yurkevich et al. *(54)* prepared two boronated

dextran polymers derived from commercial dextrans, DEAE-Sephadex A-25 and DEAE-Sephadex LH-20. Direct alkylation of Sephadex beads with tris[4-(bromomethyl) benzene] boroxine in *N,N*-dimethylformamide produced [2-{[4-bromobenzene) methyl] diethylammonio} ethyl] Sephadex A-25 or LH-20. Derivatization to the extent of 60–70% of the original amine groups was achieved for both Sephadex A-25 and Sephadex LH-20. In the case of Sephadex A-25, 1.52 mmol of boron was linked/dry g of Sephadex. For Sephadex LH-20, 0.45 mmol of boron/dry g of Sephadex was incorporated. The actual operational pK_a values for these gels were not determined or reported in this work. In all likelihood, the large amount of residual, underivatized amine groups would have interfered with an attempt to titrate the boronate groups. An approximate model compound, 4-methylbenzeneboronic acid, has been reported to have pK_a of 10.0. However, the chromatographic performance of the boronated Sephadexes suggested that the boronate anion is stabilized by the bipolar, zwitterionic polymeric structure and, thus, may exist even in weakly acid medium as low as pH 5.0. For example, D-glucitol and D-fructose, even at pH 5, were strongly complexed with the borono groups of the Sephadex A-25 polymer. The behavior of adenosine on boronated Sephadex A-25 revealed that interaction of nucleosides with these polymers was not restricted merely to complex formation with 2',3'-hydroxyl groups. Thus, the polymer capacity to retain adenosine was almost constant over a broad range of pH from 2.5–8.2. This effect could not have been owing to complexation between *cis*-diols and the boronate anion. Paradoxically, results obtained on the chromatography of seven different carbohydrates were in good agreement with the stability constants observed earlier for complexes of various polyols with benzeneboronic acid *(55)*.

In summary, the overall performance of boronated Sephadexes points to a need for more extensive efforts to define their true structures. Workers at Pierce Chemical Company (Rockford, IL) have evidence that Sephadex itself has groups that apparently can bind to the boronate residues (Paul Smith, personal communication). Therefore, the somewhat unpredictable behavior toward certain *cis*-diol compounds detracts from broad use of Sephadexes for boronate matrices.

c. *Acrylic polymer matrices.* An acrylic-based, boronated matrix has been synthesized from the commercial cation-exchanger, Bio-Rex 70, by coupling 3-aminobenzeneboronic acid to the resin carboxyl groups with carbodiimide treatment *(56)*. Bio-Rex 70 is a cation exchanger containing carboxylic acid groups on a macroreticular acrylic polymer lattice. It has a large capacity for high-mol wt solutes and does not denature proteins as do the styrene-based resins. Covalent boronate chromatography with this resin proved to be particularly effective for purifying deproteinized poly(ADP-ribose) from nuclear extracts of Ehrlich ascites tumor cells. This particular matrix is attractive for future development of boronate resins, because Bio-Rex 70 is marketed with functional carboxyl-group capacities typically in excess of 10 mmol/dry g of resin. The potential for extensive derivatization with boronate ligands is therefore high. Unfortunately, the extent of incorporation of 3-aminobenzeneboronate ligand into Bio-Rex 70 resins has apparently not been determined or reported *(56)*. One possible limitation in the use of this matrix is that the resin exhibits large vol changes (swelling) upon displacement of the hydrogen form with other cations during use as a cation exchanger. Whether similar shrinkage and swelling accompanies pH changes during its use as a derivatized boronate matrix has not been reported.

d. *Polyacrylamide matrices.* A particularly attractive method to prepare a boronated matrix has exploited aminoethylpolyacrylamide for coupling to boronic acid ligands *(3,10)*. Quantitative succinylation and quantitative coupling steps are the salient advantages of this approach. Consequently, attending problems of residual charges left on a particular derivatized matrix are avoided by complete coupling. Aminoethylpolyacrylamide is now commercially available in two bead forms, namely, aminoethyl Bio-Gel P-2 and aminoethyl Bio-Gel P-6 from Bio-Rad Laboratories, Richmond, CA. Both of these supports are prepared from Bio-Gel P-2 and Bio-Gel P-6 according to the methods of Inman *(57)*. The P-150 aminoethyl derivative used originally *(10)* is no longer sold. These beads, made up of crosslinked copolymers of acrylamide and N,N'-methylenebisacrylamide, are spherical in shape and are commercially available in several size ranges. When contacted by water or aqueous solutions, the beads swell to form spherical gel particles having pores of molecular dimensions. Pore size is controlled by the degree of crosslinkage. Typically, the amino group capacity for aminoethylpolyacrylamide gels is 1–2 mmol/dry g. By exhaustively succinylating both of these gels with succinic anhydride followed by condensation of the resulting succinamyl carboxylate with 3-aminobenzeneboronic acid using a carbodiimide, a matrix was prepared that was virtually free of residual, underivatized amine groups. Different preparations of P-150 polyacrylamide-boronate contained 0.9–1.38 mmol of bound boronate/dry g. P-2 polyacrylamide-boronate was prepared with 2.46 mmol of covalently bound boronate/dry. The gels synthesized by these methods represented nearly quantitative coupling, and yielded a boron content at least three or four times higher than that for previously described column supports. A wide variety of small biomolecules was found to bind to the P-150 boronated gel (3). At saturation, 2.5–80 mmol of selected biomolecules were bound/mL of wet packed gel. Further evaluation of the P-150 boronated polyacrylamide confirmed its high capacity to bind mononucleotides or short oligonucleotides specifically (53). However, for reasons that are yet unclear, boronate polyacrylamides prepared by the method of Hageman and Kuehn (10) using 3-aminobenzeneboronic acid do not bind polyribonucleotides (53). This behavior does not seem to be dependent on the polyacrylamide matrix, because nitrobenzeneboronic acid substituted P-150 polyacrylamide was found suitable for purifying isoaccepting transfer RNAs at a pH as low as 6.15 (58). Immobilized nitrobenzeneboronic acids were found to be particularly suited for separating aminoacyl-tRNA from unacylated species. Nitration of benzeneboronic acids lowers their pK_a values, so that they ionize to form the boronate anion at a lower pH, one compatible with the stability of the aminoacyl bond. Aminoacyl-tRNA eluted at the column void vol and unacylated tRNA, with a free 2'3'-*cis*-diol moiety, remained bound by the boronate groups.

Polyacrylamide matrices exhibit a number of problems that may limit their applicability. Polyacrylamide beads tend to shrink in buffers of high ionic strength or pH values below 6 (59). They swell in buffers of high pH. Polyacrylamide exhibits instability under alkaline conditions, and therefore, prolonged exposure to buffers above pH 8–8.5 should be avoided. Finally, polyacrylamide boronated gels that are most effective for binding of ribonucleosides, and especially of ribonucleotides, require derivatization of gels that have a high amount of crosslinking (i.e., low-mol-wt exclusion beads, such as Bio-Gel P-2). The preparation of gel with a high content of available boronyl groups, generally above 2.0 mmol/dry g, is important for selective adsorption of nucleotides. The capacity to couple boronate ligands to high-mol-wt exclusion beads, which contain a low amount of crosslinking, is generally lower, and the resulting boronated gels have low binding capacities. The reason for these differences between low and high crosslinked gels is unclear, but it may be related to the hydrated vol of the gel. A high degree of crosslinking in the polyacrylamide bead and substitutions on the gels both tend to diminish the hydrated volumes.

Boronate acrylamide gels have also been prepared from polyacrylylhydrazide beads marketed as Bio-Gel P-2 hydrazide, Bio-Gel P-4 hydrazide, or Bio-Gel P-6 hydrazide by Bio-Rad Laboratories, Richmond, CA *(59)*. The hydrazide derivatives of polyacrylamide beads can be prepared according to the method of Inman and Dintzis *(60)*. Boronated gels were prepared by coupling a succinic acid group to the hydrazine-substituted polyacrylamide followed by use of a water-soluble carbodiimide to link 3-aminobenzeneboronic acid to the free succinyl carboxylate. The boronated gel product contained an extraordinarily high 4.9 mmol of boronate/dry g of starting hydrazide gel. Nucleosides from urine samples, containing a free diol group on the ribosyl moiety, were retained by this gel at pH 8.8 in $0.25 M$ NaCl solution with a retention vol of 15 column vol or larger. Nucleosides were effectively separated from free purine and pyrimidine bases by this gel.

e. *Agarose matrices.* Failures of boronated polyacrylamide gels to bind polyribonucleotides with sufficient tenacity for preparative use provided the major impetus to develop boronate-substituted agarose. This support matrix has several advantages over other types of gels. Agarose gels have fractionation ranges at considerably higher mol wt than expected by comparison with dextran or polyacrylamide gels. Moreover, on agarose gels, there tends to be less interaction between solute and matrix than with other types of gels. Hjerten *(61)* reported that even crystal violet, which is notorious for its tendency to stick to chromatographic materials, is not attracted to agarose. Columns of agarose do not significantly change in bed vol with high salt concentration or with pH changes in an operating range of 4–13. At least four different types of boronated-agarose gels have been prepared, and commercial preparations are also marketed.

Akparov and Stepanov *(62)* condensed 4-(ω-amino-methyl)benzeneboronic acid with CH-Sepharose 4B using the water-soluble carbodiimide, N-cyclohexyl-N′[-N-2-(4-morpholinyl) ethyl]-carbodiimide *p*-toluenesulfonate. CH-Sepharose 4B, a prod-

uct of Pharmacia Fine Chemicals Co., Piscataway, NJ, is a derivative of crosslinked carboxymethylagarose that has been coupled
to ε-aminocaproic acid yielding free carboxyl groups on the agarose bead. This sorbent was prepared with a very low bound
boronate content of 5–10 μmol of ligand/mL of swollen gel. This
represented about 50% coupling to the available carboxyl functions. The sorbent reportedly was effective for the covalent chromatography of serine proteinases from *Bacillus subtilis*

Singhal et al. *(63)* also synthesized a Sepharose 4B boronate gel
with an aminohexanoic acid spacer molecule using 3-
aminobenzeneboronic acid and the carbodiimide, 1-ethyl-3-(3-
dimethylaminopropyl)carbodiimide. Analysis indicated 0.29 mmol
of boryl group coupled/dry g of agarose. However, since the starting material contained only 0.05 mmol of free carboxyl groups/
dry g, the boronate ligand must have reacted at other sites on the
agarose matrix. Columns of this gel proved unsatisfactory in the
fractionation and recovery (about 30%) of aminoacyl-tRNA from
uncharged tRNA. The same authors also linked 3-aminobenzene-
boronic acid to Affi-Gel 10, purchased from Bio-Rad Laboratories, Richmond, CA. Affi-Gel is an *N*-hydroxysuccinimide ester of
a succinylated diethylamine spacer arm condensed to Bio- Gel A,
a 4% crosslinked agarose gel bead support. Analysis indicated that
0.072 mmol of boronate was bound/ g of Affi-Gel when coupling
was carried out at pH 7.5. This boronated matrix also failed to
complex with adenosine, 5'-adenylate, or uncharged tRNAs. In
general, boronate matrices containing spacer arms longer than
1.0 nm were not found useful for chromatography of tRNAs in
this study *(63)*.

In contrast, Pace and Pace *(53)* were able to prepare boronated
agarose gels from CM-Sepharose 4B, condensed with 3-
aminobenzeneboronic acid using carbodiimide, that were effective in the fractionation of uncharged tRNA. CM-Sepharose 4B is
a carboxymethyl derivative of agarose. Linkage of this matrix directly to 3-aminobenzeneboronic acid forms a boronated support
essentially without a lengthy spacer arm between the carboxylated
agarose bead and the boronate ligand. The boronate content in
this gel was quite low, about 15 μmol/mL. Boronated agarose prepared in this way proved suitable for fractionation of trace amounts
of transfer RNA, 5S ribosomal RNA and 23S ribosomal RNA. 5'-
Monoribonucleotides were retarded by this gel, but not strongly
bound. This contrasts with the performance of boronate polyac-

rylamide gels described earlier to which monoribonucleotides bound well but to which polyribonucleotides did not bind. The claim has been made that boronated polyacrylamide will bind tRNA *(64)*, but supportive data were not provided.

Wilk et al. *(65)* prepared a gel containing 3-aminbenzeneboronate groups that were coupled directly to the carboxyl groups of CM-Sepharose CL-6B, a product of Pharmacia Fine Chemicals Co., Piscataway, NJ, using the carbodiimide coupling procedure described by Pace and Pace *(53)*. This gel demonstrated efficient binding of 5'-capped small nuclear RNA (snRNA) and 5'-capped cytoplasmic mRNA. Noncapped RNA, such as rRNA, or decapped snRNA was not bound. The presence of a positive charge on the 3'-terminal nucleotide, which occurs on all mature caps that bear a positively charged, methylated m^7G residue, proved to be the discriminating factor for the binding of capped RNA species to boronate gels.

f. *Silica matrices.* Linkage of boronate ligands to solid inorganic supports has extended the range of applications of boronate chromatography to high-performance liquid chromatography. Early attempts to derivatize controlled pore glass *(65,66)* and passivated silica *(15,66)* produced matrices with very low bound ligand concentrations ranging from 0.063–0.14 mmol of boryl derivative/ dry g of support matrix. More recently, methods have been improved to yield microparticulate silicas with over 0.5 mmol of boryl groups /dry g of silica *(67)*. Boronated silica has been characterized most extensively among these types of supports *(15,67)*. The matrix is prepared by first linking α-glycidoxy-propyltrimethoxysilane to porous silica, thus forming an epoxy-substituted silica; 3-aminobenzeneboronic acid is coupled to the epoxy–substituted silica at neutral pH over a 20-h period at room temperature *(67)*. Boronated silica prepared in this manner could tolerate pressures of 2100 psi (14 mPa) and was operational with flow rates of 1 mL/ min. A single column was reportedly used for about 150 different separations over a period of several months without any noticeable deterioration of the column *(67)*. Catechols, compounds with related catecholic structure, ribonucleosides, ribonucleotides, and carbohydrates with *cis*-1,2-diol moieties were efficiently separated on this type of support.

g. *Reversed-phase matrices.* A reversed-phase boronate matrix was prepared by Singhal et al. *(63)* that proved effective for separation of mammalian and bacterial aminoacyl-tRNAs from uncharged tRNAs and *O*-methyl nucleosides from ribose-unsubstituted nucleosides in one chromatographic step. The matrix was prepared by acylating 3-aminobenzeneboronic acid with decanoyl chloride. This produced a nine-carbon aliphatic benzeneboronate, which was coated in chloroform on inert 10-μm solid beads of macroreticular poly(chlorotrifluoroethylene). The acylated boronate is bound (coated) to the beads by hydrophobic interactions. A maximum of 0.34 mmol of boronate group/dry g wt of beads was bound to this support. The nine-carbon aliphatic "tail" of the boronate ligand provided a side chain about 1 nm long. Complex formation between the boronate group and ribofuranoses appeared to be enhanced by the hydrophobic "tail," by a high ionic solvent environment of 0.5–1.0M NaCl, and by the hydrophobic nature of the inert support.

h. *Other matrices.* In an unusual synthetic route to produce substituted boronate matrices, several types of benzeneboronate monomers have been incorporated into polymerized supports. For example, Letsinger and Hamilton *(68)* first reported making boronic acid-containing polymers prepared from vinyl monomers. Both homopolymers and copolymers were produced with either 4-chlorostyrene or acrylamide. Further reports of polymers of this type were made by Pellon et al. *(69)*, Hoffman and Thomas *(70)*, and Lennarz and Snyder *(71)*. Elliger et al. *(72)* investigated four systems for polymerizing 4-vinylbenzeneboronic acid. These included:

i. Copolymerization with styrene,

ii. Copolymerization with *N,N*-methylene-bis(acrylamide), and

iii. Copolymerization with acrylamide and *N,N*-methylene-bis (acrylamide) and interstitial homopolymerization of porous polystyrene beads.

Schott et al. *(73–75)* treated 3-aminobenzeneboronic acid with methacryloyl chloride to yield a polymerizable boronic acid derivative, 3-aminobenzeneboronic acid methacrylamide. This monomer unit underwent radical polymerization with tetramethylene dimethacrylate to produce a crosslinked boronic acid. Although many of these boronated matrices were found to be selective in chromatography experiments, in general, their binding capacities were very low, and the elution peaks for compounds with vicinal diols were very broad. These supports have not gained wide utility for these reasons.

i. *Summary.* In summary, the insoluble polymers that have been used as the matrices for boronate immobilization are cellulose,

Sephadex, polystyrene, polyacrylamide, agarose, silica, acrylic acid polymers, and reversed-phase matrices. In the reported applications of immobilized boronates, few comparisons of the properties of different supports have been described *(76)*. Those matrices most likely to gain wide acceptance in laboratory practice will be those that demonstrate high equivalents of bound boronate ligand, and good flow rates in column applications, and those that perform in a manner that is explicable in terms of the boronate group interaction with vicinal diols.

References

1. Bergold, A. and Scouten, W. H. (1983) Boronate chromatography, in *Solid Phase Chromatography* (Scouten,W. H., ed) Wiley, New York, pp. 149–187.
2. Mazzeo, J. R. and Krull, I. S. (1989) Immoblized boronates for the isolation and separation of bioanalytes. *Biochromatogr.* **4,** 124–130.
3. Maestas, R. R., Prieto, J. R., Kuehn, G. D., and Hageman, J. H. (1980) Polyacrylamide-boronate beads saturated with biomolecules: A new general support for affinity chromatography of enzymes. *J. Chromatog.* **189,** 225–231.
4. Bouriotis, V., Galpin, I. J., and Dean, P. D. G. (1981) Applications of immobilized phenylboronic acids as supports for group-specific ligands in the affinity chromatography of enzymes. *J. Chromatog.* **210,** 267–278.
5. Solms, J. and Deuel, H. (1957) Ion-exchange resins, XI. Ion-exchange resins with boric acid groups. *Chimis (Swik)* **11,** 311–318.
6. Letsinger, R. L. and Hamilton, S. B. (1959) Organoboron compounds, X. Popcorn polymers and highly cross-linked vinyl polymers containing boron. *J. Amer. Chem. Soc.* **81,** 3009–3012.
7. Elliger, C. A., Chan, G. B., and Stanley, W. L. (1975) p-Vinylbenzeneboronic acid polymers for separation of vicinal diols. *J. Chromatog.* **104,** 57–61.
8. Weith, H., Wiebers, J., and Gilham, P. (1970) Synthesis of cellulose derivatives containing the dihydroxyboryl group and a study of their capacity to form specific complexes with sugars and nucleic acid components. *Biochemistry* **9,** 4396–4401.
9. Yurkevich, A. M., Kolodkina, I. I., Ivanova, E. A., and Pichuzhkina, E. I. (1975) Study of the interaction of polyols with polymers containing *N*-substituted [(4–boronophenyl) methyl]-ammonio groups. *Carbohyd. Res.* **43,** 214–224.
10. Hageman, J. H. and Kuehn, G. D. (1977) Assay of adenylate cyclase by use of polyacrylamide-boronate gel columns. *Anal. Biochem.* **80,** 547–554.
11. Akparov, V. Kh. and Stepanov, V. M. (1978) Phenylboronic acid as a ligand for biospecific chromatography of serine proteinases. *J. Chromatog.* **155,** 329–336.
12. Pace, B. and Pace, N. R. (1980) The chromatography of RNA and oligoribonucleotides on boronate-substituted agarose and polyacrylamide. *Anal. Biochem.* **107,** 128–135.

13. Duncan, R. and Gilham, P. (1975) Isolation of transfer RNA isoacceptors by chromatography on dihydroxyboryl-substituted cellulose, polyacrylamide, and glass. *Anal. Biochem.* **66,** 532–539.

14. Singhal, R. P., Bajaj, R. K., Buess, C. M., Small, D. B., and Vakharia, V. N. (1980) Reversed-phase boronate chromatography for the separation of o-methylribose nucleosides and aminoacyl-tRNAs. *Anal. Biochem.* **109,** 1–11.

15. Glad, M., Ohlson, S., Hansson, L., Månsson, M. O., and Mosbach, K. (1980) High performance liquid affinity chromatography of nucleosides, nucleotides and carbohydrates with boronic acid-substituted microparticulate silica. *J. Chromatog.* **200,** 254–260.

16. Johnson, B. J. B. (1981) Synthesis of nitrobenzeneboronic acid substituted polyacrylamide and its use in purifying isoaccepting transfer ribonucleic acids. *Biochemistry* **20,** 6103–6108.

17. Wielckens, K., Bredehorst, R., Adamiek, P., and Hilz, H. (1981) Protein-bound polymeric and monomeric ADP-ribose residues in hepatic tissues. Preparative analyses using a new procedure for the quantification of poly(ADP-ribose). *Eur. J. Biochem.* **117,** 69–74.

18. Fulton, S. (1981) *Boronate Ligands in Biochemical Separations.* Amicon Corporation, Danvers, MA.

19. Soundararajan, S., Badawi, M., Kohrust, C. M., and Hageman, J. H. (1989) Boronic acids for affinity chromatography: Spectral methods for determinations of ionization and diol-binding constants. *Anal. Chem.* **178,** 125–134.

20. McCutchan, T. F., Gilham, P. T., and Söll, D. (1975) An improved method for the purification of RNA by chromatography on dihydroxyboryl substituted cellulose. *Nucleic Acids Res.* **2,** 853–864.

21. Uziel, M., Smith, L. H., and Taylor, S. A. (1976) Modified nucleosides in urine: Selective removal and analysis. *Clin. Chem.* **22,** 1451–1455.

22. Hansson, L., Glad, M., and Hansson, C. (1983) Boronic acid-silica: A new tool for the purification of catecholic compounds on-line with reversed-phase high performance liquid chromatography. *J. Chromatog.* **265,** 37–44.

23. Kennedy, G. R. and How, M. J. (1973) Interaction of sugars with borate: An nmr spectroscopic study. *Carbohydrate Res.* **28,**13–19.

24. Lorand, J. P. and Edwards, J. O. (1959) Polyol complexes and structure of the benzeneboronate ion. *J. Org. Chem.,* **24,** 769–774.

25. Boeseken, J. (1949) The use of boric acid for the determination of the configuration of carbohydrates. *Adv. in Carb. Chem.* **4,** 189–210.

26. Torssell, K. (1957) Zurkenntris der arylborsäuren. VII. Komplexbildung zwischen phenylborsaure und fruktose. *Arkiv. Kemi.* **10,** 541–547.

27. Foster, A. B. (1957) Zone electrophoresis of carbohydrates. *Adv. Carbohydr. Chem.* **12,** 81–115.

28. Ivanova, E. A., Kolodkina, I. I., and Yurkevich, A. M. (1971) Anionic complexes of nucleosides and nucleotides with areneboronic acids. Translated from *Zhurnal Obshchei Khimii* **41,** 455–459.

29. Angyal, S. J. and McHugh, D. J. (1957) Cyclitols. Part V. Paper ionophoresis, complex formation with borate, and the rate of periodic acid oxidation. *J. Chem. Soc.* 1423–1431.

30. Pizer, R. and Babcock, L. (1977) Mechanisms of the complexation of boron acids with catechol and substituted catechols. *Inorg. Chem.* **16,** 1677–1681.
31. Philipp, M. and Bender, M. L. (1971) Inhibition of serine proteases by arylboronic acids. *Proc. Nat. Acad. Sci. USA* **68,** 478–480.
32. Lindquist, R. N. and Terry, C. (1974) Inhibition of subtilisin by boronic acids, potential analogs of tetrahedral reaction intermediates. *Arch. Biochem. Biophys.* **160,** 135–144.
33. Bone, R., Shenvi, A. B., Kettnes, C. A., and Agard, D. A. (1987) Serine protease mechanisms: Structure of an inhibitory complex of α-lytic protease and a tightly bound peptide boronic acid. *Biochemistry* **26,** 7609–7614.
34. Breitenbach, J. M. and Hausinger, R. P. (1988) *Proteus mirabilis* urease. Partial purification and inhibition by boric acid and boronic acids. *Biochem. J.* **250,** 917–920.
35. Crompton, I. E., Cuthbert, B. K., Lowe, G., and Waley, S. G. (1988) β-Lactamase inhibitors. The inhibition of serine β-lactamase by specific boronic acids. *Biochem. J.* **251,** 453–459.
36. Kinder, D. H. and Katzenellenbogen, J. A. (1985) Acylamino boronic acids and difluoroborane analogues of amino acids: Potent inhibitors of chymotrypsin and elastase. *J. Medicinal Chemistry* **28,** 1917–1925.
37. Duncan, K., Faraci, W. S., Matteson, D. S., and Walsh, C. T. (1989) (1 Aminoethyl)boronic acid: A novel inhibitor for *Bacillus stearothermophilus* alanine racemase and *Salmonella typhimurium* D-alanine:D-alanine ligase (ADP forming). *Biochemistry* **28,** 3541–3549.
38. Garner, C. W. (1980) Boronic acid inhibitors of porcine pancreatic lipase. *J. Biol. Chem.* **255,** 5064–5068.
39. Garner, C. W., Little, G. H., and Pelley, J. W. (1984) Serum cholinesterase inhibition by boronic acid. *Biochim. Biophys. Acta* **790,** 91–93.
40. Baker, J. O., Wilkes, S. H., Bayliss, M. E., and Prescott, J. M. (1983) Hydroxamates and aliphatic boronic acids: Marker inhibitors for aminopeptidase. *Biochemistry* **22,** 2098–2103.
41. Kato, Y., Kido, H., Fukusen, N., and Katunuma, N. (1988) Peptide boronic acids, substrate analogs, inhibit chymase, histamine release from rat mast cells. *J. Biochem.* **103,** 820–822.
42. Middle, F. A., Bannister, A., Bellingham, A. J., and Dean, P. D. G. (1983) Separation of glycosylated haemoglobins using immobilized phenylboronic acid. *Biochem. J.* **209,** 771–779.
43. Mallia, A. K., Hermanson, G. T., Krohn, R.I., Fujimoto, E. K., and Smith, P. K. (1981) Preparation and use of a boronic acid affinity support for separation and quantitation of glycosylated hemoglobins. *Anal. Lett.,* **14,** 649–661.
44. Klenk, D. C., Hermanson, G. T., Krohn, R. l., Fujimoto, E. K., Mallia, A. K., Smith, P. K., England, J. D., Wiemeyer, H. M., Little, R. R., and Goldstein, D. E. (1982) Determination of glycosylated hemoglobin by affinity chromatography: Comparison with colorimetric and ion-exchange methods, and effects of common interferences. *Clin. Chem.* **28,** 2088–2094.
45. Ainley, A. D. and Challenger, F. (1930) Studies of the boron-carbon linkage. Part 1. The oxidation and nitration of phenyl(boric acid). *J. Chem. Soc.* **1930,** 2171–2180.

46. Ferrier, R. J. (1978) Carbohydrate boronates. *Adv. Carb. Chem. Biochem.* **35,** 31–80.

47. Yatscoff, R. W., Tevaarwerk, G. J. M., and MacDonals, J. C. (1984) Quantitation of nonenzymically glycated albumin and total serum protein. *Clin. Chem.* **30,** 446–449.

48. Dean, P. D. G. (1986) Affinity tails and ligands, in *Bioactive Microbial Products.* (Stowell, J. D., Bailey, P. J., and Winstanley, D. J. eds.), Academic, London, pp. 147–160.

49. Anspach, F., Wirth, H. J., Unger, K. K., Stanton, P., and Davies, J. R. (1989) High performance liquid chromatography with phenylboronic acid, benzamidine, tri-L-alanine, and concanavalin A immobilized on 3-isothiocyanatopropyltriethoxysilane-activated nonporous monodisperse silicas. *Anal. Biochem.* **179,** 171–181.

50. Okayama, H., Kunihiro, U., and Hayaishi, O. (1978) Purification of ADP-ribosylated nuclear proteins by covalent chromatography on dihydroxyboryl polyacrylamide beads and their characterization. *Proc. Nat. Acad. Sci.* USA **75,** 111–115.

51. Weith, H., Wiebers, J., and Gilham, P. (1970) Synthesis of cellulose derivatives containing the dihydroxyboryl group and a study of their capacity to form specific complexes with sugars and nucleic acid components. *Biochemistry* **9,** 4396–4401.

52. McCutchan, T. F., Gilham, P. T., and Soll, D. (1975) An improved method for the purification of RNA by chromatography on dihydroxyboryl substituted cellulose. *Nucleic Acids Res.* **2,** 853–864.

53. Pace, B. and Pace, N. R. (1980) The chromatography of RNA and oligoribonucleotides on boronate-substituted agarose and polyacrylamide. *Anal. Biochem.,* **107,** 128–135.

54. Yurkevich, A. M., Kolodkina, I. I., Ivanova, E. A., and Pichuzhkina, E. 1. (1975) Study of the interaction of polyols with polymers containing N-substituted [(4-boronophenyl) methyl]ammonio groups. *Carbohyd. Res.* **43,** 214–224.

55. Lorand, J. P. and Edwards, J. O. (1959) Polyol complexes and structure of the benzeneboronate ion. *J. Org. Chem.* **24,** 769-774.

56. Wielckens, K., Bredehorst, R., Adamietz, P., and Hilz, H. (1981) Protein-bound polymeric and monomeric ADP-ribose residues in hepatic tissues. Preparative analyses using a new procedure for the quantification of poly(ADP-ribose). *Eur. J. Biochem.* **117,** 69–74.

57. Inman, J. K. (1974) Covalent linkage of functional groups, ligands, and proteins to polyacrylamide beads. *Methods Enzymol.* **34,** 30–58.

58. Johnson, B. J. B. (1981) Synthesis of nitrobenzeneboronic acid substituted polyacrylamide and its use in purifying isoaccepting transfer ribonucleic acids. *Biochemistry* **20,** 6103–6108.

59. Uziel, M., Smith, L. H., and Taylor, S A. (1976) Modified nucleosides in urine: Selective removal and analysis. *Clin. Chem.* **22** ,1451–1455.

60. Inman, J. K. and Dintzis, H. M. (1969) The derivitization of cross-linked polyacrylamide beads. Controlled introduction of function groups for the preparation of special-purposes, biochemical absorbents. *Biochemistry* **8,** 4074–4082.

61. Hjerten, S. (1961) Agarose as an anticonvection agent in zone electrophoresis. *Biochim. Biophys. Acta.* **53**, 514–517.
62. Akparov, V. Kh. and Stepanov, V. M. (1978) Phenylboronic acid as a ligand for biospecific chromatography of serine proteinases. *J. Chromatog.* **155**, 329–336.
63. Singhal, R. P., Bajaj, R. K., Buess, C. M., Small, D. B., and Vakharia, V. N. (1980) Reversed-phase boronate chromatography for the separation of o-methylribose nucleosides and aminoacyl-tRNAs. *Anal. Biochem.* **109**, 1–11.
64. Duncan, R. and Gilham, P. (1975) Isolation of transfer RNA isoacceptors by chromatography on dihydroxyboryl-substituted cellulose, polyacrylamide, and glass. *Anal. Biochem.* **66**, 532–539.
65. Wilk, H. E., Kecakemethy, N., and Schafer, K. P. (1982) *m*-Aminophenylboronate agarose specifically binds capped snRNA and mRNA. *Nuc. Acids. Res.* **10**, 7621–7633.
66. Akparov, V., Nutsubdize, N., and Potanova, T. (1980) Purification of pancreatic lipase on an affinity sorbent containing a phenylboronic acid derivative. *Bioorg. Khim.* **6**, 609–614.
67. Hansson, L, Glad, M., and Hansson, C. (1983) Boronic acid-silica: A new tool for the purification of catecholic compounds on-line with reversed-phase high performance liquid chromatography. *J. Chromatog.* **265**, 37–44.
68. Letsinger, R. L. and Hamilton, S. B. (1959) Organoboron compounds, X. Popcorn polymers and highly cross-linked vinyl polymers containing boron. *J. Amer. Chem. Soc.* **81**, 3009–3012.
69. Pellon, J., Schwind, L. H., Guinard, M. J., and Thomas, W. M. (1961) Polymerization of vinyl monomers containing boron. II. *p*-Vinylphenylboronic acid. *J. Polymer Sci.* **55**, 161–167.
70. Hoffman, A K. and Thomas, W. M. (1959) The synthesis of *p*-vinylphenylboronic acid and some of its derivatives. *J. Amer. Chem. Soc.* **81**, 580–582.
71. Lennarz, W. J. and Snyder, H. R. (1960) Arylboronic acids. 111. Preparation and polymerization of *p*-vinylbenzeneboronic acid. *J. Amer. Chem. Soc.* **82**, 2169–2171.
72. Elliger, C., Chan, B., and Stanley, W. (1975) *p*-Vinylbenzeneboronic acid polymers for separation of vicinal diols. *J. Chromatog.* **104**, 57–61.
73. Reske, K. and Schott, H. (1973) Column-chromatographic separation of neutral sugars on a dihydroxyboryl-substituted polymer. *Angew. Chem.* **12**, 417–418.
74. Schott, H. (1972) New dihydroxyboryl-substituted polymers for column-chromatographic separation of ribonucleoside-deoxyribonucleoside mixtures. *Angew. Chem. Int. Ed.* **11**, 824–825.
75. Schott, H., Rudloff, E., Schmidt, P., Roychoudhury, R., and Kössel, H. (1973) A dihydroxylborylsubstituted methacrylic polymer for the column chromatographic separation of mononucleotides, oligonucleotides and transfer ribonucleic cid. *Biochemistry* **12**, 932–938.
76. Alvarez-Gonzalez, R., Juarez-Salinas, H., Jacobson, E. L., and Jacobson, M. K. (1983) Evaluation of immobilized boronates for studies of adenine and pyridine nucleotide metabolism. *Anal. Biochem.* **135**, 69–77.

CHAPTER 5

Calcium-Dependent Hydrophobic Interaction Chromatography

Nicholas H. Battey and Michael A. Venis

1. Introduction

Calcium-dependent hydrophobic interaction chromatography has been widely used for the purification of calcium-binding proteins, following the report that calmodulin could be purified using this procedure *(1)*. The method makes use of the fact that proteins such as calmodulin, undergo a conformational change and expose a hydrophobic region on binding calcium *(2)*. This means that they bind to a hydrophobic resin, such as phenyl Sepharose, in the presence of calcium, and can be eluted with the calcium chelator EGTA. The procedure has been developed to allow separation of calmodulin from other calcium-binding proteins, exploiting differences in affinity for calcium and in hydrophobicity, and hence elution time in EGTA *(3,4)*. Changes in pH in conjunction with EGTA elution have also been used for fractionation of calcium-regulated proteins on phenyl Sepharose *(5)*.

The procedure described here allows the separation of calcium-activated protein kinase from calmodulin on a column of phenyl Sepharose. The method has been deliberately worked out so that a crude soluble protein fraction (from maize coleoptiles) can be loaded directly on to the column, and fractions containing calmodulin and calcium-activated protein kinase obtained by differential EGTA and pH elution. Thus, although the fractions obtained are not pure, the method is simple and may be useful as an initial step in the purification of calmodulin, or calcium-activated protein kinase, as well as other calcium-binding proteins.

From: *Methods in Molecular Biology, Vol. 11: Practical Protein Chromatography*
Edited by: A. Kenney and S. Fowell Copyright © 1992 The Humana Press Inc., Totowa, NJ

2. Materials

1. All chromatography buffers contain 0.25M sucrose and so should not be stored for more than 5 d at 4–5°C. Dithiothreitol (DTT) and phenylmethylsulfonyfluoride (PMSF) are added immediately before each buffer is used. PMSF is made up as a stock solution of 100 mM in methanol and stored at 4–5°C. All reagents are Analar grade, except sucrose, which is FSA Specified Laboratory Reagent grade. It is important that water of the highest possible purity is used—for instance, water produced by the Barnstead Nanopure II system (supplied by Fisons Scientific Apparatus, Leicestershire, UK) or similar.

   Buffer A: 0.25M sucrose, 50 mM Tris-acetate (pH 8.0), 0.1 mM MgCl$_2$, 1 mM EDTA, 1 mM DTT, 0.5 mM PMSF.

   Buffer B: 0.25M sucrose, 50 mM Tris-acetate (pH 8.0), 0.1 mM MgCl$_2$, 1 mM CaCl$_2$, 0.5 mM DTT, 0.25 mM PMSF.

   Buffer C: 0.25M sucrose, 50 mM Tris-acetate (pH 8.0), 0.1 mM MgCl$_2$, 1 mM CaCl$_2$, 0.5 mM DTT, 0.25 mM PMSF, 0.5M NaCl.

   Buffer D: 0.25M sucrose, 20 mM Mes-NaOH (pH 6.8), 0.1 mM CaCl$_2$, 0.5 mM DTT, 0.25 mM PMSF.

   Buffer E: 0.25M sucrose, 20 mM Mes-NaOH (pH 6.8), 1 mM EGTA, 0.5 mM DTT, 0.25 mM PMSF.

   Buffer F: 0.25M sucrose, 20 mM glycine-NaOH (pH 8.5), 0.1 mM CaCl$_2$, 0.5 mM DTT, 0.25 mM PMSF.

   Buffer G: 0.25M sucrose, 20 mM glycine-NaOH (pH 8.5), 1 mM EGTA, 0.5 mM DTT, 0.25 mM PMSF.

   Buffer H: 10 mM MgCl$_2$, 50 mM Hepes-NaOH (ph 7.0), 0.2 mM EGTA, 0.1 mg/mL histone (Sigma type IIIS).

   Buffer K: 20% (w/v) glycerol, 125 mM Tris-HCl (pH 6.8), 4.7% (w/v) sodium dodecyl sulfate, bromophenol blue marker dye.

   Buffer L: 25% (w/v) trichloroacetic acid, 20 mM sodium pyrophosphate, 10 mM EDTA.

   Buffer M: 25% (w/v) tricholoroacetic acid.

   Buffer N: 100% (w/v) tricholoroacetic acid.

2. Phenyl Sepharose CL 4B (ligand density 40 μmol/mL gel) is obtained from Pharmacia LKB. The column (8 cm × 1 cm in diameter, total bed vol 6.3 mL) is stored in 20% ethanol and can be reused.

3. For sodium dodecyl sulfate polyacrylamide gel electrophoresis (SDS-PAGE), Analar-grade chemicals are used with the buffer system of Laemmli *(6)* and a 12% resolving gel. (*See* Chapter 20.)

4. (γ-^{32}P) ATP for protein kinase assays is obtained carrier-free at a specific activity of 3000 Ci/mmol (Amersham Int., product code PB10168).

3. Method

3.1. Protein Kinase Assay

Activity is assayed as described by Hetherington et al. (7). Aliquots of test samples are diluted to a final vol of 100 µL in an assay medium (Buffer H) containing $CaCl_2$ at a concentration calculated to give a final free-Ca^{2+} level of 0 or 300 µM (8), and (γ-^{32}P) ATP. (γ-^{32}P) ATP is used at a specific activity of 400–500 dpm/pmol ATP; the ATP concentration is approx 50 µM. After 1 min at 15°C, 40-µL aliquots are pipeted onto 2-cm filter squares, each of which has been impregnated with 100 µL of buffer L. These squares are then allowed to dry, transferred to a beaker containing buffer L (approx 5 mL/filter), and left overnight at 4°C. Unincorporated label is removed from the filters by washing in buffer M, boiling for 5 min in buffer L, and washing again in buffer M. They are then rinsed in acetone and allowed to air-dry. Incorporated $^{32}PO_4$ (peak fractions incorporate about 150,000–200,000 cpm/mL) is measured by liquid scintillation counting.

3.2. Electrophoresis

Samples are prepared for electrophoresis by adding buffer N to give a final concentration of 10% tricholoroacetic acid. After precipitation overnight at 4–5°C, protein is pelleted by centrifugation in a microcentrifuge at 11,600g (max) for 10 min. The supernatant is discarded, and any remaining drops of solution are removed from the tube and around the pellet using cotton-wool buds and filter-paper wicks. The pellet is then resuspended in 25 µL of buffer K. If the tracker dye turns yellow, indicating that some remaining tricholoroacetic acid has caused the pH to fall, sufficient 1M NaOH (usually not more than 2 µL) is added to neutralize the acid. The samples are made 1mM with respect to $CaCl_2$ or EGTA, and then diluted to a final vol of 50 µL. They are heated for 5 min in a boiling-water bath, cooled, briefly centrifuged, and loaded onto the gel. (*See* Chapter 20.)

3.3. Protein Extraction

All steps are carried out at 4–5°C. We have chosen the simplest method for rapid preparation of a crude mixture of Ca^{2+}-binding (and other) proteins. Seeds of *Zea mays* (cv. Clipper) are soaked overnight in running water, sown in vermiculite, and grown for 5 d in darkness. The etiolated coleoptiles and enclosed leaf rolls are harvested, frozen in liquid N_2, and stored frozed at –18°C until required. Then 25g of tissue is homogenized in a mortar and pestle with 50 mL of buffer A, strained through muslin (two layers), and centrifuged at 36,900g (av) for 45 min. To the supernatant $1M$ $CaCl_2$ is added to give a final concentration of 5 mM, and hence a free-Ca^{2+} concentration of approx 4 mM. Addition of $CaCl_2$ causes the supernatant to become cloudy (*see* Note 3); the solution is therefore centrifuged at 36,900g (av) for 10 min. The clear supernatant is then ready for chromatography.

3.4. Chromatography on Phenyl Sepharose

The following step-elution procedure allows EGTA elution of calmodulin at low pH, followed by elution of Ca^{2+}-activated protein kinase at high pH.

1. The column is equilibrated in at least 10 column vol of buffer B. The sample (vol approx 60mL) is then loaded at a flow rate of approx 0.8 mL/min. The column is washed with buffer B until the absorbance at 276 nm is approximately constant.
2. Nonspecifically bound protein is eluted with a salt wash of about 15 mL in buffer C (Fig. 1).
3. The pH and ionic strength are lowered by washing the column to approximately constant absorbance in buffer D (Step 1/2, Fig. 1 and 2).
4. The column is washed in buffer E, containing EGTA (Step 3). This causes calmodulin to elute in about 10 mL; the presence of calmodulin is most easily confirmed by its distinctive Ca^{2+}-dependent mobility shift on SDS-PAGE (Fig. 2).
5. The column is again washed with about 2 column vol of buffer D, to replace Ca^{2+}. The pH is then raised by washing with buffer F. A large amount of protein elutes in a vol of about 20 mL at this step, including most of the Ca^{2+}-activated protein kinase (Step 5, Figs. 1 and 2). The kinase is purified approx 20-fold by this procedure.
6. The column is washed to approximately constant absorbance with buffer G, containing EGTA at high pH. No calmodulin is visible on SDS-PAGE of this fraction (Step 6, Fig. 2), and only a little Ca^{2+}-activated protein kinase is present (Fig. 1). This step is therefore not needed for this maize extract however, it may be necessary with preparations from other tissues (*see* Note 5).
7. To clean the phenyl Sepharose, wash the column successively in water, 8M urea, water, and 20% ethanol (at least 2 column vol of each).

4. Notes

1. With plant extracts, particularly crude extracts, such as those used here, many phenolics are present and bind to phenyl Sepharose.

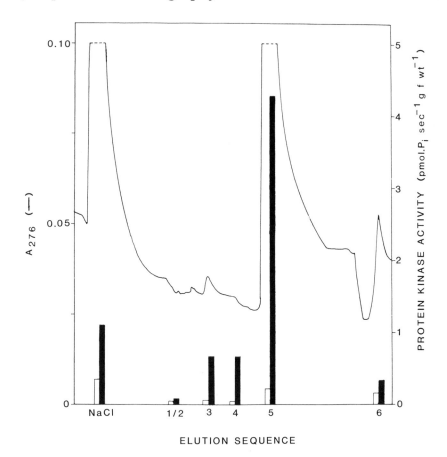

Fig. 1. Elution profile and protein kinase activity of fractions from a phenyl Sepharose column. Protein in the eluate was detected by measuring the absorbance at 276 nm; protein kinase activity was measured in the absence of Ca^{2+} (open histogram) or in the presence of 300 µM free Ca^{2+} (closed histogram). Elution sequence: NaCl; wash in 500 µM NaCl; 1/2: buffer strength reduced from 50 to 20 mM, pH reduced from 8.0 to 6.8; 3: 1 mM EGTA instead of 0.1 mM $CaCl_2$ (pH 6.8); 4: 0.1 mM $CaCl_2$ instead of 1 mM EGTA; 5: pH increased from 6.8 to 8.5; 6: 1 mM EGTA instead of 0.1 mM $CaCl_2$ (pH 8.5).

Although the washing procedure removes most of these, the gel does become gradually discolored with repeated use. This may mean that the effective life of the phenyl Sepharose is less with plant extracts than, for example, with animal extracts.

2. According to the manufacturer, phenyl Sepharose has a capacity of approx 20 mg protein/mL gel. However, the capacity is very dependent

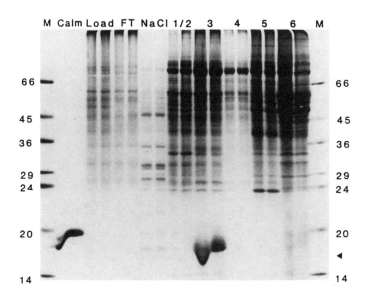

Fig. 2. SDS-PAGE of fractions from a phenyl Sepharose column. All lanes except marker lanes were electrophoresed in the presence of $CaCl_2$ (left of pair) or EGTA (right of pair). Proteins were stained with Coomassie blue. M: mol-wt markers—mol wt $(\times 10^{-3})$ are indicated alongside; Calm: calmodulin from bovine brain (Calbiochem); Load: crude soluble fraction loaded on phenyl Sepharose; FT: flow-through (nonretained) fraction from column; NaCl, and 1/2–6: see legend to Fig. 1. Arrowhead: maize calmodulin.

on the protein tested and on the loading conditions. For crude maize protein we have found that when approx 18 mg protein/mL gel is loaded, about 75% passes through in the start-buffer fraction (containing little or no Ca^{2+}-activated protein kinase or calmodulin), so in our experiments there is presumably considerable excess binding capacity available. This needs to be verified for each tissue type or crude-protein mix.

3. Before loading the 36,900g supernatant on to the phenyl Sepharose, $CaCl_2$ is added to give a final concentration of 5 mM. This induces a precipitate, which is removed by centrifugation. This precipitate forms even if the preceding centrifugation step is at more than 100,000g; it contains no calmodulin visible on SDS-PAGE, but some protein kinase. We presume that it is composed of membrane fragments and vesicles that aggregate in the presence of Ca^{2+}. It is important to remove this material, as it can partially block the column. An additional

problem with some plant extracts is precipitation of pectin on addition of Ca^{2+}: this can also slow the flow rate through the column unless removed.

4. Clearly the calmodulin fraction obtained is not pure (Fig. 2). It also gives a deceptively small peak when monitored at 276 nm, because of the low aromatic amino acid content of calmodulin. More representative elution profiles may be obtained by monitoring at 235 nm or below (*see* ref. *9*). The purity of this postphenyl Sepharose fraction could be improved by an initial purification step, e.g., low-pH precipitation with ammonium sulfate *(10)*, or heat treatment *(11)*. However, both methods have the disadvantage that, at least in maize, they result in loss of protein kinase activity. By contrast, recovery of protein kinase activity is very good using phenyl Sepharose as the first step in purification.

5. Under the conditions described here, soluble maize Ca^{2+}-activated protein kinase elutes as the pH is raised to 8.5 after EGTA elution of calmodulin at pH 6.8. When the same technique is used with protein solubilized from apple membranes, Ca^{2+}-activated protein kinase elutes at pH 8.5 only in the presence of EGTA *(5)*. We are currently investigating the basis for this difference.

6. We have described a column method, but the procedure could be modified for batchwise preparation of calmodulin and protein kinase; this might considerably reduce the overall procedure time (approx 4 h from the start of sample loading to elution of the last fraction).

References

1. Gopalakrishna, R. and Anderson, W. B. (1982) Ca^{2+}-induced hydrophobic site on calmodulin: Application for purification of calmodulin by phenyl Sepharose affinity chromatography. *Biochem. Biophys. Res. Commun.* **104,** 830–836.
2. LaPorte, D. C., Wierman, B. M., and Storm, D. R. (1980) Calcium-induced exposure of a hydrophobic surface on calmodulin. *Biochemistry* **19,** 3814–3819.
3. Walsh, M. P., Valentine, K. A., Ngai, P. K., Carruthers, C. A., and Hollenberg, M. D. (1984) Ca^{2+}-dependent hydrophobic interaction chromatography: Isolation of a novel Ca^{2+}-binding protein and protein kinase C from bovine brain. *Biochem. J.* **224,** 117–127.
4. McDonald, J. R., Groschel-Stewart, U., and Walsh, M. P. (1987) Properties and distribution of the protein inhibitor (M_r 17000) of protein kinase C. *Biochem J.* **242,** 695–705.
5. Battey, N. H. and Venis, M. A. (1988) Separation of calmodulin from calcium-activated protein kinase using calcium-dependent hydrophobic interaction chromatography. *Anal. Biochem.* **170,** 116–122.

6. Laemmli, U. K. (1970) Cleavage of structural proteins during the assembly of the head of bacteriophage T4. *Nature* **227,** 680–685.

7. Hetherington, A. M. Battey, N. H., and Millner, P. A. (1990) Protein kinase, in *Methods in Plant Biochemistry* vol. 3, (Lea, P. J., ed.), Academic, London, pp. 371–383.

8. Robertson, S. and Potter, J. D. (1984) The regulation of free Ca^{2+} ion concentration by metal chelators. *Methods Pharmacol.* **5,** 63–75.

9. Klee, C. B. and Vanaman, T. C. (1982) Calmodulin. *Adv. Protein Chem.* **35,** 213–321.

10. Burgess, W. H., Jemiolo, D. K., and Kretsinger, R. H. (1980) Interaction of calcium and calmodulin in the presence of sodium dodecyl suphate. *Biochim. Biophys. Acta* **623,** 257–270.

11. Dedman, J. R. and Kaetzel, M. A. (1983) Calmodulin purification and fluorescent labelling. *Methods Enzymol.* **102,** 1–8.

CHAPTER 6

Lectin Affinity Chromatography

Iris West and Owen Goldring

1. Introduction

Lectins are glycoproteins or proteins that have a selective affinity for a carbohydrate, or a group of carbohydrates. Many purified lectins are readily available and these may be immobilized to a variety of chromatography supports.

Immobilized lectins are powerful tools that can be used to separate and isolate glycoconjugates *(1–3)*, polysaccharides, soluble cell components, and cells *(4)* that contain specific carbohydrate structures. Glycopeptide mixtures can be fractionated by sequential chromatography on different lectins *(5)*. Immobilized lectins can also be used to remove glycoprotein contaminants from partially purified proteins. In addition, immobilized lectins can be used to probe the composition and structure of surface carbohydrates *(6)*. Subtle changes in glycoprotein structure caused by disease may be reflected in altered affinity for lectins *(7)*. The predominant lectins used for affinity chromatography are listed in Table 1.

In principle, the mixture of glycoconjugates to be resolved is applied to the immobilized lectin column. The glycoconjugates are selectively adsorbed to the lectin and components without affinity for the lectin are then washed away. The adsorbed carbohydrate containing components are dissociated from the lectin by competitive elution with the "hapten sugar," i.e., the best saccharide inhibitor. This may be a simple monosaccharide, or an oligosaccharide *(8,9)*. There are many matrices suitable for coupling to lectins, and a number of

From: *Methods in Molecular Biology, Vol. 11: Practical Protein Chromatography*
Edited by: A. Kenney and S. Fowell Copyright © 1992 The Humana Press Inc., Totowa, NJ

Table 1
Common Lectins Used in Affinity Chromatography

Lectin	Specificity	Useful eluants	Uses	Ref.
Concanavalin A (Con A)	α-D-mannopyranosyl with free hydroxyl groups at C3, C4, and C6	0.01–0.5 M Methyl-α-D-mannoside D-mannose D-glucose	Separation of glycoproteins Purification of glyco-protein enzymes Partial purification of IgM Separation of lipoproteins	10 11 12 13
Lens culinaris (Lentil lectin)	α–D-glucopyranosyl residues α-D-mannopyranosyl α-D-glucopyranosyl residues	Methyl-a-D-glucoside 0.1 M Na borate, pH 6.5 0.15 M Methyl-α-D-mannoside	Purification of gonadotrophins Purification of He-La cells Isolation of mouse H₂ anti-gens	14 15 16
	Binds less strongly than Con A		Purification of detergent-solubilized glycoproteins Purification of lymphocyte membranes glycoproteins	17 18
Tritum Vulgaris	N-acetyl-D-glucosamine	0.1 M N-acetyl gluco-samine	Biochemical characterization of the H4G4 antigen from HOON pre B leukemic cell line Purification and analysis of RNA polymerase transcription factors	19 20
Ricinus communis RCA 1 Jacalin (29)	α-D-galactopyranosyl residues D-galactopyranosyl residues	0.15 M D-galactose 0.1 M melibiose in PBS	Fractionation of glycopeptides binding glycoproteins Separating IgA, IgA₂ Purification of C1 inhibitor	21 22 23
Bandeiraea simplicifolia	α-D-galactopyranosyl and N-acetyl-D-galactosamyl	PBS	Resolving mixtures of nucleotide sugars	24

Table 2
Commonly Available Activated Matrices

Commercial name	Supplier
Affi 10, Affi 15	Bio-Rad
Affi Prep 10, and 15 *(25)*	Bio-Rad
CNBr Agarose	Sigma
Epoxy activated agarose	Sigma
CDI agarose	Sigma
Reacti 6X	Pierce
Polyacrylhydrazide agarose	Sigma

reliable coupling methodologies have been developed. Many supports are now supplied preactivated. These are simple to use. Table 2 lists a few of the most popular activated matrices.

The lectin must be chosen for its affinity for a carbohydrate or sequence of sugars. Many of the lectins are not specific for a single sugar and may react with many different glycoproteins. Concanavalin A (Con A) has a very broad specificity and will bind many serum glycoproteins, lysosomal hydrolases, and polysaccharides. The binding efficiency of con A is reduced by the presence of detergents: 0.1% sodium dodecyl sulfate (SDS), 0.1% sodium deoxycholate, and 0.1% cetyltrimethyl-ammonium bromide (CTAB) *(26)*. *Lens culinaris* lectin has similar carbohydrate specificities to Con A, but binds less strongly. The lectin retains its binding efficiency in the presence of sodium deoxycholate. It is therefore useful for the purification of solubilized membrane glycoconjugates. Wheat germ lectin can interact with mucins, which contain *N*-acetylglucosamine residues. It retains its binding efficiency in the presence of 1% sodium deoxycholate. Jacalin is a lectin isolated from the seeds of *Artocarpus integrifolia,* which has an affinity for *D*-galactose residues *(29)*. It is reported to separate IgA1 from IgA2 subclass antibodies. It is therefore useful for removing contaminating IgA from preparations of purified IgG.

Two simple procedures (Sections 3.1. and 3.2.) for the immobilization of lectins are given below, one for Con A onto carbonyldiimidazole (CDI)-activated agarose, the other for Con A onto Affi-Gel 15. Once the Con A has been immobilized to the matrices, the purification and elution procedures are common. A single protocol (Section 3.3.) is given detailing this procedure.

2. Materials

2.1. Immobilization of Con A on CDI-Agarose

1. 1,1' carbonyldiimidazole-activated agarose (Sigma). The activated gel is stable in the acetone slurry for 1–2 y at 4°C.
2. Affinity-purified Con A.
3. Protective sugar: Methyl-α-D-mannoside.
4. Coupling buffer: $0.1M$ Na_2CO_3, pH 9.5.
5. $0.1M$ Tris HCl, pH 9.5.
6. Coupling buffer (PBS): 8.0 g NaCl, 0.2 g KH_2PO_4, 2.9 g Na_2HPO_4· $12H_2O$, 0.2 g KCl, made up to 1 L with distilled water.
7. $0.1M$ Sodium acetate, pH 4.0.
8. $0.1M$ $NaHCO_3$, pH 8.0.

2.2. Immobilization of Con A on Affi-Gel 15

1. Affi-gel 15.
2. Affinity-purified Con A.
3. Protective sugar methyl-α-D-mannoside.
4. Phosphate buffered saline (PBS).
5. Equilibration buffer: 5 mM sodium acetate buffer, pH 5.5, containing $0.1M$ NaCl, 1 mM $MnCl_2$, 1 mM $CaCl_2$, 1 mM $MgCl_2$, and 0.02% sodium azide.

2.3. Purification of Glycoproteins on Immobilized Con A

1. Equilibration buffer: 5 mM sodium acetate buffer, pH 5.5, containing $0.1M$ NaCl, 1 mM $CaCl_2$, and 0.02% sodium azide.
2. $0.01M$ Methyl-α-mannoside in equilibration buffer.
3. $0.3M$ Methyl-α-mannoside in equilibration buffer.
4. 5 mM Sodium acetate buffer, pH 5.5, containing $1M$ NaCl.

3. Methods

3.1. Immobilization of Con A on 1,1' Carbonyldiimidazole Activated Agarose

1. Place the gel in a sintered glass funnel. Apply gentle suction (a water pump is adequate).
2. Wash the gel free of acetone with 10 vol of ice-cold distilled water (*see* Note 2). Do NOT dry cake.
3. Dissolve Con A at 20 mg/mL in $0.1M$ Na_2CO_3 buffer, pH 9.5, containing $0.5M$ methyl-α-D-mannoside, the protective sugar (*see* Note 3). Keep at 4°C.

4. Gently break up the gel cake, then add the cake directly to the coupling buffer containing the ligand, 1 vol of gel to 1 vol of buffer.
5. Incubate for 48 h at 4°C with gentle mixing.
6. Transfer the gel to a column, then wash the gel with 0.1M Tris HCl, pH 9.5, to block any remaining unreacted sites.
7. Wash the gel with 3 vol of 0.1M sodium acetate buffer, pH 4.0, then with 3 vol of 0.1M NaHCO$_3$, pH 8.0.
8. Repeat Step 7.
9. Wash the gel with PBS containing 0.02% sodium azide, then store the gel at 4°C in buffer containing 0.02% sodium azide.

3.2. Immobilization of Con A on Affi-Gel 15

1. Place the gel in a sintered glass funnel. Apply gentle suction to remove the isopropyl alcohol. Do NOT let the gel dry out. (*See* Notes 2,3, and 8.)
2. Wash the gel with 3 vol of cold distilled water (*see* Note 2).
3. Drain off excess water.
4. Dissolve Con A at 10 mg/mL in PBS containing 0.3M methyl-α-D-mannoside (*see* Note 3).
5. Transfer the gel to the coupling buffer, 1 vol gel:1 vol coupling buffer (*see* Note 6).
6. Mix gently for 2 h at 4°C.
7. Transfer the gel to a column and wash with 5 vol of PBS.
8. Wash with equilibration buffer, then store the column at 4°C in the presence of 0.2% sodium azide.

3.3. Purification of Glycoproteins on Immobilized Con A

1. Equilibrate the column with 5 mM sodium acetate buffer containing 0.1M NaCl, 1 mM MnCl$_2$, 1 mM CaCl$_2$, 1 mM MgCl$_2$, 0.02% sodium azide (*see* Note 10), at a flow rate of 0.1 mL/min.
2. Either dissolve solid glycoprotein in a minimum volume of equilibration buffer, or if the glycoprotein is in solution, dilute 1:1 with equilibration buffer. Remove any particulate matter.
3. Apply to the column and wash with equilibration buffer to remove nonbound components. Monitor the effluent at 280 nm until the absorbance is zero (*see* Note 11).
4. Elute with 0.01M methyl-α-D-mannoside in acetate buffer (5 vol). Collect 1 mL fractions.
5. Elute with 0.3M methyl-α-D-mannoside in acetate buffer. Collect 1 mL fractions.

6. Wash the column with acetate buffer containing $1M$ NaCl, then regenerate the column by extensive washing with equilibration buffer.
7. Pooled bound fractions can either be concentrated by ultrafiltration or dialyzed against a buffer to remove the hapten sugar.

4. Notes

1. When using CNBr-activated matrices, charged groups may be introduced during the coupling step. This may lead to nonspecific adsorption from ion-exchange effects.
2. Adequate washing of the gel prior to coupling is required. Many proteins are sensitive to the acetone or alcohols used to preserve the activated gels.
3. Coupling methods are similar for different lectins; however, the lectin binding site must be protected by the presence of the hapten monosaccharide. Table 3 lists the common lectins and the saccharides used to protect the active site during immobilization.
4. Con A requires Mn^{2+} and Ca^{2+} to preserve the activity of the binding site.
5. Affi-Gel 10 and 15 are N-hydroxysuccinimide esters of crosslinked agarose. Both are supplied as a slurry in isopropyl alcohol. The gels can be stored without loss of activity for up to 1 yr at $-20°C$.
6. Buffers containing amino groups should not be used for coupling. Suitable buffers include HEPES, PBS, and bicarbonate.
7. Affi-Gel 10 couples neutral or basic proteins optimally. The coupling buffers used should be at a pH at, or below, the pI of the ligand. To couple acidic proteins coupling buffers should contain 80 mM CaCl$_2$. Affi-Gel 15 couples acidic proteins optimally. The pH of coupling buffers should be above or near the pI of the ligand. To couple basic proteins, include $0.3M$ NaCl in the coupling buffer.
8. Steps 1–4 should be completed in 20 min to ensure optimum coupling.
9. The size of column depends on the amount of sample to be purified. Preliminary experiments can easily be carried out using Pasteur pipets, with supports of washed glass wool. Alternatively, the barrels of disposable plastic syringes can be used. Long, thin columns will give the best resolution of several components. However, where the requirement is to remove only contaminants from the protein, a short, fat column will elute the proteins in a smaller volume.
10. Equilibration buffers used will depend on the stability of the glycoconjugate being purified. Some glycoconjugates are sensitive to acetate *(21)*. Phosphate buffers will precipitate out any Ca^{2+} added as calcium phosphate. The optimal pH for the equilibration buffer should be established to ensure maximum adsorption to the affinity

Table 3
Common Lectins and Their Protective Saccharides

Lectin	Hapten saccharide	Ref.
Con A	Methyl-α-D-mannoside	27
Lens culinaris	Methyl-α-D-mannoside	
Tritium vulgaris	Chitin oligosaccharides	28
Ricinus communis	Methyl-β-galactoside	21
Jacalin	D-Galactose	29

column. Binding may be enhanced in certain cases by the inclusion of NaCl in the buffer *(30)*.

11. Sparingly reactive glycoconjugates may be retarded rather than bound by the column *(31)*. Washing should continue for at least 5 bed vol. The nonbound fractions can be retained and reapplied to the column on a subsequent occasion, when it is suspected that the load applied to the column is saturating.

12. Elution can be achieved either as a "step," i.e., a change to a buffer containing the hapten sugar. Alternatively, one can use a gradient of the hapten sugar. Different inhibiting sugars can be used for elution, e.g., D-glucose, α-D-mannose, methyl-α-D-glucoside.

13. Occasionally, glycoconjugates may bind very tightly to the matrix, and not elute even with $0.5M$ hapten sugar. Sodium borate buffers offer an alternative: these may elute components with high affinity for the lectin without denaturing it. The eluant temperature can be raised above 4°C, depending on the protein stability. Detergents, e.g., 0–5% SDS, together with $6M$ urea will often strip components from the lectin, but may irreversibly denature the lectin. Occasionally, it may be expedient to use stringent denaturing conditions to desorb bound fractions with a high yield. Glycoproteins have been eluted from Con A Sepharose by heating in media containing 5% SDS and $8M$ urea *(32)*. NH_4OH (1M) has also been used as an eluant *(14)* with high yield, but irreversibly denaturing the lectin.

References

1. Kennedy, J. F. and Rosevear, A. (1973) An assessment of the fractionation of carbohydrates on Concanavalin A Sepharose by affinity chromatography. *J. Chem. Soc. Perkin Trans.* **19,** 2041–2046.
2. Fidler, M. B., Ben-Yoseph, Y., and Nadler, H. L. (1979) Binding of human liver hydrolases by immobilised lectins. *Biochem. J.* **177,** 175.
3. Kawai, Y. and Spiro, R. G. (1980) Isolation of major glycoprotein from fat cell plasma membranes. *Arch. Biochem. Biophys.* **199,** 84–91.

4. Sharma, S. K. and Mahendroo, P. P. (1980) Affinity chromatography of cells and cell membranes. *J.Chromatography* **184**, 471–499.
5. Cummings, R. D. and Kornfeld, S. (1982) Fractionation of asparagine linked oligosaccharides by serial lectin affinity chromatography. *J. Biol.Chem.* **257**, 11235–11240.
6. Boyle, F. A. and Peters, T. J. (1988) Characterisation of galactosyltransferase isoforms by ion exchange and lectin affinity chromatography. *Clin. Chim. Acta* **178**, 289–296.
7. Pos, O., Drechoe, A., Durand, G., Bierhuizen, M. F. A., Van der Stelt, M. E., and Van Dik, W. (1989) Con A affinity of rat α-1-acid glycoprotein (rAGP); Changes during inflmmation, dexamethasone or phenobarbital immuno-electrophoresis(CAIE) are not only a reflection of biantennary glycan content. *Clin. Chim. Acta* **184**, 121–132.
8. Debray, H., Pierce-Cretel, A., Spik, G., and Montreuil, J. (1983) Affinity of ten insolubilised lectins towards various glycopeptides with the N-glycosylamine linkage and related oligosaccharides, in *Lectins in Biology, Biochemistry, Clinical Biochemistry*, (Bog-Hanson, T. C. and Spengler, G. A., eds.) De Gruyter, Berlin, vol. 3, p. 335.
9. Debray, H., Descout, D., Strecker, G., Spik, G., and Montreuil, J. (1981) Specificity of twelve lectins towards oligosaccharides and glycopeptides related to *N*-glycosylglycoproteins. *Eur. J. Biochem.***117**, 41–55.
10. Banerjee, D. K. and Basu, D. (1975) Purification of normal urinary *N*-acetyl-β-hexosaminidase by affinity chromatography. *Biochem. J.* **145**, 113–118.
11. Esmann, M. (1980) Concanavalin A Sepharose purification of soluble Na,K-ATPase from rectal glands of the spiny dogfish, *Anal. Biochem.* **108**, 83–85.
12. Weinstein, Y., Givol, D., and Strausbauch, D. (1972) The fractionation of immunoglobulins with insolubilised concanavalin A. *J. Immunol.* **109**, 1402–1404.
13. Yamaguchi, N., Kawai, K., and Ashihara, T. (1986) Discrimination of gamma-glutamyltranspeptidase from normal and carcinomatous pancreas. *Clin. Chim.Acta* **154**, 133–140.
14. Matsuura, S. and Chen, H. C. (1980) Simple and effective solvent system for elution of gonadotropins from concanavalin A affinity chromatography. *Anal. Biochem.* **106**, 402–410.
15. Kinzel, V., Kübler, D., Richards, J., and Stöhr, M. (1976) *Lens Culinaris* lectin immobilised on sepharose: Binding and sugar specific release of intact tissue culture cells. *Science* **192**, 487–489.
16. Kvist, S., Sandberg-Trägardh, L., Östberg, L. (1977) Isolation and partial characterisation of papain solubilised murine H-2 antigens. *Biochemistry* **16**, 4415–4420.
17. Dawson, J. R., Silver, J., Shepherd, L. B., and Amos, B. D. (1974) The purification of detergent solubilised HL-A antigen by affinity chromatography with the haemagglutin from *Lens culinaris. J. Immunol.* **112**, 1190–1193.
18. Hayman, M. J. and Crumpton, M. J. (1973) Isolation of glycoproteins from pig lymphocyte plasma membrane using *Lens culinaris* phaetohemagglutin. *Biochem. Biophys. Res. Commun.* **47**, 923–930.

19. Gougos, A. and Letarte, M. (1988) Biochemical characterisation of the 44g4 antigen from the HOON preB leukemic cell line. *J. Immunol.* **141(6)**, 1934–1940.

20. Jackson, S. F. and Tjian, R. (1989) Purification and analysis of RNA polymerase 11 transcription factors by using wheat germ agglutinin affinity chromatography. P. N. A. S. **86(6)**, 1781–1785.

21. Dulaney, J. T. (1979) Factors affecting the binding of glycoproteins to concanavalin A and *Ricinus communis* agglutin and the dissociation of their complexes. *Anal. Biochem.* **99**, 254.

22. Gregory, R. L., Rundergren, J., and Arnold, R. R. (1987) Separation of human IgA1 and IgA2 using jacalin-agarose chromatography. *J. Immunol. Methods* **99**, 101–106.

23. Pilatte, Y., Hammer, C. H., and Frank, M. M. (1983) A new simplified procedure for Cl inhibitor purification. *J. Immunol. Methods* **120**, 37–43.

24. Blake, D. A. and Goldstein, I. J. (1982) Resolution of carbohydrates by lectin affinity chromatography. *Methods Enzymol.* **83**, 127–132.

25. Matson, R. S. and Siebert, C. J. (1988) Evaluation of a new N-hydroxysuccinimide activated support for fast flow immunoaffinity chromatography. *Prep. Chromatogr.* **1**, 67–91.

26. Lotan, R. and Nicolson, G. L. (1979) Purification of cell membrane glycoproteins by lectin affinity chromatography. *Biochem et Biophys Acta* **559**, 329–376.

27. Cook, J. H. (1984) Turnover and orientation of the major neural retina cell surface protein protected from tryptic cleavage. *Biochemistry* **23**, 899–904.

28. Mintz, G. and Glaser, L. (1979) Glycoprotein purification on a high capacity wheat germ lectin affinity column. *Anal. Biochem.* **97**, 423–427.

29. Roque-Barreira, M. C. and Campos-Neto, A. (1985) Jacalin: An IgG-binding lectin. *J. Immunol.* **134**, 1740.

30. Lotan, R., Beattie, G., Hubbell, W., and Nicolson, G. L. (1977) Activities of lectins and their immobilised derivatives in detergent solutions. Implications on the use of lectin affinity chromatography for the purification of membrane glycoproteins. *Biochemistry* **16(9)**, 1787.

31. Narasimhan, S., Wilson, J. R., Martin, E., and Schacter, H. (1979) A structural basis for four distinct elution profiles on concanavalin A Sepharose affinity chromatography of glycopeptides. *Can. J. Biochem.* **57**, 83–97.

32. Poliquin, L. and Shore, G. C. (1980) A method for efficient and selective recovery of membrane glycoproteins from concanavalin A Sepharose using media containing sodium dodecyl sulphate and urea. *Anal. Biochem.* **109**, 460–465.

CHAPTER 7

Dye–Ligand Affinity Chromatography

Steven J. Burton

1. Introduction

Dye–ligand affinity chromatography is performed with an adsorbent consisting of a solid support matrix to which a dye has been covalently bonded. The protein mixture is passed through a packed bed of adsorbent that selectively binds proteins that are able to interact with the immobilized dye. Nonbound proteins are washed from the adsorbent bed and bound protein(s) are subsequently eluted by adjusting the composition of the eluant.

A class of textile dyes, known as reactive dyes, are commonly used for protein purification purposes since they bind a wide variety of proteins in a selective and reversible manner, and are easily immobilized to polysaccharide-based support matrices, such as beaded agarose, dextran, and cellulose. From a chromatographic viewpoint, synthetic affinity ligands, such as reactive dyes, are preferable to biological ligands (e.g., antibodies) since the former are relatively inexpensive and are more resistant to chemical and biological degradation. The use of immobilized reactive dyes for protein purification has been the subject of intense research effort, the results of which have been summarized in numerous review articles *(1–4)*.

Unlike the relatively specific affinity adsorbents described in Chapter 9, immobilized dye adsorbents are able to bind a wide range of seemingly unrelated proteins with varying degrees of affinity. In some instances, protein binding is highly selective, enabling very high degrees of purification in a single step *(5,6)*. More usually, a cross-section of proteins are bound from a mixture, though relatively high

From: *Methods in Molecular Biology, Vol. 11: Practical Protein Chromatography*
Edited by: A. Kenney and S. Fowell Copyright © 1992 The Humana Press Inc., Totowa, NJ

degrees of purification can still be achieved by careful adjustment of the conditions used for binding and elution. Dye–ligand affinity chromatography generally gives higher degrees of purification compared to low-specificity techniques, such as ion exchange (Chapter 16), hydrophobic interaction (Chapter 5), and gel permeation (Chapter 14) chromatography.

Purification can be achieved either in a positive mode where the target protein is bound to the exclusion of contaminants, or in a negative mode, where contaminants are bound and the target protein passes through the adsorbent bed unretarded. Ultimately it is the structure of the dye that determines which proteins are bound. Thus, by screening a suitably diverse range of dye adsorbents for protein binding, purification strategies can be developed even though relatively little is known about the structure and properties of the protein in question (7,8). Generally, some form of adsorbent screening exercise is performed as the first step in the development of a purification protocol.

The reactive dyes of interest to biochemists usually consist of a chromophore (either azo, anthraquinone, or phthalocyanine) linked to a reactive group (often a mono or dichlorotriazine ring). Sulfonic acid groups are commonly added to confer solubility in aqueous solution. Most dyes are produced for the textile industry (ICI Procion™ range and Ciba-Geigy Cibacron™ range), though a new range of adsorbents with ligands developed specifically for protein purification work are now available (Affinity Chromatography Limited MIMETIC™ range). The structures of two commonly encountered reactive triazine dyes are shown in Fig. 1.

The overall size, shape, and distribution of charged and hydrophobic groups enables reactive dyes to interact with the ligand binding sites of proteins. Protein binding is often dominated by electrostatic interactions; however, dye–ligand affinity chromatography must not be compared to simple ion exchange chromatography (Chapter 16) since binding is frequently possible at pHs that are considerably greater than the pI of the target protein (9). In addition, bound proteins are commonly eluted with a soluble competing ligand, which suggests that the dyes in question normally interact with proteins at discrete sites rather than in an indiscriminate fashion.

2. Materials

All buffers should be freshly made with analytical grade buffer salts and, ideally, passed through a 0.2 μm pore-sized filter prior to use.

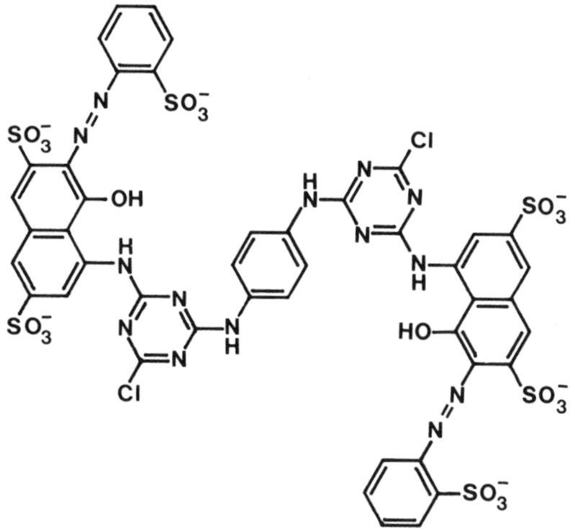

Fig. 1. Triazine dye structures. **(A)** C.I. Reactive Blue 2 (Procion Blue H-B/Cibacron Blue 3G-A); **(B)** C.I. Reactive Red 120 (Procion Blue HE-3B).

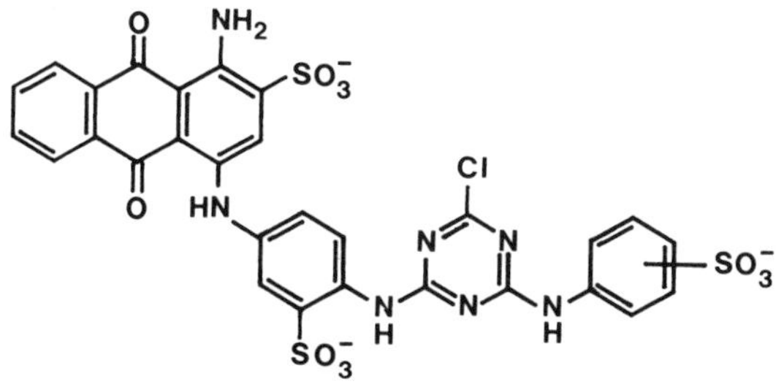

1. Dye–ligand affinity columns: A selection of immobilized dye adsorbents with diverse ligand structures are required for initial screening trials. Preswollen ready-to-use adsorbents that have undergone extensive washing during manufacture are preferable, otherwise dye leakage problems may be encountered. Dye–ligand adsorbents are generally supplied as slurries in bacteriostat-containing solution (e.g., 25% (v/v)ethanol; 0.02% (v/v) sodium azide) and should be stored at 4°C.

   Small chromatographic columns (1–2 mL bed vol) are sufficient for adsorbent screening, although larger columns are generally required for preparative work; the column size being dictated by the amount of adsorbent required to process a given volume of protein solution. Adsorbent screening kits with prepacked columns are available commercially.

2. Equilibration buffer: 25 mM Sodium phosphate buffer, pH 6.0 (3.9 g NaH$_2$PO$_4$·2H$_2$O dissolved in 1 L of water and adjusted to pH 6.0 with 1M NaOH).
3. Elution buffer: 50 mM Sodium phosphate buffer, pH 8.0, containing 1M potassium chloride (3.55 g Na$_2$HPO$_4$ and 37.25 g KCl dissolved in 500 mL of water and adjusted to pH 8.0 with 1M HCl).
4. Column regenerating solution: 1M Sodium hydroxide solution (40 g NaOH dissolved in 1 L of water).

3. Methods

3.1. Dye Screening

1. If prepacked columns are unavailable, slurry-pack small columns with adsorbent to give a bed vol of 1–2 mL. Ideally, degass the adsorbent suspension prior to column packing, especially if stored at 4°C immediately before use (otherwise air bubbles may form in the adsorbent bed and disrupt liquid flow through the column). A porous disk (frit) placed on the top of the adsorbent bed is useful for preventing disturbance of the column packing during chromatography. Flush the packed columns with approx 10 vol of equilibration buffer in preparation for protein binding.
2. The protein sample applied to the columns must contain soluble protein and be of an equivalent ionic strength and pH to the equilibration buffer. This is generally achieved by dialyzing the protein extract against 50 vol of equilibration buffer. Alternatively, buffers may be exchanged using a desalting column. A number of additives may be incorporated into the equilibration buffer to ensure protein stabilization (*see* Note 4). The sample solution must be filtered (0.2 μm pore-sized filter) or centrifuged to remove particulate material prior to chromatography (excessive particle contamination will lead to rapid adsorbent fouling, recognizable as a progressive reduction in flow rate and increase in column back pressure). The level of contamination and the total protein concentration of the applied sample may vary enormously. Ideally, 1–5 mg of target protein should be applied to each column in a vol of 1–5 mL. Thus, assuming the target protein constitutes 5% of the total protein, a minimum column loading of 20 mg total protein/mL of adsorbent is required.
3. Screen the adsorbent columns for protein binding simultaneously. Separations can be performed at room temperature, though it is advisable to carry out all chromatographic procedures at 4°C if the target protein is particularly labile. Run the sample solutions through

the equilibrated columns at a linear flow rate of 10–30 cm/h. Most macroporous dye–ligand adsorbents allow such flow rates under gravity, otherwise use a multichannel peristaltic pump for buffer delivery. Flush nonbound proteins from the column with 5 bed vol of equilibration buffer at the same flow rate. Eluted protein is collected as a single fraction for each column; consequently, automated fraction collection devices are not required at this stage. The majority of bound proteins are desorbed by flushing the columns with 5 mL of elution buffer. Again, collect the eluted protein as one fraction.

4. Assay collected fractions for both total protein (the Bradford protein assay is recommended—*see* vol. 4 and ref. *10*) and the desired protein. This latter assay will vary depending on the nature of the protein in question; enzyme assay, immunoassay, or electrophoresis are suitable for quantification purposes.

5. Rank the dye–ligand adsorbents according to their ability to bind the desired protein. Adsorbents that bind the target protein but allow contaminants to pass through should be selected for development of a positive binding purification step. Conversely, adsorbents that bind contaminants but not the desired protein are selected for a negative binding step. Note the effectiveness of the eluting buffer in removing bound proteins. Calculation of the total protein recovery for each column shows how much (if any) protein remains bound to the column after the desorption step. After use, flush all columns with 10 bed vol of regenerating solution, equilibrate with preservative solution (follow manufacturer's recommendations), and store at 4°C. Dye–ligand adsorbents are reusable many times.

3.2. Optimization of a Dye–Ligand Purification Step

1. An optimized purification protocol can be developed relatively simply once suitable dye–ligand adsorbents have been identified. In general, small scale purification trials (using 1–2 mL columns) are performed prior to scaling up the separation. Several optimization experiments can be performed simultaneously by the use of multiple columns. Use a peristaltic pump for buffer delivery during optimization to ensure reproducible results on scale-up. In-line detector(s) (UV, pH, conductivity) chart recorder and fraction collector may also be included at this stage.

For a positive binding step, the pH and ionic strength of the equilibration buffer are adjusted to maximize protein binding. Ideally, few contaminating proteins should bind, otherwise both the capacity of the adsorbent for the desired protein, and the potential degree of purification will be reduced. The reverse applies to a negative binding step, where contaminant binding is maximized. Generally speaking, increasing either the ionic strength or pH of the equilibration buffer serves to reduce protein binding and vice versa. Certain di- and trivalent metal ions may also be added to increase the affinities of proteins for immobilized dyes (3) (see Note 4).

2. Optimal column loading is determined by frontal analysis, where the sample solution is pumped continuously onto the column until the desired protein is detectable in the eluate. The optimal loading is equivalent to the volume of sample solution that passes through the adsorbent bed up to the point where 5–10% of the initial concentration of the desired protein is detected in the column eluate. Breakthrough concentrations of 5–10% give rise to loading losses in the region of 2–5%. In the case of a negative binding step, the optimal load volume is determined by the breakthrough of contaminants in the column eluate. The desired level of purification dictates the maximum breakthrough concentration of contaminants.

3. The composition, volume, and flow rate of the buffer used to wash the adsorbent prior to protein elution may be varied to effect the removal of contaminating proteins that may have bound in addition to the desired protein (Fig. 2). Elution of contaminants is performed in exactly the same manner as elution of the target protein. The buffer flow rate of the washing step can often be increased to speed up process times.

4. Elution of bound proteins is performed in a selective or nonselective manner. Selective elution generally gives higher degrees of purification as compared to nonselective techniques, such as changes of pH, ionic strength, or polarity. If the elution buffer used for screening trials effected partial desorption of the target protein, increase either the pH, salt concentration, or both to improve the recovery of bound protein. Ideally, the desired protein is completely eluted with a minimum volume of buffer (normally 2–3 column vol). However, reduce the pH and/or ionic strength of the elution buffer if complete desorption was observed during screening trials. Excessive salt concentration or pH enhances the desorption of the majority of bound proteins, consequently, higher degrees of purification are achieved using the minimum ionic strength and pH necessary for sharp elution of the target protein. In instances where increases of ionic strength fail to promote elution, reduce the polarity of the elution buffer by adding ethylene glycol or glycerol at concentrations of about 10–50% (v/v).

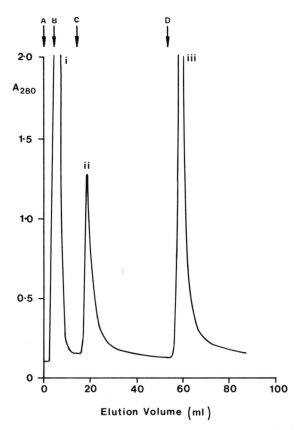

Fig. 2. Purification of human serum albumin from whole plasma by affinity chromatography on MIMETIC Blue 1 A6XL. Column: 1 cm dia. × 5 cm (4 mL bed vol); flow rate 0.33 mL/min. Applied solutions: (**A**) pooled buffered plasma, pH 6.0 (65 mg protein); (**B**) 25 mM MOPS buffer, pH 6.0; (**C**) 50 mM sodium phosphate buffer, pH 8.0; (**D**) 50 mM sodium phosphate buffer, pH 8.0, containing 2.0M NaCl. Eluted peaks: (**i**) nonbound proteins; (**ii**) eluted contaminants; (**iii**) purified human serum albumin (>96% pure).

Selective elution is often possible using a soluble ligand that competes with the dye for the same ligand-binding site on the protein. Any competing ligand may be used providing that it has an affinity greater than the dye, otherwise excessive concentrations of competing ligand will be required for elution. Oxidoreductases and phosphotransferases are commonly eluted by the addition of nucleotide cofactors, such as $NAD(P)^+$ and ATP, respectively. In these instances, relatively low concentrations of competing ligand (0.1–10 mM) are required for elution. Selection of a competing ligand requires some knowledge of the ligand-binding properties of the protein in question. Substrates, products, cofactors, inhibitors, or allosteric effectors are all potential candidates. Perform trial elutions to identify which ligand is competitive with dye binding as this is difficult to predict in advance. Selective elution may be used in combination with increases of pH and ionic strength to reduce the concentration (and cost) of the competing ligand required for elution.

5. Dye–ligand adsorbents are conveniently regenerated with 1M NaOH, although alternative chemical cleaning agents can also be used. Chaotropic solutions of urea or guanidine hydrochloride may be employed at 6–8M concentration to remove residual protein. Chaotropes can also be used to elute bound proteins as a last resort; sodium thiocyanate (2–4M) is often very effective in this role. Hypochlorite solutions must not be used for regenerating dye–ligand adsorbents.

6. Scaling up an optimized dye–ligand purification step is relatively straightforward, provided that certain precautions are observed. The column bed volume is increased to accommodate larger volumes of protein mixture, with the loading per mL of adsorbent (determined by frontal analysis) remaining constant. Larger bed volumes are achieved by increasing the column diameter rather than by adjusting column length. Increases in bed length are often inevitable, but keep these to a minimum whenever possible. Increase buffer volumes in direct proportion to the column volume while keeping the linear flow rate for each step constant (the overall flow rate increases in proportion to the cross-sectional area of the column). The column temperature and the composition of all buffers remain constant.

4. Notes

1. Dye–ligand affinity adsorbents can be made in the laboratory according to established methods *(11)*. However, suitable reactive dyes are not widely available and home-made adsorbents tend to give irreproducible results from batch to batch unless a great deal of care and attention is paid to their syntheses. Batch inconsistencies are partly attributable to the quality of commercial textile dyes, which are often highly heterogeneous *(12)*. In general, it is better to use defined ready-made packings.

2. Early dye–ligand adsorbents often shed dye during chromatography, particularly during protein elution. There are several causes of dye

leakage, including loss of noncovalently bound dye, degradation of the ligand–matrix bond, or degradation of the support matrix. Although reactive dyes are mostly harmless, their presence in purified protein preparations is undesirable. Recently introduced adsorbents (MIMETIC™ Ligand adsorbents) that incorporate advanced bonding chemistry and washing procedures have largely eliminated this problem *(13)*.

3. It is important that crude protein, not purified protein, is used in screening studies and for the development of an optimized dye–ligand purification step. The binding and elution characteristics of contaminating proteins cannot be predicted in advance, therefore they must be present during preliminary purification trials. Otherwise what appears to be an optimized purification protocol will often fail in practice owing to low column capacities and coelution of bound proteins.

4. As with all forms of preparative adsorption chromatography, wide, squat columns are preferable to long, thin columns. The use of columns with small length to diameter ratios (>1.0) enables the use of higher bulk flow rates and lessens bed compression problems. However, long thin columns can be useful in instances where the target protein has a low affinity for the immobilized dye. If the immobilized ligand only weakly retards the protein, it will soon appear in the effluent following column loading. Provision of a longer pathlength column in these circumstances affords a greater adsorbent–protein interaction time which, in turn, improves the resolution of nonbound contaminants from the weakly bound target protein.

5. The composition of buffers used for dye–ligand affinity chromatography is highly variable; the examples given are only intended as a guide. Most buffer ions may be used, though phosphate salts are particularly favored for reducing nonselective protein–dye interactions. Similarly, sodium or potassium chloride may be added to increase the ionic strength of the buffer. When low conductivity buffers are required, MOPS, MES, HEPES, or Tris are often used. Metal ions, such as Mg^{2+}, Ca^{2+}, Zn^{2+}, Mn^{2+}, Cu^{2+}, Co^{2+}, Fe^{3+}, and Al^{3+}, may be added to promote protein binding *(3)*, normally at concentrations in the region of 0.1–10 mM. The buffer system must be considered when multivalent metal ions are added, otherwise precipitation problems may arise (e.g., calcium phosphate formation). Desorption of proteins whose binding is enhanced by metal ions is often achieved by adding EDTA to the elution buffer. In some instances, metal ions may also be added to stabilize proteins.

Similarly, thiols (e.g., glutathione), EDTA, and protease inhibitors (e.g., phenylmethylsulfonylfluoride (PMSF)) are often included in buffers to prevent protein degradation and preserve enzyme activity. Most dye–ligand adsorbents are compatible with these additives, which are generally added to final concentrations of 1 mM.

6. The pH of some buffer systems (especially Tris buffers) can vary appreciably with changes of temperature. Buffer solutions should be equilibrated at the desired temperature (usually 4°C or 25°C) before final pH adjustment.

7. The affinity of proteins for immobilized dyes is frequently influenced by temperature. Changes of temperature have a variable effect on protein binding and may increase or decrease the dissociation constant of a protein for an immobilized dye. To avoid inconsistencies, screening and optimization experiments should be performed at the same temperature as the final purification procedure. If ambient temperatures are likely to fluctuate, jacketed columns should be used.

8. Buffer flow rates must be carefully controlled, especially during protein adsorption and desorption. High flow rates reduce protein binding capacities during adsorption and cause peak-tailing during desorption. Linear flow rates of 10–30 cm/h are suitable for binding and elution, though higher flow rates can often be used to speed up washing and column regeneration steps. Maximum flow rates are determined by the mechanical properties of the support matrix. Protein binding may be promoted by halting buffer flow for 10 min prior to commencement of the washing stage *(11)*.

9. Throughout this chapter it is assumed that the protein requiring purification is present in a cell-free extract contaminated with several other proteins. However, dye–ligand affinity chromatography may be used to remove nonproteinaceous contaminants (e.g., DNA or pigments) from otherwise pure protein preparations. In this case, determination of purity will be based on criteria different from those described in Section 3.1.

10. As an alternative to specific protein detection assays, SDS-polyacrylamide gel electrophoresis (SDS-PAGE) may be used to analyze fractions from adsorbent screening experiments. Miniaturized electrophoresis systems allow the simultaneous analysis of several samples in just a few hours. Electrophoresis is advantageous in that specific assays are not required for sample analysis, however, a purified protein standard is required for identification purposes.

11. An initial screen of adsorbent-binding properties may not give a clear result, especially when all of the columns either bound little protein

or most of the protein present in the applied sample. In these instances, the composition of the equilibration buffer should be changed and the experiment repeated. If little protein binding was observed in the initial screen, the ionic strength and the pH of the buffer should be lowered (e.g., to 10 mM sodium acetate buffer, pH 5.0). Metal ions may also be added to increase protein binding avidity (e.g., 10 mM magnesium chloride). If the majority of proteins present in the applied sample (including the desired protein!) were bound in the initial screen, the pH and ionic strength of the buffer should be increased (e.g., to 50 mM sodium phosphate buffer, pH 8.0).

12. A "frontal" screening procedure may be used instead of the "zonal" method described here. Dye screening by frontal analysis (where the column is continuously loaded with sample until the protein concentration of the eluate is equivalent to the applied protein concentration) gives additional information on the relative affinities of different proteins in the mixture *(8)*. A frontal approach to screening is particularly useful in instances where the target protein is present in relatively low concentration compared to the total protein concentration of a multicomponent mixture. In such cases, it is advantageous if the target protein binds to the adsorbent more tightly than contaminating proteins. Otherwise, target protein that is bound in the initial phase of column loading may be displaced later on by contaminating proteins with a greater affinity for the immobilized dye.

13. Purification by a negative binding step is particularly advantageous when large volumes of labile protein solution are to be processed. Column residence times are very much reduced for a negative binding step, since the adsorbent bed merely acts as a contaminant removing filter. Consequently, negative binding steps often yield higher recoveries of active protein as compared to conventional positive binding protocols. Highly effective two-step purification procedures can be developed by combining a negative binding step and a positive binding step in tandem. Such an arrangement is particularly attractive if the conditions allow direct application of the effluent from the negative binding column onto the positive binding column.

14. Whereas most immobilized dyes are relatively inert, the chemical resistance of the matrix and the dye–matrix bond is variable. The chemical compatibility of commercially available adsorbents should be checked before regenerating dye–ligand columns with sodium hydroxide.

15. Elution of bound proteins by reducing the polarity of eluant generally gives poor results (broad peak profiles) as compared to salt/pH

elution. Consequently, elution with low polarity solvents should only be contemplated if satisfactory results cannot be obtained by other means.

16. As an alternative to stepwise elution of bound proteins, elution gradients may be used to resolve coeluting proteins or sharpen trailing peak profiles.

17. A chaotrope (literally chaos-causing) is a compound that disrupts hydrogen-bond formation in aqueous solution. Proteins that are not stabilized by disulfide bonds are completely denatured in strongly chaotropic solutions. Furthermore, hydrophobic and hydrogen bonding interactions between proteins and ligands are also neutralized in the presence of concentrated chaotropes. Consequently, solutions of chaotropic agents, such as urea, guanidine hydrochloride, or sodium thiocyanate, are highly effective in stripping bound proteins from affinity columns. One advantage of using chaotropic solutions for column regeneration is that, residual protein contaminants may be identified by electrophoresis if so required. However, chaotropes are not very useful for elution purposes since some degree of protein denaturation (and complete or partial loss of function) inevitably occurs.

18. On scaling up a purification protocol involving immobilized affinity ligands, it is important that the column loading per mL of adsorbent and column residence time remain constant *(7)*. Protein-binding capacities are variable depending on the purity of the applied sample and the affinity of the protein for the immobilized dye. Capacities in excess of 40 mg per mL are common.

References

1. Lowe, C. R.(1984) Applications of reactive dyes in biotechnology, in *Topics in Enzyme and Fermentation Biotechnology* (Wiseman, A., ed.) Ellis Horwood, Chichester, UK, pp. 78–161.
2. Kopperschlager, G., Bohme, H.-J., and Hofmann, E. (1982) Cibacron Blue F3G-A and related dyes as ligands in affinity chromatography, in *Advances in Biochemical Engineering* (Fiechter, A., ed.), Springer-Verlag, Berlin-Heidelberg, pp.101-138
3. Clonis, Y. D., Atkinson, A., Bruton, C. J., and Lowe, C. R. (1987) *Reactive Dyes in Protein and Enzyme Technology.* Stockton, New York.
4. Lowe, C. R. and Pearson, J. C. (1984) Affinity chromatography on immobilised dyes. *Methods Enzymol.* **104,** 97–113.
5. Ghiggeri, G. M., Candiano, G., Delfino, G., and Queirolo, C. (1985) Highly selective one-step chromatography of serum and urinary albumin on immobilised Cibacron Blue F3G-A: Studies on normal and glycosylated albumin. *Clin. Chem. Acta* **145,** 205–211.

6. Smith, K., Sundram, T. K., Kernick, M., and Wilkinson, A. E. (1982) Purification of bacterial malate dehydrogenases by selective elution from a triazinyl dye column. *Biochim. Biophys. Acta* **708,** 17–25.

7. Scopes, R. K. (1986) Strategies for enzyme isolation using dye–ligand and related adsorbents. *J. Chromatogr.* **376,** 131–140.

8. Kroviarski, Y., Cochet, S., Vadon, C., Truskolaski, A., Boivin, P., and Bertrand, O. (1988) New strategies for the screening of a large number of immobilised dyes for the purification of enzymes. *J. Chromatogr.* **449,** 403–412.

9. Lascu, I., Porumb, H., Porumb, T., Abrudan, I., Tarmure, C., Petrescu, I., Presecan, E., Proionov, I., and Telia, M. (1984) Ion-exchange properties of Cibacron Blue 3G-A Sepharose (Blue Sepharose) and the interaction of proteins with Cibacron Blue 3G-A. *J. Chromatogr.* **283,** 199–210.

10. Bradford, M. M. (1976) A rapid and sensitive method for the quantitation of microgram quantities of protein using the principle of protein-dye binding. *Anal. Biochem.* **72,** 248–254.

11. Lowe, C. R., Hans, M., Spibey, N., and Drabble, W. T. (1980) The purification of inosine 5'-monophosphate dehydrogenase from *Escherichia coli* by affinity chromatography on immobilised procion dyes. *Anal. Biochem.* **104,** 23–28.

12. Burton, S. J., McLoughlin, S. B., Stead, C. V. and Lowe, C. R. (1988) Design and applications of biomimetic anthraquinone dyes: Synthesis and characterisation of terminal ring isomers of C.I. Reactive Blue 2. *J. Chromatogr.* **435,** 127–137.

13. Stewart, D. L., Purvis, D., and Lowe, C. R. (1990) Comparison of ligand leakage and human serum albumin binding capacity of nine different C.I. Reactive Blue 2 affinity adsorbents. *Anal. Biochem.,* in press.

CHAPTER 8

High-Performance Affinity Chromatography for Protein Separation and Purification

Yannis D. Clonis

1. Introduction

1.1. Affinity Chromatography

Affinity chromatography *(1,2)*, the most powerful of all protein-fractionation techniques, relies on the formation of reversible specific complexes between a ligand immobilized on an insoluble polymer support, termed affinity adsorbent, and the species to be isolated free in solution. The modern support materials (natural, synthetic, or inorganic), consist of macroporous hydrophilic beaded particles, usually bearing free hydroxyl groups available for ligand immobilization. When the support is made of noncompressible particles of small diameter (e.g., 5–20 µm) and narrow size distribution (e.g., 0.2–2 µm) the technique is termed high-performance affinity chromatography (HPAC) *(3–5)*.

The following questions would be interesting to deal with briefly:

1. Why do the particles need to be small? The solutes present in the mobile liquid phase establish an equilibrium with those in the liquid phase contained in the particles. This mass-transfer phenomenon is achieved by means of diffusion between the two phases. Obviously, the larger the contact area of the two phases, the larger the fraction of solute molecules passing from one phase to the other per unit of time; therefore, equilibrium is reached faster. Consequently, reduc-

From: *Methods in Molecular Biology, Vol. 11: Practical Protein Chromatography*
Edited by: A. Kenney and S. Fowell Copyright © 1992 The Humana Press Inc., Totowa, NJ

tion in the size of the particle leads to an increase in the surface area per unit of volume and shorter diffusion paths into the interior of pores. This, in turn, permits faster equilibrium, hence use of higher flow rates. Establishment of faster equilibrium would also improve the shape of the eluting peaks (make them sharper) because the solutes at the rear and the front of the eluting zones would no longer lag behind or advance further, respectively.

2. Why do the particle need to be noncompressible? The use of high flow rates, in combination with the large surface area (small particle radius), leads to a high pressure drop along the bed. Therefore, the particles need to be rigid, otherwise they will be deformed.

3. Why do the particles need to be beaded and with a narrow particle-size distribution? There is a type of diffusion, not to be confused with the type described above, termed *eddy diffusion,* associated entirely with the physical properties of the stationary phase and the quality of the packing within the column. When the packing consists of irregular, nonuniform particles and the bed is inhomogeneous, a path length for the sample may vary considerably in different parts of the column, leading to broad asymmetric zones. Minimization of eddy diffusion would result in narrow and sharp zones (peaks).

4. Does gradient elution influence zone broadening? Because the concentration of a gradient is higher at the rear, the rear of the eluting zone moves at a higher speed than the front. Consequently, the zone will be sharper with than without a gradient; in fact, the steeper the gradient, the sharper the zone. On the other hand, selectivity is reduced with increasing gradient slope; thus, the gradient slope selected would be a compromise.

It should be stressed that the length of the column must never exceed that needed for full desorption of all species. This is because a longer bed does not produce higher resolution and can lead to zone broadening as a result of diffusion.

1.2. Column-Support Materials Used in HPAC

1.2.1. Matrix Requirements

An ideal matrix suitable for HPAC should be:

1. Noncompressible, so that it can be used under high pressure and flow rate;
2. Made of spherical, rigid particles of a typical diameter between 5 and 30 μm, and with narrow size distribution;
3. Macroporous, so that the macromolecules can diffuse freely in the interior of the particles;

4. Hydrophilic and uncharged, in order to minimize nonspecific interactions;
5. Chemically stable, in order to undergo ligand immobilization chemistry and hygiene maintainance, as well as to resist biological degradation; and
6. Functionable, by means of surface free groups, usually hydroxyls, or other appropriately modified groups suitable for ligand immobilization.

Table 1 summarizes some commercially available *beaded* support materials suitable for HPAC.

1.2.2. Inorganic Matrices—Silica

Among inorganic matrices, such as silica, titania, alumina, and zirconia, silica has received most attention. The surface of unmodified silica consists of siloxane bridges and silanol groups; the latter occur both as free silanol functions and hydrogen-bridged ones at a concentration of about $4.8/nm^2$. A 5–15% ratio of adsorbed to absorbed water is expected for silica; physically bound water can be removed by heating at 110–150°C under reduced pressure, whereas chemically bonded water requires more than 200°C, which leads to the formation of siloxane bridges between silanol groups. Because of the employment of high temperatures in the silica manufacturing process, it may be useful, prior to other derivatizations, to treat silica with acid (e.g., 10 mM HCl) in order to open the siloxane bridges to free silanol groups. It is these free, negatively charged silanol functions that endow unmodified silica with significant nonspecific adsorption. Coating the silica's surface with organofunctional silanes circumvents this deleterious characteristic by providing a neutral hydrophilic surface suitable for ligand attachment. During derivatization with trimethoxy-organosilanes under aqueous conditions, new silanols are formed on the coated surface as a result of partial hydrolysis of the silane, thus leading to multilayer formation and narrowing of the pore size of the silica particles. Capping of the remaining free silanol functions after coating with a silane can be achieved by reacting with trimethyl-chlorosilane; however, this compromises the hydrophilicity of the silica's surface.

A major and practically unsolved problem with silica is its poor stability at pH 8.0 and above, although pretreatment with mixtures of zirconium salts improves its resistance in alkaline conditions.

The last main drawback of silica matrices (the relatively small pore size [up to 500Å]) is now solved with the introduction of wide-pore

Table 1
Beaded Support Materials Suitable for Ligand Coupling and Use in HPAC

Support	Surface group	Manufacturer
Silica, unmodified		
Hypersil WP 300	Silanol	Shandon, Runcorn, UK
LiChrospher®	Silanol	Merck, Darmstadt, FRG
Spherisorb®-WPS	Silanol	Phase Sep, Cewyd, UK
Silica, coated		
Aminopropyl silica	Amino	J. T. Baker, Phillipsburg, NJ
Diol silica	Diol	J. T. Baker
Hexamethylenediamine silica	Amino	J. T. Baker
Succinoylaminopropyl silica	Carboxyl	J. T. Baker
Silica, coated and activated		
Bakerbond™ diazofluoborate	RN_2^+	J. T. Baker
Bakerbond epoxypropyl	Epoxy	J. T. Baker
Bakerbond glycidoxypropyl	Epoxy	J. T. Baker
Bakerbond glutaraldehyde	Aldehyl	J. T. Baker
FMP-Activated	1-Methylpyridinium toluene-4-sulfonate	Bioprobe, Tustin, CA
HyPAC Carbonite N	4-Nitrophenyl carbonate	Barspec, Rehovot, Israel
SelectiSpher™-10	Tresyl	Pierce, Oud-Beijerland, Holland
Ultraffinity™-EP	Epoxy	Beckman, Berkeley, CA
Synthetic, unmodified		
Dynospheres® XP3507	Hydroxyl	Dyno Particles, Lillestrøm, Norway
Separon H100	Hydroxyl	LIW, Prague, Czechoslovakia
TSK® PW	Hydroxyl	Toyo Soda, Tokyo, Japan
Synthetic, activated		
Affi-Prep™ 10	Succinimide ester	BioRad, Richmond, CA
Eupergit® 30N C	Epoxy	Röhm Pharma, Darmstadt, FRG
Natural, crosslinked		
Superose® 6 and 12	Hydroxyl	Pharmacia, Uppsala, Sweden

silica, e.g., 500–2000 Å. However, unfortunately this has been achieved at the expense of surface area, a rather important characteristic for the adsorption-type packings.

Table 1 includes silica supports suitable for use in HPAC after appropriate derivatization. These supports have been classified into three commercially available categories: (1) unmodified silicas, which, prior to any use in HPAC, need coating; (2) coated silicas bearing different free functionable groups; and (3) silicas that have been coated and subsequently activated, and are therefore suitable for direct ligand coupling through their active moieties.

1.2.3. Synthetic Matrices

Rigid synthetic hydrophilic macroporous beaded supports represent a breakthrough in HPAC technology, as they overcome successfully most problems encountered with silica supports. Dynospheres® XP-3507 are true monodisperse (monosized) acrylate-copolymer particles. Separon H1000 beads are made by copolymerization of hydroxyalkylmethacrylate with alkene dimethacrylate, followed by aggregation of the heavily crosslinked microparticles. TSK®-PW is a vinyl-based hydroxyl-bearing support available in a wide range of pore sizes and particle diameters. The following two supports are available in preactivated form: (1) Eupergit® C 30N, which bears epoxy groups through which affinity ligands possessing nucleophilic groups can be coupled (This support is obtained by copolymerization of methacrylamide, methylen-bis-methacrylamide, glycidylmethacrylate, and/or allyl-glycidylether), and (2) Affi-Prep™ 10, a synthetic macroporous support activated with the *N*-hydroxy succinimide active ester method. Likewise, a wide range of affinity ligands bearing nucleophilic groups can be immobilized directly on Affi-Prep 10.

1.2.4. Natural Matrices

Rigid natural matrices should attract considerable technological interest because of their commercial potential, which can be easily appreciated (cf the advantages that agarose supports offer in classical affinity chromatography). Unfortunately, the rigid agarose supports available today for HPLC require further development if they also are to be used for HPAC. This is because they possess only few free-hydroxyl functions for ligand coupling, since a large proportion of them are involved in the crosslinking.

2. Materials

2.1. Coating and Activation

2.1.1. Preparation of Glycidoxypropyl Silica (Epoxy-Activated Silica) (3,5)

1. Macroporous silica (5 g, 10 μm, 550 m^2/g).
2. γ-Glycidoxypropyltrimethoxysilane solution (1% v/v) in 0.1 M sodium acetate buffer (pH 5.5). This solution should be prepared fresh just prior to use.
3. Acetone.
4. Sintered-glass funnel No. 4.

2.1.2. Preparation of Glycidoxypropyl Silica Under Dry Conditions (5,6)

1. Macroporous silica (10 g, 10 μm, 250 m^2/g).
2. Glycidoxypropyltrimethoxysilane.
3. Triethylamine.
4. Dry toluene (sodium or molecular sieve, 0.4 nm).
5. Toluene.
6. Acetone.
7. Drying tube (CaCl$_2$ or CaSO$_4$).
8. Sintered-glass funnel No. 4.

2.1.3. Preparation of Glycerylpropyl Silica (Diol Silica) (3)

1. Macroporous silica (5 g, 10 μm, 550 m^2/g).
2. γ-Glycidoxypropyltrimethoxysilane aqueous solution (1% v/v, pH 5.5) made fresh just prior to use.
3. HCl (10 mM).
4. Acetone.
5. Sintered-glass funnel No. 4.

2.1.4. Preparation of Aldehyde Silica (5,6)

1. Macroporous diol silica (5 g, 10 μm, 250 m^2/g).
2. Acetic acid/water solution (9/1, v/v).
3. Sodium periodate.
4. Methanol.
5. Diethyl ether.
6. Sintered-glass funnel No. 4.

2.1.5. Preparation of Tresyl Silica
(Sulfonyl Chloride Activated Silica) (5–7)

1. Macroporous diol silica (2.5 g, 5 μm, approx 500 m²/g).
2. Tresyl chloride.
3. Dry acetone (molecular sieve, 0.4 nm or commercial).
4. Acetone.
5. Dry pyridine (molecular sieve, 0.4 nm or commercial).
6. HCl solution (5 mM).
7. Sintered-glass funnel No. 4.

2.1.6. Preparation of Imidazol Carbamate–Silica
(Carbonyldiimidazole-Activated Silica) (5)

1. Macroporous diol silica (5 g, 10 μm, 50 m²/g).
2. Carbonyldiimidazole (CDI).
3. Dry acetonitrile (molecular sieve or commercial).
4. Dry acetone.
5. Drying tube.
6. Sintered-glass funnel No. 4.

2.1.7. Preparation of Chloroformate-Activated
Silica (8,9)

1. Macroporous silica with primary hydroxyl groups (1 g silica, 10 μm, 250–550 m²/g).
2. Dry acetone.
3. Chloroformate (*p*-nitrophenyl- or *N*-hydroxysuccinimide chloroformates).
4. Dry 4-dimethylaminopyridine.
5. Acetone.
6. Acetic acid/dioxane (5% acid).
7. Diethyl ether.
8. Sintered-glass funnel No. 4.

2.2. Immobilization

2.2.1. Immobilization of 3-Aminobenzene Boronic Acid
to Glycidoxypropyl Silica (10)

1. Glycidoxypropyl silica.
2. 3-Aminobenzene boronic acid.
3. Sodium hydroxide (3M).
4. Sodium chloride (0.5M).
5. Sodium bicarbonate (0.1M).

6. Hydrochloric acid (1 m*M*).
7. Acetone.
8. Acetone/water 1/1 (v/v).
9. Sintered-glass funnel No. 4.

2.2.2. Immobilization of N^6-(6-Aminohexyl)-AMP to Aldehyde Silica (3)

1. Aldehyde silica.
2. N^6-(6-Aminohexyl)-AMP.
3. Sodium bicarbonate (0.1*M*).
4. Sodium borohydride.
5. Acetone.
6. Acetone/water 1/1 (v/v).
7. Sintered-glass funnel.

2.2.3. Immobilization of Antihuman Serum Albumin (Anti-HSA) to Aldehyde Silica

1. Aldehyde silica.
2. Anti-HSA.
3. Sodium bicarbonate buffer, 0.1*M*, pH 7.9 (buffer A).
4. Buffer A containing 0.5*M* NaCl (buffer B).
5. Sodium borohydride.
6. Sintered-glass funnel.

2.2.4. Immobilization of N^6-(6-Aminohexyl)-AMP to Tresyl Silica (7)

1. Tresyl silica.
2. N^6-(6-Aminohexyl)-AMP.
3. Sodium bicarbonate (0.7*M*)
4. Tris-HCl buffer, 0.1*M*, pH 7.5.
5. Acetone.
6. Acetone/water 1/1 (v/v).
7. Sintered-glass funnel No. 4.

2.2.5. Immobilization of Trimethyl (p-Aminophenyl) Ammonium Chloride to ACA-Separon (TAPA) (11)

1. ACA-Separon (Separon substituted with a spacer molecule carrying a terminal free-carboxyl group. This product is available from LIW, Prague, Czechoslovakia).
2. Trimethyl (*p*-aminophenyl)ammonium chloride.

3. 1-Ethyl-3-(3-dimethylaminopropyl)carbodimide hydrochloride.
4. Sodium hydroxide/sodium chloride solution ($0.05M/1M$).
5. Acetone.
6. Acetone/water 1/1 (v/v).
7. Sintered-glass funnel No. 4.

2.2.6. Immobilization
of Procion Blue MX-R to TSK PW (5,12)

1. TSK PW-type, e.g., TSK G5000PW (All PW types contain surface free-hydroxyl groups. These materials are available from Toyo Soda, Tokyo, Japan).
2. Procion blue MX-R (A dichlorotriazine reactive dye, ICI, Manchester, UK).
3. Sodium carbonate solution (1% w/v).
4. Acetone/water 1/1 (v/v).
5. Acetone.
6. Sintered-glass funnel No. 4.

2.2.7. Immobilization to Hydrophobic Supports
by Adsorption of Surfactant-Affinity Ligands (13)

1. HPLC column of Davisil™ octadecyl-bonded silica (30–40 μm, 300 Å, 0.2 × 2 cm, 21.3 mg packing; available from Alltech Associates, Inc.).
2. Ligand: octaethylene glycol *n*-hexadecyl ether pyridinium.
3. Methanol.
4. Deionized water.
5. Tris-HCl buffer, $0.05M$, pH 8.0.

2.3. Applications

2.3.1. Resolution of Lactate Dehydrogenase (LDH)
Isoenzymes on N^6-(6- Aminohexyl)-AMP
Immobilized to Aldehyde-Silica

1. HPAC silica-based matrix (LiChrosorb® Si 60, 10 μm, about 1.2 g) synthesized as in Sections 3.1.4. and 3.2.2. and packed with the slurry-packing technique *(5,12)* in a stainless steel column (100 × 5 mm id).
2. Sodium phosphate buffer, $0.1M$, pH 7.5.
3. Assay reagents for LDH *(12,14,15)*.
4. Sample: 1 mg of each bovine serum albumin (BSA), pig-heart LDH (H_4) and rabbit-muscle LDH (M_4) in 1 mL of buffer.

2.3.2. Separation of BSA, LDH,
and Alcohol Dehydrogenase (ADH)
on N^6-(6- Aminohexyl)-AMP Immobilized to Tresyl Silica

1. HPAC silica-based matrix (LiChrospher Si 500, Merck) synthesized as described in Sections 3.1.5. and 3.2.4. and packed in a stainless steel column to a vol of 0.5 × 10 cm. It contains 29 μmol AMP-analog/g dry support.
2. Sodium phosphate buffer, 0.05M, pH 7.5.
3. Assay reagents for LDH *(12)* and ADH *(14,15)*.
4. Sample: 25 μg of BSA, 5 μg of LDH, and 25 μg of ADH in 450 μL of buffer.

2.3.3. Purification of Rabbit Muscle LDH
on Procion Blue MX- R Immobilized
to TSK G5000 PW

1. HPAC synthetic matrix (TSK, Tokyo, Japan, PW type) synthesized as in Section 3.2.6. and packed in a stainless steel column (0.45 × 5.0 cm) with the slurry method. Immobilized dye concentration at 18 μmol/g dry adsorbent.
2. Sodium phosphate buffer, 0.05M, pH 7.0.
3. Sodium chloride solution, 4M, in buffer.
4. Assay reagents for LDH *(12,14)*.
5. Sample: commercial (Sigma) crude extract from rabbit muscle dialyzed for 15 h at 4°C against 500-fold excess buffer and filtered through a membrane filter (Durapore® Millipore, GVWP type, 0.2 μm). The sample contains, typically, 430 U of enzyme activity and 12.3 mg of protein/mL.

2.3.4. Purification of Human Cholinesterase
on Surfactant-Affinity Ligands Adsorbed
to Hydrophobic Silica

1. HPAC matrix synthesized as in Section 3.2.7. (30–40 μm, 300 Å) and packed in a stainless steel column (0.2 × 2.0 cm, 21.3 mg of packing).
2. Tris-HCl buffer, 0.05M, pH 8.0.
3. Sodium chloride, 1M in 0.05M Tris HCl, pH 8.0.
4. Assay reagents for cholinesterase *(13–15)*.
5. Sample: human serum (0.066 U cholinesterase/mg protein, 56.8 mg/mL by absorption at 280 nm).

3. Methods

3.1. Coating and Activation

3.1.1. Preparation of Glycidoxypropyl Silica (3,5)

In a 500-mL flask add 5 g of silica and 200 mL of the silane solution (1% v/v). Sonicate the suspension for 10 min under vacuum and then heat it at 90–95°C for 3 h with gentle mixing using an overhead device or shaking in a bath. Allow the reaction to cool and wash the glycidoxypropyl silica with 200 mL each of water and acetone, and dry it under reduced pressure. Store in a desiccator.

3.1.2. Preparation of Glycidoxypropyl Silica Under Dry Conditions (5,6)

1. Preparation of dry silica: In a 500-mL three-necked flask add 10 g silica, and heat the flask at 150°C for 6 h under vacuum (<0.01 bar), using a vacuum pump to remove moisture. The drying tube should be inserted between the flask and the pump.

2. Silica coating in dry toluene: Cool the flask (<70°C), disconnect the vacuum, fit with an overhead stirring device, and immediately introduce a total of 150 mL of dry toluene containing 10 mL of silane and 0.25 mL of triethylamine (see Note 1). Sonicate the suspension for 5 min under reduced pressure. (Insert the drying tube between the flask and the pump.) Disconnect the pump and immediately fit the flask with a reflux condenser carrying a drying tube at the end. (Both items should have been previously dried at 140°C for 90 min.) Reflux the suspension for 16–18 h with gentle stirring. Allow the reaction to cool and then wash (sintered funnel) with 500 mL each of toluene and acetone. Dry under reduced pressure and store in a desiccator.

3.1.3. Preparation of Glycerylpropyl Silica (3)

In a 500-mL flask add 5 g silica and 200 mL of the silane solution (1% v/v, pH 5.5). Sonicate the suspension under vacuum for 10 min and then heat it at 90–95°C for 2 h in a shaking bath. Adjust the pH to 3.0 using HCl (10 mM) and heat the suspension for a further hour to convert the epoxy groups to diol groups. Allow the reaction to cool and wash (sintered funnel) with 200 mL each of water and acetone. Dry diol silica in an oven overnight (typically 50°C) and store in a desiccator.

3.1.4. Preparation of Aldhyde Silica (5,6)

In a 250-mL flask add 5 g of diol silica (Section 2.1.3.) and 100 mL of a freshly prepared solution of acetic acid/water (9/1, v/v) containing 25 g of sodium periodate. Sonicate the suspension for 5 min under reduced pressure before shaking it for 2 h at 25°C. Wash (sintered funnel) the aldehyde silica with 250 mL each of water, methanol, and diethyl ether. Dry the silica under reduced pressure and store it in a desiccator.

3.1.5. Preparation of Tresyl Silica (5–7)

Wash 2.5 g of diol silica with 3 × 80 mL of dry acetone, sucked dry, and add it in a dried 50-mL flask (washed with 10 mL of dry acetone), followed by 10 mL of dry acetone and 0.5 mL of dry pyridine (*see* Note 2). Replace the stopper immediately and cool the suspension to 0°C while stirring gently. Add dropwise, over a period of 1 min, 500 μL of tresyl chloride with vigorous stirring. Replace stopper immediately and gently stir the reaction for 20 min. Wash tresyl silica with 150 mL each of acetone, acetone/HCl (5 mM) (1/1, v/v), HCl (5 mM) and acetone. Dry silica under reduced pressure and store in a desiccator.

3.1.6. Preparation of
Imidazol Carbamate–Silica (5)

Wash 5 g of diol silica with 3 × 150 mL of dry acetonitrile, suck dry, and add it to a dried 100-mL flask, followed by 40 mL of a solution of dry acetonitrile containing 2.4 g of CDI. Sonicate the suspension for 5 min under reduced pressure (insert the drying tube between the flask and the pump) and replace the stopper in the flask. Leave the reaction to proceed for 30 min at room temperature with gentle shaking. Wash (sintered funnel) the imidazole carbamate–silica with 250 mL of dry acetone, dry it under reduced pressure, and store in a desiccator.

3.1.7. Preparation of
Chloroformate-Activated Silica (8,9)

Suspend 1 g of silica bearing primary hydroxyl groups (*see* Note 3) in 5 mL of dry acetone and add, while stirring, 10 mM chloroformate; replace the stopper in the flask. Cool the suspension in an ice bath and, while stirring, introduce dropwise 10 mL of a solution of dry acetone containing 15 mmol 4-dimethylaminopyridine, and replace the stopper in the flask. Allow the activation to proceed for 30 min. Wash (sintered funnel) the activated silica with acetone, acetic acid/dioxane, acetone, and diethyl ether. Dry under reduced pressure and store in a desiccator.

3.2. Immobilization

3.2.1. Immobilization of 3-Aminobenzene Boronic Acid to Glycidoxypropyl Silica (10)

In a 50-mL flask add 15 mL of water and 2 g epoxy silica, and mix well before introducing 600 mg of 3-aminobenzene boronic acid. Adjust the pH to 8.5 with $3M$ NaOH and sonicate the suspension for 10 min under reduced pressure (*see* Note 4). Readjust the pH to 8.5 and incubate the reaction for 24 h at 20°C with gentle shaking. Wash (sintered funnel) the silica product with, in the following order, 50 mL each of NaCl, NaHCO$_3$, HCl, water, water/acetone (1/1 v/v), and acetone, and dry it under reduced pressure in a desiccator.

3.2.2. Immobilization of N^6-(6-Aminohexyl)-AMP to Aldehyde Silica (3)

In a 25-mL flask add 5 mL of NaHCO$_3$, 2 g of aldehyde silica, and mix well before introducing 400 mg of nucleotide ligand. Sonicate for 5 min under reduced pressure and allow the reaction to proceed for 24 h at 20°C with gentle shaking. Afterward, over a period of 1 h and while stirrring gently, add in portions, a solution of 80 mg of NaBH$_4$ in 1 mL of water. Wash (sintered funnel) the silica product with 150 mL each of water, water/acetone (1/1 v/v), and acetone, and dry it under reduced pressure in a desiccator.

3.2.3. Immobilization of Antihuman Serum Albumin (Anti-HSA) to Aldehyde Silica

In a 50-mL flask add, in the following order, 10 mL of buffer A, 80 mg of anti-HSA in 4 mL of buffer A, and 5 g of aldehyde silica. Degas the suspension under reduced pressure and allow the coupling reaction to proceed for 50 min at 4°C. Wash (sintered funnel) the silica product with 500 mL of buffer B and, if to be used immediately, with the appropriate solvent system. Store the affinity adsorbent wet at 4°C. For prolonged storage add 0.02% sodium azide.

3.2.4. Immobilization of N⁶-(6-Aminohexyl)-AMP to Tresyl Silica (7)

In a 25-mL flask add 8 mL of NaHCO$_3$ and 2.5 g of tresyl silica (Section 2.1.5.), and mix well before introducing 40 mg of nucleotide ligand. Sonicate the suspension under reduced pressure for 5 min and allow the reaction to proceed for 24 h at 20°C with gentle shaking. Wash (sintered funnel) the silica product with 100 mL of water before treating it with 10 mL of Tris-HCl buffer for 4 h at 20°C. Wash the silica adsorbent with water, water/acetone (1/1, v/v), and acetone, and dry it under reduced pressure in a desiccator.

3.2.5. Immobilization of Trimethyl (p-Aminophenyl) Ammonium Chloride (TAPA) to ACA-Sparon (11)

In a 100-mL flask add, in the following order, a freshly prepared solution made of 500 mg of carbodiimide in 35 mL of water (pH 5), 270 mg of TAPA, and 5 g of ACA-Separon (*see* Note 5). Sonicate the suspension for 5 min under reduced pressure and allow the reaction to proceed for 24 h at 25°C with shaking. Wash (sintered funnel) the TAPA–ACA-Separon gel with 500 mL each of NaOH–NaCl solution, water, water/acetone (1/1, v/v), and acetone, and dry it at 40°C overnight. It can also be dried under reduced pressure in a desiccator.

3.2.6. Immobilization of Procion Blue MX-R to TSK-PW (5,12)

In a 50 mL-flask add 100 mg of dye and 25 mL Na$_2$CO$_3$, and immediately sonicate the suspension for 30 s. Add 1 g of dry TSK gel, sonicate again for 5 min under reduced pressure, and leave the reaction for 14 h at 40°C with shaking. Wash (sintered funnel) with water until efluents are dye-free, then with 100 mL acetone/water (1/1, v/v), 200 mL of acetone, and dry it overnight at 40°C or under reduced pressure in a desiccator (*see* Note 6).

3.2.7. Immobilization to Hydrophobic Supports by Adsorption of Surfactant-Affinity Ligands (13)

Wash the reverse-phase silica column (*see* Note 7) with 20 mL each of methanol and deionized water at a flow rate of 1 mL/min. Then apply continuously 50 mL of an aqueous solution of 0.1 mM of the affinity ligand until the absorbance at 259 nm is constant. Wash the affinity column with the appropriate buffer system prior to application of the protein sample (e.g., with 100 mL of Tris-HCl buffer).

3.3. Applications

1. It is important that all buffers and biological samples employed in HPAC runs should be filtered through 0.2 μm of hydrophilic membrane prior to use (e.g., Millipore GV or WP type). All the buffers used for HPAC runs, but not the protein samples, should also be degassed after filtration.

2. Resolution of LDH isoenzymes on N^6-(6-aminohexyl)- AMP immobilized to aldehyde silica (*see* Section 2.3.1.): The chromatographic run should be performed at 20–22°C, at a pressure of 400 psi and at a flow rate of 1.6 mL/min. Fig. 1 shows the expected events. Inject a 10-μL sample of a mixture of BSA, pig-heart LDH, and rabbit-muscle LDH onto the affinity column. After the appearance of unbound BSA (followed at 215 nm) in the washings, elute by affinity elution the bound LDH isoenzymes with a NADH gradient (8 mL, 0–0.15 mM) made in phosphate buffer. Detection of eluted LDH isoenzymes should be performed by following the enzyme activity in a spectrophotometer (absorbance decrease at 340 nm) as described before *(12)*.

3. Separation of BSA, LDH, and ADH on N^6-(6-aminohexyl)- AMP immobilized to tresyl silica (*see* Section 2.3.2.): The chromatographic run should be performed at room temperature, at a pressure of 1000 psi and at a flow rate of 1 mL/min. Fig. 2 shows the expected events. Apply a sample (450 μL) of the mixture of the three proteins onto the affinity column. Unbound BSA passes through (100%). Then elute by affinity elution the two bound enzymes with short pulses of NAD+ plus oxamate (elution of LDH) and NAD+ plus pyrazole (elution of ADH). Follow the column effluents at 280 nm (protein monitoring) and 340 nm (enzyme activity monitoring) *(12,14,15)*. An 80% recovery is expected for the two enzymes.

4. Purification of rabbit-muscle LDH on Procion blue MX-R immobilized to TSK G5000 PW *(5,12)* (*see* Section 2.3.3.): The chromatographic run should be performed at room temperature and at a flow rate of 1 mL/min, which should result in a pressure of about 360 psi. Equilibrate the affinity column with phosphate buffer and apply 2 mL of dialyzed sample (860 LDH units, 94.4 mg protein). Wash away nonadsorbed protein with buffer (12–14 mL) and subsequently elute LDH activity with a linear gradient of sodium chloride (0–4M, total vol, 10 mL). Collect 2-mL fractions and assay each for LDH activity and protein content (absorbance). Pool those fractions containing LDH activity and measure the total LDH activity (U) and protein (mg) present. Calculate the specific activity of purified LDH, the purification achieved (fold) over the starting material, and the yield (%).

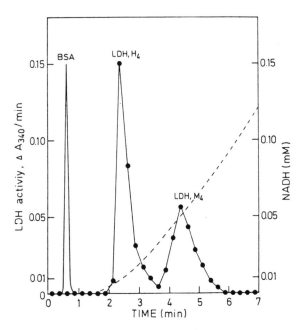

Fig. 1. Resolution of pig-heart LDH and rabbit-muscle LDH isoenzymes on N^6-(6-aminohexyl)-AMP immobilized on aldehyde silica.

The protein content of crude enzyme should be determined by the $A_{280/260}$ absorbance method, whereas that of pure enzyme may be from the absorbance at 280 nm, using the formula mg/mL = $1.44 \times A_{280}$ *(12,14)*.

5. Purification of human cholinesterase on surfactant-affinity ligand adsorbed to hydrophobic silica *(13)* (*see* Section 2.3.4.): The chromatographic run should be performed at a flow rate of 1 mL/min. Apply 200 µL of sample on the column and wash with buffer until A_{214} goes down to about 0.1. Apply a linear gradient of 0–1 M NaCl for 2 min to elute bound cholinesterase. An enzyme recovery of about 80% should be obtained with a specific activity for the purified protein of 5.2 U/mg, corresponding to a 79-fold purification.

Determine protein amounts from the integrated peak areas and a BSA standard curve, using as a column a 40-cm piece of stainless steel tubing (0.17 mm ID) connected to an HPLC system (214 nm).

4. Notes

1. The amount of silane and base used should be approximately proportional to the surface area of the silica. Therefore, if a silica of 50 m^2/g is used, then five times less reagents should be used.

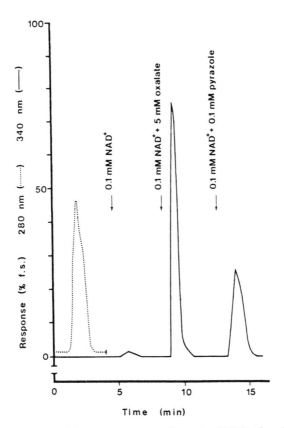

Fig. 2. Separation of bovine serum albumin (BSA), beef-heart lactate dehydrogenase (LDH), and horse-liver alcohol dehydrogenase (ADH) on N^6-(6-aminohexyl)-AMP immobilized on tresyl silica.

2. Tresyl chloride (2,2,2-trifluoroethanesulfonyl chloride) is readily hydrolyzed by water; therefore it is imperative to use dry solvents during the activation reaction. However, the tresyl-activated diol silica is stable in an aqueous environment at low pH (e.g., pH 3.0).

3. Silica with primary hydroxyl groups is obtained by reducing aldehyde silica (Section 3.1.4.) as follows: 4.5 g of aldehyde silica is suspended in 100 mL of sodium acetate, pH 4.0, and 3.8 mL (100 mM) of $NaBH_4$ is added in portions over 1 h, while stirring. The silica is finally washed with water, acetone, and ether, and dried under reduced pressure in a desiccator.

4. Sonication facilitates dissolution of the boronic acid and removal of the air from the silica pores.

5. The carbodiimide-promoted condensation between TAPA and ACA-Separon can also take place in an aqueous buffer of $0.2M$ 2-(morpholino)ethenesulphonic acid–NaOH (Mes-NaOH), pH 4.7, with periodic adjustment of the pH with $1M$ HCl or $1M$ NaOH.
6. This is a general procedure for immobilizing highly reactive dichlorotriazine dyes to synthetic gels bearing free hydroxyl groups *(5,12)*. However, immobilization of the less-reactive monochlorotriazine dyes on synthetic supports, as above, leads to a low ligand-substitution level. In this case it is recommended that the monochlorortiazine dye first be converted to the corresponding 6-aminohexyl dye by reacting the dye with 1,6-diaminohexane *(5)*. Subsequently the dye analog can readily be immobilized to activated supports (*see* Sections 3.1. and 3.2.).
7. The surfactant from which the affinity ligand is synthesized may be obtained from Nikkol Chemicals Co., Japan. The affinity ligand is readily synthesized *(14)* by tresylation of the surfactant and nucleophilic substitution with the pyridine. Pyridine was chosen because it is an inhibitor, thus an affinity ligand, for the enzyme to be purified, horse serum cholinesterase.

References

1. Clonis, Y. D. and Lowe, C. R. (1988) Affinity chromatography, in *Principles of Clinical Biochemistry* (William, D. L. and Marks, V., eds.), Heinemann, Oxford, UK, pp. 383–390.
2. Clonis, Y. D. (1987) Large-scale affinity chromatography. *Bio/technology* **5**, 1290–1293.
3. Ohlson, S., Hansson, L., Larsson, P.-O., and Mosbach, K. (1978) High-performance liquid affinity chromatography and its applications to the separation of enzymes and antigens. *FEBS Lett.* **93**, 5–9.
4. Clonis, Y. D. and Small, D. A. P. (1987) High performance dye-ligand chromatography, in *Reactive Dyes in Protein and Enzyme Technology* (Clonis, Y. D., Atkinson, T., Bruton, C., and Lowe, C. R., eds.),Macmillan, Basingstoke, UK, pp. 87–100.
5. Clonis, Y. D. (1989) High performance affinity chromatography, in *HPLC of Macromolecules - A Practical Approach* (Oliver, R. W., ed.), IRL, Oxford, UK, pp. 157–182.
6. Larsson, P. -O. (1984) High performance liquid affinity chromatography. *Methods Enzymol.* **104**, 212–223.
7. Nilsson, K. and Mosbach, K. (1981) Immobilization of enzymes and affinity ligands to various hydroxyl group carrying supports using highly reactive sulfonyl chlorides. *Biochem. Biophys. Res. Commun.* **102**, 449–457.

8. Ernst-Cabrera, K. and Wilchek, M. (1986) Silica containing primary hydroxyl groups for high performance affinity chromatography. *Anal. Biochem.* **159**, 267–272.
9. Ernst-Cabrera, K. and Wilchek, M. (1987) Coupling of ligands to primary hydroxyl-containing silica for high performance affinity chromatography, optimization conditions. *J. Chromatogr.* **397**, 187–196.
10. Glad, M., Ohlson, S., Hansson, L., Mansson, M. -O., and Mosbach, K. (1980) High performance liquid affinity chromatography of nucleogides, nucleotides and carbohydrates with boronic acid-substituted microparticulate silica. *J. Chromatogr.* **200**, 254–260.
11. Taylor, R. F. and Marenchic, I. G. (1984) Comparison of low- and high-pressure affinity chromatography for the purification of serine and sulfhydryl esterases. *J. Chromatogr.* **317**, 193–200.
12. Clonis, Y. D. (1987) Matrix evaluation for preparative high performance affinity chromatography. *J. Chromatogr.* **407**, 179–187.
13. Torres, J. T., Guzman, R., Carbonell, R. G., and Kilpatrick, P. K. 1988) Affinity surfactants as reversible bound ligands for high performance affinity chromatography. *Anal. Biochem.* **171**, 411–418.
14. Decker, L. A. (1977) *Worthington Enzyme Manual*, (Worthington Biochemical, Freehold, NJ)
15. Boehringer Mannheim GmbH, West Germany (1975) *Biochemica Information I and II*.

CHAPTER 9

Immunoaffinity Chromatography

George W. Jack

1. Introduction

Immunoaffinity chromatography (IAC) harnesses the specificity and avidity of the interaction between an antigen and its antibody to purify the antigen. The technique may be used with either polyclonal or monoclonal antibodies (MAb) *(1)*, but MAbs are preferable for a number of reasons. Polyclonal sera are never specific for a single antibody but reflect the variety of immunological challenges sustained by an animal since birth. Furthermore, the best polyclonal sera are generated using highly purified antigen, so if a purification method already exists, why raise sera to develop a second purification method? Perhaps the main disadvantage of polyclonal sera lies in the wide range of avidity of antibodies they contain; although this ensures that such antibodies will bind their antigen, the dissociation of the antigen–antibody complex may prove impossible using conditions that retain the activity of both.

On the other hand, MAbs may be generated even when pure antigen is not available, and then produced on whatever scale is required, either by growing as ascites or in tissue culture. A unique and consistent product may be obtained by this means, whereas every animal immunized with a given antigen will produce a different set of antibodies, each with its own properties. Each MAb exhibits a unique association constant (K_a) for its antigen, so from a panel of MAbs one can be chosen for use in IAC with sufficient avidity to ensure good binding

From: *Methods in Molecular Biology, Vol. 11: Practical Protein Chromatography*
Edited by: A. Kenney and S. Fowell Copyright © 1992 The Humana Press Inc., Totowa, NJ

of the antigen while still allowing the MAb–antigen complex to be dissociated under relatively mild conditions.

The technique of IAC involves the immobilization of a MAb on an insoluble support matrix, followed by packing into a column on which to perform adsorption/desorption chromatography. A wide range of matrices for antibody immobilization is available commercially, but probably the most commonly used are the Sepharoses produced by Pharmacia and the Affi-Gel® series from Bio-Rad (Richmond, CA). The Affi-Gel series are crosslinked agaroses derivatized with *N*-hydroxysuccinimide esters to give spontaneous coupling to proteins through free amino groups via a 10- or 15-atom-long spacer arm (*see* Fig. 1). Pharmacia (Uppsala, Sweden) markets a range of activated and derivatized Sepharoses, also agarose-based, the most commonly used of which is CNBr-activated Sepharose (*see* Fig. 2 for the coupling reaction). CNBr-activated agarose may be synthesized in house (*2*) but since this involves the handling of highly toxic chemicals and may result in varying degrees of activation between batches, it is probably best bought as a prepared derivative.

The degree of substitution of the support matrix influences the capacity of the immunoadsorbent for the antigen (*3*), but in general substitution levels in excess of 10 mg MAb/mL of matrix result in inefficient utilization of the MAb. The choice of MAb to be used in IAC may be determined by trial and error, i.e., by immobilizing a number of MAbs and identifying those that will both bind antigen and allow subsequent elution. If the affinity constants of a panel of MAbs are known, the number to be tested can be restricted to those with K_as in the range 5×10^7–5×10^8, since this has been shown to be optimal for IAC (*4*). Pure MAb should be used to minimize the amount of nonfunctioning immobilized protein. This will also reduce the nonspecific binding capacity of the sorbent since by virtue of their amphoteric nature, immobilized proteins may act as ion exchangers, while any exposed hydrophobic sequences may result in nonspecific hydrophobic interactions. The selection of an appropriate buffer for the binding stage followed by adequate washing of the MAb–antigen complex will minimize contamination of the eluted product.

Among the advantages of IAC is its ability to produce virtually pure protein in a single step. Complicated multistep purification procedures, which inevitably result in low overall yields, can be avoided. Because of the high specificity and avidity of the adsorption stage, very dilute antigen solutions in the presence of large amounts of extraneous pro-

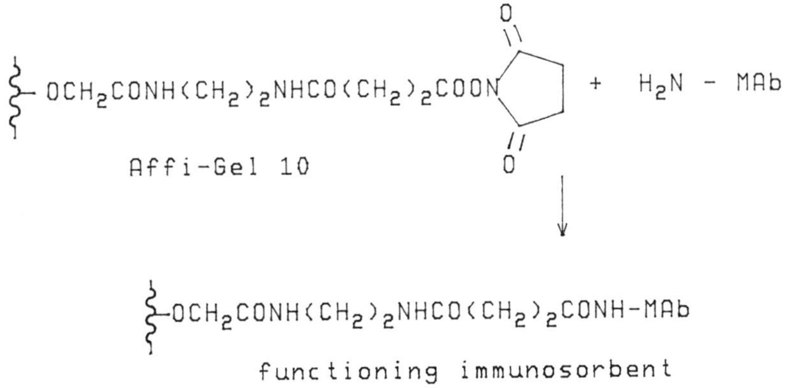

$$\xi-OCH_2CONH(CH_2)_2NHCO(CH_2)_2COON \underset{O}{\overset{O}{\diagdown}} + H_2N - MAb$$

Affi-Gel 10

↓

$$\xi-OCH_2CONH(CH_2)_2NHCO(CH_2)_2CONH-MAb$$

functioning immunosorbent

Fig. 1. MAb coupling to Affi-Gel 10.

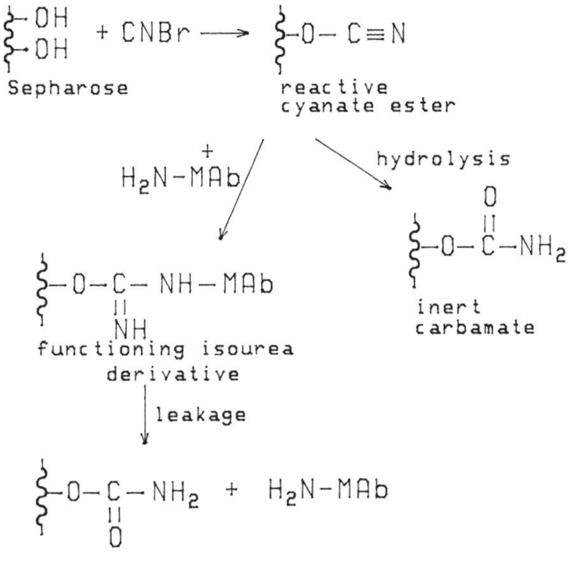

$$\xi\begin{matrix}-OH \\ -OH\end{matrix} + CNBr \longrightarrow \xi-O-C\equiv N$$

Sepharose

reactive
cyanate ester

$$H_2N-MAb \diagup \qquad \diagdown hydrolysis$$

$$\xi-O-C-NH-MAb \atop \| \atop NH$$

functioning isourea
derivative

$$\xi-O-\overset{O}{\overset{\|}{C}}-NH_2$$

inert
carbamate

↓ leakage

$$\xi-O-C-NH_2 \atop \| \atop O + H_2N-MAb$$

Fig. 2. MAb couping to CNBr-Sepharose.

tein may be processed successfully, since the adsorbent will bind only the antigen of the immobilized MAb. The capacity of immunosorbents for antigen is, however, low, with capacities in excess of 1 mg antigen/ mL gel something of a rarity. This factor, together with the high cost of producing IAC columns, has encouraged the use of relatively small columns operated in a cycling mode and automated by microprocessor or computer control of pumps, multiway solenoid valves, and fraction collectors *(3–6)*. The use of short, fat columns at relatively high flow rates enables several cycles to be completed each day. The inclusion of protective devices, such as pressure and level sensors in automated systems, allows for unattended operation 24 h/d, ensuring high productivity from the system despite a low yield per cycle. Furthermore, should the column become contaminated and lose activity, the replacement cost is not as high for a short column as for a large column.

For small-scale laboratory investigative use, several cycles of chromatography may be run daily using relatively unsophisticated equipment; the high resolving power of the technique ensures high product quality in a single step.

2. Materials

Use Analar- or even Aristar-grade reagents and distilled water for all solutions. All solutions must be sterilized either by filtration (0.2 μm filter) or by autoclaving (122°C for 30 min), as appropriate.

1. The apparatus required for running IAC is available commercially and includes a glass column with adjustable flow adaptors, peristaltic pump, a flow-through UV-absorbance monitor, and a fraction collector. For synthesis of the IAC sorbent, a sintered funnel of porosity 2 to British Standard 1752, i.e., 40–90 μm, with a Buchner flask attached to a water vacuum pump is required.
2. Coupling buffer: 0.2*M* bicarbonate plus 0.5*M* NaCl, pH 8.6. Dissolve NaHCO$_3$ (16.8 g) and NaCl (29.2 g) in water, adjust the pH to 8.6 by the addition of solid Na$_2$CO$_3$, and make up to 1 L. Sterilize by filtration.
3. Acid wash: 1 m*M* HCl. Dilute 85.9 mL of conc. HCl to 1 L with water to produce 1*N* HCl, and then dilute 1 mL of 1*N* HCl to 1 *L* for use.
4. Blocking buffer: 0.5*M* Tris-HCl, pH 8.5. Dissolve 60.5 g of Tris in water, and adjust to pH 8.5 with conc. HCl and then to 1 L with water. Sterilize by autoclaving.
5. Binding buffer: borate/salt, pH 8.5. Dissolve boric acid (60 g), NaCl (140 g), and NaOH (10 g) in water and make to 1 L. Sterilize by autoclaving. For use, dilute 200 mL of buffer with 800 mL of sterile water.
6. Wash buffer (acid): acetate/salt, pH 4. Dissolve sodium acetate trihydrate (13.6 g) and NaCl (29.2 g) in water and adjust to pH 4.0 with glacial acetic acid and then with water to 1 L. Sterilize by autoclaving.
7. Column-cleaning buffer: 8*M* urea–acetate, pH 4.0. Dissolve urea of Aristar grade (480 g) and sodium acetate trihydrate (13.6 g) in water, and adjust to pH 4.0 with glacial acetic acid and then to 1 L with water. Sterilize by filtration.

3. Method

1. Take 2 g of CNBr-activated Sepharose (Pharmacia) and suspend in 1mM HCl in a sintered-glass funnel. Stir the suspension gently with a glass rod until the freeze-dried powder has swollen, and then suck the liquid through.
2. Wash the gel with 200–300 mL of 1 mM HCl/g gel with gentle stirring in several aliquots to remove protective additives.
3. Suck the liquid through between successive additions of HCl. Suck the gel dry and then wash it rapidly with 100 mL of coupling buffer before sucking dry again; immediately add the gel to a flask containing up to 70 mg of MAb in 30 mL of coupling buffer.
4. Stir the gel gently to obtain an even suspension; then leave at room temperature for 3 h with occasional mixing or shaking, and then at 4°C overnight.
5. Take a sample of the supernatant above the gel beads and assay for protein. If more than 80% of the MAb is no longer in solution, proceed; if not, allow the reaction to continue longer at room temperature.
6. When coupling is complete, wash the gel on the sintered funnel with 200 mL of coupling buffer, suck dry, and suspend for 30 min at room temperature in 50 mL of blocking buffer.
7. Return the gel to the sintered funnel and wash it with alternate 100 mL vol of binding buffer and wash buffer. Five cycles should be adequate.
8. Pack the gel in a glass column with a diameter of either 1.6 or 2.2 cm, with the flow adaptors in contact with the gel bed. At flow rates of 40 or 70 mL/h, depending on the column diameter, wash the gel with 3 column vol (20 mL) of binding buffer followed by the same vol of wash buffer. Repeat this process three times; then wash the gel with 1 column vol (7 mL) of urea-acetate before equilibrating the gel with 3 column vol of binding buffer.
9. Equilibrate the extract to be chromatographed with the binding buffer, either by dialysis or by buffer exchange through Sephadex G-25. Filter-sterilize the extract.
10. Run the extract through the IAC sorbent at room temperature at the above flow rates, and wash the column with binding buffer until the UV-absorbance of the column effluent falls back to the baseline level. Use a sensitive range on the absorbance monitor, such as 0–0.1 absorbance units, for full-scale recorder deflection.
11. Wash the column with 2 column vol of a solution of low ionic strength. This can be the binding buffer with the salt omitted, or even water will sufffce.

12. Stop the flow through the columns and alter the configuration of the plumbing to reverse the direction of flow through the column. Elute the column with 3 column vol of a solution (*see* Note 4), which will dissociate the MAb–antigen complex.

13. Wash the column with 1 column vol of urea–acetate followed by 2 column vol of binding buffer. Again reverse the direction of flow through the column and equilibrate the gel with binding buffer (3 column vol) before applying the next sample of extract. The entire chromatographic procedure should take between 2 and 3 h.

4. Notes

1. Provided the manufacturer's recommendations regarding the storage and shelf-life of CNBr-activated Sepharose are adhered to, no problems should be encountered in coupling the MAb to the support matrix. The reactive cyanate ester intermediate is relatively stable in acid conditions and does not hydrolyze to an inert carbamate until the pH is raised to neutrality. For this reason, after acid-washing the gel, it should be washed rapidly with coupling buffer and transferred immediately to buffer containing the MAb to be coupled. One gram of freeze-dried powder will give approx 3.5 mL of swollen gel. Similarly, with the Affi-Gel series, compliance with the manufacturer's recommendations should result in successful coupling of MAb to matrix, although special note should be made of the relationship between the pI of the protein to be coupled and the efficiency of coupling to Affi-Gel 10 or 15. In general, MAbs couple more efficiently to Affi-Gel 10.

2. All proteins are susceptible to proteolysis, even when immobilized. An IAC column run with nonsterile buffers or extract will rapidly lose its antigen-binding capacity as a result of bacterial of fungal proteolysis of the immobilized antibody. The use of sterile buffers and extracts is essential for the prolonged activity of an IAC column. Choice of the sterilization method is also crucial, since urea-containing and bicarbonate/carbonate buffers do not survive autoclaving. If operated carefully under sterile conditions, columns have been used for over 50 cycles of operation spread over many months, but such longevity is enhanced if the column is used under "clean room" conditions. The ease with which a column can be contaminated and inactivated reinforces the recommendation to use multiple cycles on a small column, since there is less material to lose after an accidental contamination.

3. All IAC columns will bind protein nonspecifically, although selection of the appropriate buffer for the absorption stage can reduce this phenomenon quite markedly. In the purification of human thyrotropin *(7)*, altering the binding buffer from Tris to borate and then adding NaCl to the buffer reduced the contamination of the thyrotropin by lutropin by approx 40-fold. Since protein binds nonspecifically to IAC columns by both ionic and hydrophobic interactions, washing the column with solutions of both high and low ionic strength prior to antigen elution will eliminate these opportunistic contaminants.

4. The nature of the eluting agent used in IAC must be determined empirically in each instance. A change to more acid conditions is one of the most commonly used elution strategies, but to be successful requires some knowledge of the pH stability of the antigen being purified. For example, human thyrotropin starts to lose activity below pH 3.5, but human leukocyte interferon is stable at pH 2.5 *(8)*. Immobilized MAbs have much greater pH stability than free solution antibodies, and can tolerate acid pH well down to pH 2, but appear susceptible to denaturation above pH 10 and also withstand the actions of protein denaturants. Protein denaturants, such as urea $(8M)$ or guanidine hydrochloride $(6M)$, particularly at neutral pH, have been used as eluents if the antigen retains its activity on removal of the denaturant. Exceedingly robust molecules, such as human growth hormone, will survive treatment with urea at pH 2 *(9)*. High concentrations, up to $3M$, of chaotropic agents, such as thiocyanate or iodide ions, are commonly used as eluents. A frequently used ploy with chaotropes is to pump approx 1 column vol of agent into the column, and then stop pumping for about half an hour before continuing with the elution. The use of thiocyanate ion may, however, compromise material intended for pharmaceutical use. Polarity-reducing agents, such as ethylene glycol, have also been used, but in defining an effective eluting agent one must keep an open mind and try many possibilities *(10)*.

5. Whenever practical, reverse-flow elution is to be preferred, since it results in a greatly reduced elution volume. Elution volumes may be only one-third of those obtained with unidirectional flow through the gel bed. Reverse-flow elution also has the advantage of cleaning away any material held by the net of what is normally the inlet-side flow adaptor.

6. Leakage of antibody occurs from all IAC columns. The extent to which such leakage results in product contamination can be minimized by

careful column washing. A regime like that outlined earlier for washing a newly synthesized gel will produce a column from which an antigen may be eluted with low, but detectable, contamination with antibody. During thyrotropin purification, the first three cycles on a newly synthesized gel yielded thyrotropin contaminated with MAb at 24, 12, and <6 ng/mL, showing that leakage from a new gel decreases with use *(7)*. If minimum contamination levels are required from the first cycle of operation, the column should be precycled with the full purification protocol several times before running any extract through the column. Similarly, if the column has been unused for some time, it should be precycled before recommencing antigen purification.

Leakage results from hydrolysis of the isourea linkage between matrix and ligand. However, since the antibody is bound to the matrix by multipoint attachment, several bonds must hydrolyze before ligand leakage occurs. Leakage may also result from mechanical fragmentation of matrix beads and may be minimized by using crosslinked supports. Several methods have been described to minimize ligand leakage, but perhaps the most commonly used is glutaraldehyde crosslinkage between antibody molecules after immobilization *(11)*. A wide range of alternate coupling chemistries have been devised, for which reduced ligand leakage is claimed *(12* and this vol., Chapter 11) as a result of utilizing chemical bonds more stable than the isourea linkage. The reality, however, is that leakage occurs to some extent in all cases, if a sufficiently sensitive assay for MAb is used to detect it.

7. MAbs are produced by cell lines derived from a cancerous cell, and, in consequence, oncogenic nucleic acid may be present in crude MAb preparations. Before use in IAC sorbents, the MAb should be purified to eliminate potential oncogenic material and minimize the nonfunctional protein on the sorbent. This may be achieved by a variety of means, but the most common methods currently in use are chromatography on either immobilized Protein A *(13)* or Protein G *(14)* and by fast protein liquid chromatography (FPLC). Both Protein A and Protein G are receptors for the Fc region of antibodies from staphylococcal and streptococcal strains, respectively. Both may be purchased already immobilized from commercial suppliers such as Pharmacia and Bio-Rad, with instructions for use. Each does not bind all subclasses of mouse and rat antibodies, but one or the other should purify almost any MAb. The product is likely to contain Protein A or G leaking off the matrix.

As an alternative, a combination of FPLC on both Mono S and Mono Q cation and anion exchangers will purify all antibodies to homogeneity. A 0–0.3M NaCl gradient in 20 mM acetate, pH 5.0, on Mono S followed by a 0–0.3M NaCl gradient in 20 mM Tris-HCl buffer, pH 7.6, on Mono Q should prove adequate.

References

1. Kohler, G. and Milstein, C. (1975) Continuous culture of fused cells secreting antibody of predefined specificity. *Nature* **250**, 495–497.
2. Cuatrecasas, P. (1970) Protein purification by affinity chromatography: Derivatisation of agarose and polyacrylamide beads. *J. Biol. Chem.* **245**, 3059–3065.
3. Chase, H. A. (1984) Affinity separations utilising immobilised monoclonal antibodies–a new tool for the biochemical engineer. *Chem. Eng. Sci.* **39**, 1099–1125.
4. Jack, G. W., Blazek, R., James, K., Boyd, J. E., and Micklem, L. R. (1987) The automated production by immunoaffinity chromatography of the human pituitary glycoprotein hormones thyrotropin, follitropin and lutropin. *J. Chem. Tech. Biotechnol.* **38**, 45–58.
5. Bazin, H. and Malache, J.-M. (1986) Rat (and mouse) monoclonal antibodies. V.A simple automated technique of antigen purification by immunoaffinity chromatography. *J. Immunol. Methods.* **88**, 19–24.
6. Eveleigh, J. W. (1982) Practical considerations in the use of immunosorbents and associated instrumentation. *Anal. Chem. Symp. Ser.* **9**, 293–303.
7. Jack, G. W. and Blazek, R. (1987) The purification of human thyroid-stimulating hormone by immunoaffinity chromatography. *J. Chem. Tech. Biotechnol.* **39**, 1–10.
8. Staehelin, T., Hobbs, D. S., Kung, H-F., Lai, C.- Y., and Pestka, S. (1981) Purification and characterisation of recombinant human leukocyte interferon (IFL + A) with monoclonal antibodies. *J. Biol. Chem.* **256**, 9750–9754.
9. Jack, G. W. and Gilbert, H. J. (1984) The purification by immunoaffinity chromatography of bacterial methionyl- (human somatatropin). *Biochem. Soc. Trans.* **12**, 246,247.
10. Hornsey, V. S., Griffin, B. D., Pepper, D. S., Micklem, L. R., and Prowse, C. V. (1987) Immunoaffinity Purification of Fractor VIII Complex. *Thromb. Haemostas.* **57**, 102–105.
11. Kowal, R. and Parsons, R.C. (1980) Stabilisation of proteins immobilised on Sepharose from leakage by glutaraldehyde crosslinking. *Anal. Biochem.* **102**, 72–76.
12. Jack, G. W. (1990) The use of immunoaffinity chromatography in the preparation of therapeutic products, in *Focus on Laboratory Methods in Immunology, Vol. 2* (Zola, H., ed.) CRC, Boca Raton, FL, pp. 127–146.
13. Goding, J. W. (1978) Use of staphylococcal Protein A as an immunological reagent. *J. Immunol. Method.* **20**, 241–253.
14. Bjorck, L. and Kronvall, G. (1984) Purification and some properties of streptococcal Protein G, a novel IgG-binding reagent. *J. Immunol.* **133**, 969–974.

CHAPTER 10

Selection of Antibodies for Immunoaffinity Chromatography

Duncan S. Pepper

1. Introduction

A broad choice is available in immunoaffinity chromatography, of using polyclonal antibodies or monoclonal antibodies, and this chapter will concentrate mainly on the latter, although many points are common to both. Another broad distinction can be made between immobilized antigens (which may include immunoglobulins) used to prepare purified antibodies, and immobilized antibodies used to prepare pure antigens; again, this chapter will concentrate mainly on the latter. Until recently there was little rational choice possible in antibody selection for immunoaffinity chromatography—one simply had to use the individual animal species bleeds as available and hope for the best *(1)*. The advent of monoclonal antibodies has, in principle, broadened our choice *(2)*, but also has shown up the need for principles and techniques to produce, select, and use antibodies on a rational basis for immunoaffinity work *(3,4)*. Although this chapter will deal mainly with the selection process *(5)*, it is also necessary to consider and control other aspects, such as immunization *(6)*, fusion and cloning, preparation of solid-phase antigens and antibodies, and selection and testing of eluants *(7)*. Unfortunately, these aspects are far from being exact sciences, and so a considerable degree of

From: *Methods in Molecular Biology, Vol. 11: Practical Protein Chromatography*
Edited by: A. Kenney and S. Fowell Copyright © 1992 The Humana Press Inc., Totowa, NJ

Table 1
Polyclonal Antibodies

Advantages	Disadvantages
Easy bulk production	Specificity/titer not reproducible
Many possible species	Unpredictable crossreaction/ response
Spectrum of affinities	Difficult or impossible GMP
Low cost	Low purity
Low effort	Unknown viral hazards
Complement fixing	Difficult to purify subclasses
Fragmentable (Fab etc.)	High affinity = hard to regenerate

empirical testing is still needed *(8)*. Much of the pain and tedium can be removed from such repetitive empirical testing by the sensible choice of automated or semiautomated assay techniques (e.g., enzyme-linked immunosorbent assay [ELISA], radioimmunoassay [RIA], immunoradiometric assay [IRMA], and the like) *(9,10),* and assay development thus rests at the core of successful antibody selection for immunoaffinity work.

Tables 1–3 summarize the good and bad points of polyclonal vs monoclonal antibodies, and a rational choice can often be made on the basis of the points summarized in Table 3. Increasingly nowadays, as monoclonal antibody availability improves and with the need for a GMP (good manufacturing practice) environment, the choice is more likely to be in favor of a monoclonal antibody, particularly if cycles of reuse and protein antigens are involved. Nevertheless, one should look first at polyclonal antibody as a practical source, as it is frequently quicker and cheaper to prepare in bulk when a good supply of pure antigen is available (≥ 1 mg). One polyclonal antibody source that is frequently overlooked is the avian egg yolk (IgY) *(11–14)*—the good and bad points of this source are summarized in Table 4. In the author's experience these antibodies can readily be isolated and immunopurified to give excellent yields of high-titer, high-affinity antibodies. Of the more traditional animal species, the sheep is probably the best choice, combining good response with modest antigen requirement, low overall costs, and high volume/mass yields per bleed. Particularly if individual bleeds need to be characterized in detail, it makes sense to use a larger animal, hence a larger bleed volume. In an ideal world one would test both polyclonal and monoclonal sources of antibody

Table 2
Monoclonal Antibodies

Advantages	Disadvantages
Defined specificity/crossreaction	Cost (labor, equipment, time)
Reproducible	Uncertain success in immune
Known subclass/complement fixation	response
Easy to purify	Limited to mice, rats, and humans
High specific activity	Generally lower affinity than
Known viral risk	polyclonal antibody
GMP possible with tissue culture	
Gene manipulation: class switching/	
humanization/enzyme addition	
High affinity can coexist	
with easy regeneration	

Table 3
Choice of Polyclonal or Monoclonal Antibody?

Polyclonal good for	Monoclonal good for
Multigram quanitities of antibody	Diagnostic assays
Fab/Fab$_2$ fragments	Protein/multivalent antigens
Drugs/low mol wt/univalent antigens	Multiple reuse
Single use/low cost	Defined epitopes
	Defined properties

Table 4
Chicken Polyclonal Antibodies

Advantages	Disadvantages
Nonmammalian	Need to wait until the bird is
No protein A/G binding	laying eggs
No complement fixation	Cannot stop the bird from
No rheumatoid factor crossreaction	laying eggs
No bleeding after screening	Limited to 18 mo productivity
Only IgY in egg yolk	No IgM/A in egg yolk
Easy to purify: 2-step	Few animal house facilities
polyethelene–glycol precipitation	Few commercial reagents
Recognizes high-frequency	
mammalian Ag's	
Yields 200 mg/egg	
(1000 mg/wk)/bird	
Commerical (broiler)	
industry low cost	

to see which functions better *(3,4)*, but in the absence of adequate supplies of both, it is probably more realistic to use limited supplies of antigen and/or specific polyclonal antibody as sources of analytical reagents with which to build up a system of assays for the production of monoclonal antibodies suitable for immunopurification. In principle, one needs only either monospecific polyclonal antibody or pure antigen, but in practice both are desirable, especially when the purity or mass available are in doubt.

Assuming that antigen procurement (natural or synthetic) is completed and an immune response is obtained *(6)*, the major sequences of operation are to titrate the polyclonal antibody (by RIA, IRMA, ELISA, or the like) to immobilize on a micro scale either antibody or antigen, and to measure the amount of antigen/antibody bound to it and then attempt to elute it with a short-list panel of eluants *(15)*. If inadequate binding is found, then further antibody sources will have to be acquired, but usually the problem is that binding is adequate, but elution is difficult. Under these circumstances, a longer list of eluants can be evaluated (bearing in mind that they have to be tested in parallel for nondestruction of the biological activity of the product) and any promising candidates can be fine-tuned or combined into synergistic mixtures *(2,7)*. Finally, tertiary screening is carried out in a minicolumn format to ensure that the whole system works *(7)*.

In both polyclonal and monoclonal work, time is an important variable: if the characterization assays cannot be completed in e.g., a 2-wk period, then the polyclonal response may have changed to a less desirable titer, affinity, or specificity, in the case of hybridoma cultures, it is likely that cell overgrowth and/or death will eliminate promising candidates before they can be recloned. Another hazard not to be forgotten is the death of the donor animal, which increases in likeliness as time passes.

All these points combine to underline the need for rapid, reliable, simple assays capable of processing perhaps 250–500 samples weekly. Screening thus reduces to assay design and implementation—but in a situation in which the target antigen is likely to be a new analyte and the antibodies are likely to be changing with time. In order to stay in control in such a situation, good management of people and resources is essential, and teamwork makes for a more successful outcome. Table 5 summarizes the division of work load and sequence of events carried out in the author's group and, although some of the points are spe-

Table 5
Sequence of Monoclonal–Antibody Production

Biochemistry section	Immunology section
Antigen purification ⟶	Immunization of mice
↓	↓
Mouse serum antibody screen ⟶	Fusion
↓	↓
Primary screening ←	Growth rate
↓	↓
(ELISA, RIA, bioassay)	Cloning
	↓
	N$_2$(l) Freeze/recover
↓	
Positive 30/300 ⟶	Recloning
↓	↓
Secondary screening ⟶	Bulk Production
↓	
By use (e.g.)	Tissue Ascites
Immobilized Immobilized	Culture (100 mL = 800 mg)
Antigen Antibody	(1 L = 20 mg)
Panel n = 10	
↓	
Tertiary Screening	

cific to monoclonal-antibody technology, the major points are applicable to polyclonal antibodies as well. It is only by controlling the entire process, from the beginning of immunization through the primary, secondary, and tertiary screening processes, that the chances of getting good immunopurification reagents are maximized.

It is often assumed that ELISA or ^{125}I-assay formats are mutually exclusive. However, this is not so; antibodies that are positive in one may not be so strong in the other assay, or may even be missed. Since we eventually want good positives in an immunopurification format (i.e., column with immobilized antibody), neither ELISA nor ^{125}I-tracer will necessarily predict the right outcome (*see* Note 1) *(16)*. It pays to have both reagents and technologies in house, e.g., ^{125}I-tracer has a role in diagnosing problems in the early development stages of an ELISA (coating efficiency, capacity, and the like) and ^{125}I- tracer can be counted *in situ* in gel beads to determine the amount of noneluted antigen as well as the fluid (eluted) tracer antigen (*see* Note 9).

2. Materials

2.1. Common to All Assays

1. Antigen, preferably ≥90% pure, and 1 mg (dry wt) polyclonal antibody, preferably monospecific, of titer ≥1/1000 and ≥1 mL to act as a positive control.
2. Second antibody, e.g., sheep antimouse IgG (Sigma) (if making monoclonal antibody).
3. Second antibody, e.g., sheep antirabbit or donkey antisheep (Sigma) (if making rabbit or sheep antibodies). Both of the latter should have high titer and be devoid of significant crossreaction with IgG of other species.
4. Preactivated solid-phase gel, e.g., CNBr-Sepharose 4B® (Pharmacia, Piscataway, NJ) or a homemade activated gel, e.g., Periodate-Sephacryl S-1000® (*see* Chapter 11). Solid-phase coupled protein A, e.g., Prosep A (Bioprocessing, Consett, UK).
5. Murine/rat IgG subclass typing kits (The Binding Site, Birmingham, UK or Sigma).

2.2. ELISA Assays

1. Assay diluent: 0.05M phosphate or Tris-HCl buffer, pH 7.4, containing 1% w/v bovine serum albumin (BSA), 1% v/v Tween 20.
2. Coating buffer: 0.05M sodium bicarbonate, pH 9.0.
3. Blocking buffer: 0.05M phosphate or Tris-HCl buffer, pH 7.4, plus 10% v/v normal horse serum.
4. Peroxidase substrate: tetramethyl benzidine (Sigma) 10 mg/mL in dimethylsulfoxide (TMB-DMSO). Acetate-citrate buffer (4.1 g of sodium acetate plus 2.1 g of citric acid to pH 6.0 in 500 mL of water). Urea: hydrogen peroxide complex (BDH 30559) tablets—use one tablet in 100 mL of distilled water. Immediately prior to use, place 250 µL of TMB-DMSO and 375 µL of urea/hydrogen peroxide in 25 mL of acetate-citrate buffer.
5. Stop solution: 1M H_2SO_4 aqueous solution.
6. Eluant panel: e.g., 0.1M citric acid, pH 2.5; 0.1M NH$_4$OH, pH 11.0; 1M KI; 3M MgCl$_2$ pH 6.5; 50% v/v ethylene glycol; 8M urea; 30% v/v isopropanol; 0.2M lithium diiodosalicylate (LIS).

7. 96-well plates, plate washer, plate reader, 8- or 12-way pipet, 37°C incubator, plate shaker, plate viewer.
8. Peroxidase (or alkaline phosphatase, *see* Note 2) conjugated second antibodies (e.g., sheep antimouse or donkey antisheep, or the like). Both of these should ideally be immunopurified on the appropriate-species IgG and cross-adsorbed on irrelevant-species Ig to remove unwanted crossreaction. Peroxidase (or alkaline phosphatase) conjugated streptavidin (if biotinylated antigen or antibody are used).

2.3. RIA/IRMA Assays

1. ^{125}I-labeled antigen or antibody (specific activity = 1–10 µCi/µg) with ≥50% maximum binding should be bought or homemade.
2. ^{125}I-NaI (e.g., Amersham): 10 µL at 100 mCi/mL, pH 9, for protein iodination.
3. Phosphate buffer 0.25*M*, pH 7.4 in distilled water.
4. Chloramine T 5 mg/mL in 0.25*M* phospate buffer, pH 7.4, freshly prepared.
5. Reducing agent: sodium metabisulfite ($Na_2S_2O_5$), 0.16 mg/mL in 0.05*M* phosphate buffer, pH 7.4.
6. Carrier: potassium iodide, 1*M* in 0.05*M* phosphate, pH 7.4.
7. Sephadex G-50: 10 g, swollen in 0.05*M* phosphate, pH 7.4.
8. Nunc Cryotubes.
9. Assay tubes: polystyrene or polypropylene, 12- × 75-mm and 16- × 100-mm polystyrene tubes and suitable racks to take, e.g., 80 tubes.
10. Horizontal vibratory/rotary shaking table for above racks (e.g., Stuart SO1, Redhill, Surrey, UK, or IKA Vibrax™ VXR8, J & K, W. Germany). Manifold dispensing/aspirating system, 5- or 8-way (e.g., Novobiolabs), 12- or 16-channel multiwell gamma counter (e.g., Nuclear Enterprises NE 1600, Edinburgh, UK).

3. Methods

3.1. Assay Development and Preparation

Prior to starting immunization, it is worthwhile to spend some time obtaining suitable positive and negative controls (or a panel of these) of both polyclonal and monoclonal antibodies. This is no time for pride—such reagents should be bought from commercial suppliers whenever feasible and begged, borrowed, or stolen from your friends and colleagues wherever they are in the world. By assembling such a panel of antibodies before immunization starts, you are able to spend time developing screening assays before the animals become immune

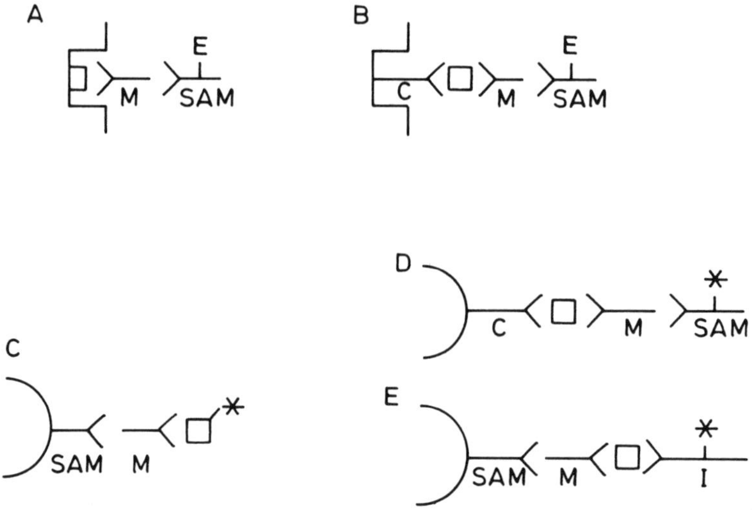

Fig. 1. ELISA **(A,B)**, RIA **(C)**, and IRMA **(D,E)** assays for primary screening of antibody response. Symbols used are □, antigen; —<, antibody; E, enzyme; ^{125}I; M, monoclonal antibody (or other test sample); SAM, sheep second antibody to mouse (or other test species) immunoglobulin; I, immunopurified specific antibody; C, polyclonal specific capture antibody. Note that antibodies C and M (or M and I) should be from different species to avoid false positive reactions with conjugated second-antibody SAM. Assays A and B use ELISA 96-well-tray formats; assays C, D, and E use porous gel-bead-immobilized antibodies in 12- × 75-mm tubes.

and you also have a set of standards against which to judge the performance of your own antibodies (*see* Notes 4, 7).

The primary screen will usually be either an ELISA- or a RIA-type assay (Fig. 1), but if possible consider running both in parallel for a while to get some idea of their relative strengths and weaknesses (*see* Note 6) during the early development phases of the assay. Many of the reagents are common to both and can easily be used later in secondary screening, so this is not wasted; likewise, if one assay "fails" during the more intensive work of hybridoma screening, it should be possible to set up an alternative backup assay quickly.

When choosing a basic format in ELISA you will need either immobilized antigen or immobilized antibody (or both) (*see* Note 1). Whichever you use needs to be available in sufficient quantity (≥100 µg) and purity (≥90%) to permit coating and yield unequivocal positives. With ^{125}I-tracer screening the choice is between ^{125}I-tracer antigen (RIA) and ^{125}I-tracer antibody (IRMA), so again the choice will depend upon which you have available in the greatest amounts and/or purity. There are, therefore, four basic formats of primary screening assay (Fig. 1):

1. ELISA with immobilized antigen captures test antibody and a signal is elicited by enzyme-conjugated second antibody (Fig. 1A);
2. ELISA with immobilized monospecific polyclonal antibody captures antigen from impure sample, and then from test antibody, which is elicited by an enzyme-conjugated second antibody (NB: The species of the coating antibody must be different from the test antibody and the enzyme-conjugated antibody should show no crossreaction with the coating antibody.) (Fig. 1B);
3. RIA with labeled antigen, complexed to soluble test antibody, and captured and separated by solid-phase second antibody (Fig. 1C); and
4. IRMA with immobilized monospecific polyclonal antibody, which captures antigen from impure sample, and then captures test antibody, which is elicited by ^{125}I-second antibody (Fig. 1D). Another variant of this assay is possible when the monospecific polyclonal antibody is in short supply, but can be immunopurified and iodinated; in this case the solid-phase bead has a "second" antibody on it, which captures the test antibody, which in turn captures an antigen from an impure sample, which in turn is elicited by the ^{125}I-tracer Ab (Fig. 1E). Other versions of these are, of course, possible; this really depends on what reagents are in short supply or plentiful *(17–19)*.

The assay should be characterized in terms of its maximum signal (OD in an ELISA or max percentage of binding of counts in RIA/IRMA) and minimum signal (i.e., background absorbance or nonspecifically bound [NSB] counts) and for this a set of good positive and negative controls are useful. Aim for a maximum absorbance of 1.0–2.0 in ELISA or 40–80% of counts bound in an RIA, and a background of ≤0.1 AU in an ELISA or ≤1% NSB in RIA. With an IRMA, aim for a positive signal of ≥3% bound counts (with an input of 300,000 counts) and a background of 2–3 times the machine counting background. At the end of the day, what matters is the distance or "window" between the strongest positive signal and the mean background. The so-called signal/noise ratio (S/N) should ideally be ≥20/1 (e.g., with a background of 1%, the maximum positive signal should be >20% bound tracer, whereas with a background noise of 2% bound a maximum bound signal of 40% is needed). You can see that ultimately, with increasing noise, the maximum positive signal cannot physically exceed 100% bound in an RIA (or OD = 2.00 AU in ELISA); therefore, an important conclusion is that in a screening assay it is better to expend effort on reducing the NSB than on increasing the maximum signal *(20–24)* (*see* Note 7).

An important question is, What constitutes a relevant positive or negative control? When screening sera, this should be nonimmune preimmunization sera, ideally from the same individual animals but otherwise a normal pool from the same species. With hybridoma screening, the positive should ideally be a known positive hybridoma in HAT medium or, failing this, ascites diluted perhaps 1/1000 in hypoxanthine–aminoprterin–thymidine (HAT) medium (*see* Note 4). For negative controls use HAT medium. If positive hybridomas are not available, use a pool of hyperimmune mouse sera prior to fusion and diluted, e.g., 1/1000 or 1/10,000 in HAT medium. However, if you have insufficient mouse positive serum samples remaining (and remember: they have to be set up in every assay batch run, *see* Note 7), as a last resort you can use another animal species as a source of positive polyclonal antibody, but remember to change to a suitable second antibody/capture antibody reagent if necessary. To summarize, time and effort spent on developing and characterizing the primary screening assays prior to immunization is time well spent and will be well rewarded later.

3.2. Primary Screening (see Note 6)

Methods of general utility for a typical ELISA and a typical RIA are given. These can be modified in the light of experience to accommodate individual cases in which the antigens' properties or supply of reagents may be unusual.

3.2.1. A Typical ELISA Primary Screen

1. Coat 96-well flat-bottomed polystyrene plates with antigen solution at 100 µL/well with 1 µg/mL of antigen in coating buffer, pH 9. Allow them to coat overnight at 4°C, and then wash with saline and block with 200 µL/well of blocking buffer for 3 h at 20°C. Wash and store the plates under maximum stability conditions for antigen (e.g., moist, 4°C, or frozen (–20°C), freeze-dried, or dried at 20°C). Prepare sufficient plates for at least a 1-wk assay workload.
2. Dispense 100-µL aliquots (in duplicate) of hybridoma supernatants (or dilutions of polyclonal antisera from 1/10 to 1/10^6) into indi-

vidual wells, together with adequate numbers of suitable positive and negative controls (the author uses four of each). Shake on a plate shaker for 2–20 h at 20°C or allow to stand undisturbed overnight at 4°C. The time required depends on the titer and sensitivity expected: High titers need shorter incubations, whereas high sensitivity requires longer contact times.

3. Wash wells three times with assay diluent, giving 2–5 min of shaking with each wash to ensure complete removal of unbound murine (or species) antibody. Add a suitable dilution of enzyme-conjugated second antibody—typically 1/1000 in assay diluent. Such a reagent can easily be bought commercially (e.g., Sigma, goat antimurine IgG–peroxidase conjugate); however, if the antigen is likely to be contaminated with IgG, then any crossreacting antibodies should have been removed prior to conjugation (e.g., Sigma). Shake or incubate the conjugate for 2–3 h and remove unbound conjugate by washing well, as in Step 2 above.

4. Add suitable substrate solution, e.g., 100 µL of peroxidase substrate (Section 2.2.4.) and shake for 1–10 min at room temperature (the speed of the reaction increases markedly over the temperature range 15–25°C, so shorter times are required at higher temperatures). Avoid azide ions as preservatives when working with peroxidase, as the former will inhibit the latter (*see* Note 2). Also avoid standing the substrate or the reaction in bright sunlight, as it is photolabile. Observe the formation of blue color by eye and when suitable density has been achieved (typically 5 min) stop the reaction with 100 µL of $1 M H_2SO_4$ added to each well with an 8- or 12-channel multipipet. It helps if the stop reagent is dispensed vigorously into the wells so as to mix thoroughly and stop the reaction rapidly and uniformly. Any color formed is yellow in consequence of the acid; wells that appear green have not been stopped properly and should be more vigorously mixed.

5. Read the optical density of the wells after blanking on suitable buffer or empty wells, and with a filter set close to 450 nm. Calculate the mean of duplicates or simply plot them in pairs.

6. The author prefers a histogram (*see* Fig. 2) in which the successive readings of numbered samples are accumulated with time and show a clear pictorial representation of the signal strength of successive samples. After 10–20 samples have been accumulated, a clear idea of the background should be emerging (and ideally should be the same as the negative controls previously incorporated). Based on the degree of stringency desired, a positive cutoff line can be set at 2, 3, or 4X the background. Provided the background is stable, these param-

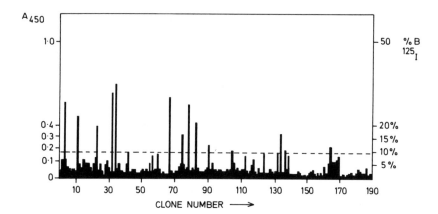

Fig. 2. Histogram showing primary screening of hybridoma supernatants in a cumulative manner following fusion, the fastest growing cells
having the lowest numbers and the slowest growing cells having the highest numbers. The screen was based on assay format 1E and the positive
cutoff was set at 10%, i.e., 5X the mean background of 2%. Representative
values for a hypothetical ELISA screen have also been added on the left-
hand axis to illustrate typical values to be expected. A total of 16 positives
clones were selected on the basis of this screen.

eters can be maintained throughout the rest of the screening. If excessive numbers of positives (e.g., ≥50) are obtained, it may be desirable to set the stringency levels higher, e.g., 10X background. However,
because of the differential growth rates of hybridomas, it should not
be assumed that the samples of highest positive signal and/or highest titer samples are necessarily the best for immuno-purification—
considerations of affinity and epitopes are also relevant. The safest
strategy is to select a range of antibodies covering high, medium, and
low signal strengths. If only low positives are obtained it is important
to repeat the screening assays in duplicate or triplicate to ensure that
the results are real, and not random noise. If they replicate on the
same sample, then later samples from the original culture should be
retested to ensure that the cells are still positive secretors and to exclude artifacts. It is unwise to attempt to discriminate for positives
with a cutoff of less than twice background. If no positives are confirmed after 300–400 wells have been screened, then a second fusion
should be contemplated (*see* Note 7).

7. Interpretation of "positives" should be unequivocal and provided in
   list form by the screening operator(s) to the cloning operator(s) as
   soon as feasible, i.e., within 6 h if possible, and certainly within 24 h.

Table 6
Reasons for Attrition–Lost Antibodies

Incubator breakdowns
Fungal/bacterial contamination
N_2(l) Freezer breakdown
Cells do not grow upon thawing
Cells will not reclone
Cells do not make good ascites
Mycoplasma contamination
Identical clones—redundancy

In this way, cell selection can proceed, overgrowth is prevented, and cloning and subcloning can proceed before valuable antibodies are lost. It is essential that selected cells are retested repeatedly throughout the cloning, recloning, and bulking-up stages to confirm that growth and secretion are proceeding satisfactorily. Any clones that cease to secrete must be ruthlessly discarded. Ultimately, because of physical limitations on the cloning process, it will be necessary to choose only about 36 positives for cloning. Initially aim for 50 positives out of 400–500 wells, i.e., at a 1/10, positive/negative ratio. Attrition occurs throughout the antibody selection process for a number of reasons (Table 6) so it is always wise to start with more than you need.

3.2.2. A Typical RIA Screen

1. Tracer preparation of good quality is the key to any RIA. Start with 10 μg of antigen of purity ≥90% in 10 μL of 50 mM phosphate buffer, pH 7.4. (If the concentration is <1 mg/mL you can increase the volume up to 100 μL, but efficiency of iodination drops off.) Add ^{125}I-NaI (5–10 μL) and 10 μL of chloramine T and, after mixing for 5–10 s, stop the reaction with 0.8 mL of reducing agent and add a carrier (Section 2.3.5.). The protein-bound ^{125}I is separated from the free ^{125}I by gel filtration on a column of Sephadex G-50 (1 cm × 10 cm) and preequilibrated in blocking buffer (*see* Section 2.2.3.). The column is eluted with this buffer and 1-mL fractions are collected and counted for ^{125}I content. Generally the first peak (fractions 3–4) contains protein-bound ^{125}I, whereas the second peak (fractions 5–10) contains free ^{125}I. Collect only the fractions of the first peak and store at 4°C. In the author's laboratory, many tracer pools have been divided into two halves with one half stored at 4°C and the other half aliquoted into 1-mL Nunc cryotubes and kept frozen in liquid nitrogen vapor at –150°C. Although frozen storage cannot slow the rate of

radioactive decay, the chemical radiolytic damage of protein, which is a consequence of radioactive decay, is largely prevented, so tracers can be used up to 6–9 mo after preparation. The specific activity upon initial iodination is often greatly in excess of what is needed; this explains why the tracer is still useful after several half-lives have elapsed.

2. Dilute an aliquot of tracer in assay diluent until it gives 10,000–20,000 counts/100 µL when counted for 100 s in a gamma counter. It helps greatly if a modern 12-well or 16-well multiplace gamma counter is available, as this speeds up screening considerably. Mix the tracer (100 µL) with serial dilutions of a known positive antibody (polyclonal or monoclonal) from 1/10 to 1/10^6 in tenfold steps using assay diluent buffer and incubate the mixture overnight without shaking at 20°C in 12- × 75-mm polystyrene tubes.

3. Separate bound from free tracer. Add an appropriate solid-phased second antibody in a 1/10 slurry using a 200-µL pipet with the plastic tip cut back to increase the lumen diameter to 1 mm. Bead slurry should be continuously mixed on a magnetic stirrer to prevent heterogeneity. If a commercially available second antibody with suitable specificity and capacity is not available, then one prepared by coupling 1 mL of raw antiserum (sheep antimouse or donkey antirabbit, or the like) to 5 mL of activated beads, either commercial CNBr-Sepharose or other homemade activated beads (25) (see Chapter 11). After shaking the slurry for 45–60 min at room temperature, separate the captured antibody–tracer complex from the free ^{125}I tracer.

4. In the author's laboratory, separation is achieved as follows: Add 300 µL of assay diluent (to wash down the insides of the tube) followed by 1.25 mL of 10% w/v sucrose in the same buffer (26). This has to be injected slowly and carefully into the bottom of the tube, so that it lifts the assay mixture, including beads, smoothly up the tube in the form of a floating cushion. Stand for 20–30 min at 20°C and 1g, by which time the beads will have settled to the bottom of the tube and have been washed free of any entrapped tracer during their passage down through the sucrose cushion. The free tracer is left in the overlaying fluid. Remove the free tracer and sucrose by means of a vacuum aspirator pump and eight-way manifold. This procedure is greatly facilitated if the tubes are held in a 5 × 16 (80-place) peg rack, e.g., Endicott No. 207, and addition of sucrose is performed with an eight-way multichannel peristaltic pump, e.g., the Watson Marlow Delta 302 is ideally suited to this work (26).

5. After sucking off the unbound tracer, count the sedimented bead pellet in about 300 µL of residual liquid for 100 s in a gamma counter,

together with antibody-free controls (i.e., tracer + buffer + solid-phase second antibody). The negative controls should give 1–2% binding: If they give more than this, consider a second wash–sedimentation step or change the additives/concentrations in the BSA/phosphate/Tween so as to reduce further the NSB. The positive control should show the maximum binding of tracer to be 40–90%, depending on the quality of the tracer. Values down to 20% maximum binding can be used with caution if the NSB is kept low (≤1%), but values below 20% are unreliable, and a better quality of tracer should be prepared. Assuming that a good quality of tracer is available, screening can start (*see* Note 6).

6. From this point on, the screening using format 1C is essentially the same as in the ELISA process (Section 3.2.1., Step 5, and Fig. 2), whereby successive samples of hybridoma supernatants are assayed, accumulated, and inspected graphically for background and positive signal levels, and ≤50 positives are selected for rescreening, cloning, recloning, and scale-up.

3.2.3. Titer, Affinity, Subclass, and the Like

1. Titer: By serially diluting the positive hybridomas at the level of scale-up corresponding to 1–20 mL, it is possible to determine a rough titer, defined as the dilution at which the maximum positive signal for that particular antibody sample is reduced by half. Serial twofold or tenfold dilutions are convenient to use, depending on the expected titer. The results can be plotted in graphical form (Fig. 3), allowing rapid comparison of the different antibody sample titers (note that these are dependent primarily on the amount of antibody in solution and therefore the growth rate of the hybridoma, as well as antibody affinity). Sometimes the situation is more complicated, as in Fig. 4: here the antibodies all have different maximum percentages of binding and differing slopes, and may also show both positive and negative slopes in the early phases of dilution. It is quite normal for RIA-type assays to show this behavior in the lower dilutions (1/10 and 1/100) because the solid-phase second antibody is unable to capture all the soluble first antibody because of its limited capacity. Roughly speaking, the shallower the slope of the dilution curve, the lower the affinity of the antibody. However, the maximum percentage of binding is also influenced by the distribution of epitope(s); some forms of the molecule (e.g., because of proteolysis or denaturation) may not express all epitopes so some antibodies will never achieve maximum tracer binding in consequence of heterogeneity and/or loss of epitopes (*see* Note 6).

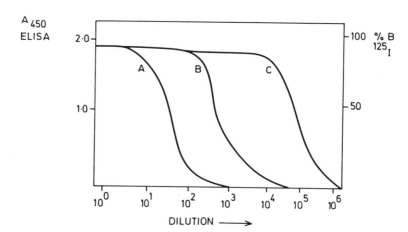

Fig. 3. Typical (hypothetical) dilution curves for individual positive hybridomas showing similar affinity (slope) but differing titers.

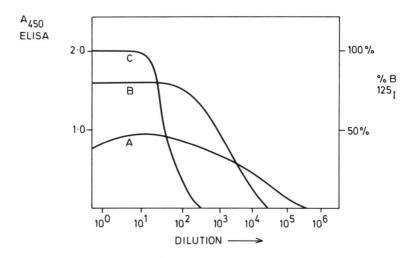

Fig. 4. Atypical (hypothetical) dilution curves for individual positive hybridomas showing differing affinity (slope) and titers. Curve A has the highest titer but the lowest affinity, curve B has intermediate titer and affinity, and curve C has the highest affinity but the lowest titer.

2. Affinity can be measured in a variety of relative or absolute methods *(27–32)*. However, there is little or no correlation between antibody affinity and ease of elution as an immunopurification reagent *(1)*, considerable efforts may be involved in purification for absolute

affinity determinations, and the quantities of antibody and antigen consumed are not practical during primary screening. It is sufficient to define an operational affinity as the relative slope of the curves, e.g., in Fig. 4, the steeper the slope (curve C), the higher the relative affinity.

3. This has important implications for protein A purification of the antibody and immunopurification applications as well as leakage (*see* Note 9). Note that in Section 3.2.1., Step 3, an IgG-chain-specific second antibody was specified. This should produce only positives with murine IgG-type antibodies. If a polyclonal antimurine Ig reagent is used, then positives could include both IgM and IgA classes, neither of which is desirable for technical reasons (i.e., they are difficult to purify and tend to have low levels of productivity and low affinity, and to be unstable in use). A simple screening method for class and subclass should be applied to all positive hybridomas at this time (i.e., after cloning $n < 36$). Various options include Ouchterlony microdiffusion (The Binding Site), hemagglutination (Serotec), and dot-ELISA (Amersham) are available commercially for murine, rat, and human immunoglobulins. In the author's laboratory the murine subclasses are also confirmed as a byproduct of the purification process of antibodies by affinity chromatography on protein A *(33,34)*. In a continuously decreasing pH gradient, murine subclass IgG_1 elutes at pH 6.5–7.0, murine IgG_{2a} elutes at pH 5.5–6.5, murine IgG_3 elutes at pH 4.5–5.5, and murine IgG_{2b} elutes at pH 3.5–4.5. Provided a single clear UV peak or simple bioassay is available, this procedure is a convenient and reliable way of assigning murine subclasses. Unfortunately neither human nor rat subclasses can readily be assigned in this way. It turns out in practice that murine IgG_1 and IgG_{2a} are the most commonly produced and the most useful. Immunoglobulins IgG_{2b} and IgG_3 are less commonly produced, are more unstable during purification, and are very sensitive to proteolysis, which makes them a bad choice for immunopurification reagents (*see* Note 9). It is usually impractical to attempt Fab fragmentation of murine subclasses IgG_{2b} and IgG_3 because of rapid destruction of biological activity *(35,36)*; furthermore, as these same subclasses are more susceptible to proteolysis, leading to severe leakage and inactivation in use, they should be avoided as solid-phase reagents if possible.

3.3. Secondary Screening (see Notes 3 and 5)

Following primary screening and depending on the rate of attrition, you will now have between 12 and 36 cloned, classed, and subclassed antibodies, which can be ranked in terms of titer and/or relative affinity as described above. A series of secondary screenings must now take place, attempting to mimic as closely as possible the final mode of use, i.e., an immobilized antibody that will reversibly bind the product antigen or vice-versa (*see* Note 1). Briefly, the immobilized Ag/Ab pair is eluted with a series of chosen eluates and the extent of elution of bound antigen or antibody is quantitated *(15)*. A profile is built up of which antibodies are readily eluted and by what reagents. If no satisfactory results are obtained, more extensive testing of eluants is performed and synergistic mixtures of eluants are studied (ref. *7,* and *see* Note 10). Finally, the chosen eluants must be tested for their ineffectiveness at damaging either the product antigen or the coupled antibody *(7)*. As before, ELISA or RIA formats are possible with labeled antigen or labeled antibody; however ELISA format 5A is convenient in that no purification or conjugation of test antibody or antigen is essential, though better results may be obtained with pure antigen-coated wells.

ELISA assays for secondary screening are available in three basic formats (Fig. 5), and which one is chosen depends on what reagents are available. Assay format 5A is very convenient in that it is physically identical to primary screening assay 1A and does not require purification of hybridoma antibody; thus it can be used earlier in the cloning process, before large volumes of tissue-culture fluid are available. Its disadvantage is that it requires more pure antigen to coat the plates. Assay format 5B differs from assay format 5A in that recognition of murine antibody Fc regions by a conjugated second antibody is not required and a more robust biotinyl–streptavidin enzyme conjugate *(24)* is used. This is useful if strong eluants have been used (e.g., pH 2.0 or 3*M* potassium thiocyanate [KSCN]), which may partially destroy epitopes, thus reducing murine Fc reactivity with the second antibody. The disadvantage of assay 5B is that it requires purification of the hybridoma antibody prior to biotinylation. Assay format 5C is more logical, in that it formally reproduces the correct orientation of an immunopurification step with an immobilized antibody and a soluble captured antigen (*see* Note 1). The disadvantages of format 5C are that both antibody and antigen have to be pure and the antigen has to be biotinylated. Note particularly in Fig. 5 that all three assays require the elution step/wash to be carried out on a single pair complex, not on a multiple sandwich, and that enzymes must be absent to avoid denaturation (and hence false positive elution) by

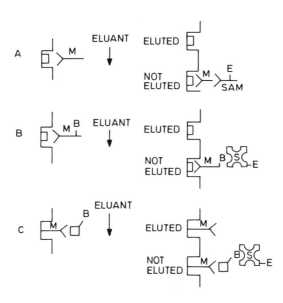

Fig. 5. ELISA assays for secondary screening of antibody/antigen elutability. Symbols are: □, antigen; —<, antibody; E, enzyme; SAM, sheep antimouse second antibody (or other species of test antibody); M, murine test antibody (or other immune species); B, biotinylated; streptavidin.

eluants. The assays are operated in general as described in Section 3.2.1., except as follows:

1. Coat the plates with 100 μL of antigen or antibody at 1 μg/mL in phosphate or bicarbonate buffer, pH 7–9, for 2–20 h at 4°C. The time, temperature, and pH can be altered to suit the stability and solubility of the proteins, as noncovalent binding of antigen/antibody to polystyrene is usually rapid and irreversible with proteins. In case of doubt, you can check for undesirable elution of antigen (or antibody) when coated directly onto polystyrene by provoking elution with the chosen eluant panel before incubation with the chosen hybridoma panel. Normally <1% of coated protein is eluted. This can most easily be demonstrated if ^{125}I-tracer is incorporated into the coating solution.

2. After coating, blocking, and washing of sufficient plates, it is convenient to reserve one coated plate for each antibody clone, so all positive and negative controls, duplicates, and representative panel of eluants can be readily compared within one plate. Add samples of hybridoma supernatant (*see* Fig. 5A) or appropriate reagents (as in Fig. 5B, C), 100 μL/well, and shake for 2–20 h at 20°C to equilibrate

binding. Wash three times, as in Section 3.2.1., Step 3, and then elute wells with 100 µL of each of the various eluants chosen to represent the various possibilities, as in Table 7 *(7)*. The eluants need not at this stage be particularly inert (Section 2.2.6.) toward the target (product) antigen, since the purpose of the experiment is to build up a "fingerprint" of the antibody in terms of what does and does not effect elution. For practical reasons the eluants should be limited in number to ≤10; additional saline controls are required to estimate maximum binding in noneluted samples. Incubate the eluants for 15 min while shaking the plates at 20°C. Then wash three times as in Section 3.2.1., Step 3, to remove all traces of eluant and to return the pH to neutrality in all wells.

3. Add second-antibody conjugate (Fig. 5A) or streptavidin-enzyme conjugate (Fig. 5C) and shake at 20°C for 1–3 h. Wash three times as in Section 3.2.1., Step 3, and add a suitable enzyme substrate, as in Section 3.2.1., Step 4. Incubate until a satisfactory color is developed in the saline (noneluted) control, stop the reaction with $1M$ sulfuric acid as in Section 3.2.1., Step 4, and read on a plate reader blanked to empty or buffered wells at 450 nm.

4. Express the optical density in each pair of duplicated eluant samples as a percentage of the optical density in the saline control, and plot these in histogram form for each antibody/eluant set (cf Fig. 6). Note that suitable choices of antigen-coating concentration and conjugated-enzyme dilution need to be made on the basis of the primary screening work, so that the optical density in the saline control lies between 1.0 and 2.0 AU. Figures 6A–E show typical antibody eluant "fingerprints" for different monoclonal antibodies directed against soybean trypsin inhibitor (SBTI). The assay format used was similar to that used in Fig. 5C, i.e., the wells were coated with monoclonal antibody to SBTI and captured biotinyl SBTI antigen, which, after elution and washing, was probed with a streptavidin–enzyme conjugate. You can readily see that each antibody has a different profile and that there is no correlation between ease of elution and affinity constant (K_d). Figure 6A shows an antibody with the highest K_d (1×10^{-10}), which is readily eluted with pH ≤3.0, but which shows potential elution with a mixture of pH 4.0 and 30% ethylene glycol—such different eluants when mixed may even be more effective by virtue of synergism *(37)*.

Figure 6B shows an antibody (CG5) with a high affinity of 4.5×10^9; however, it is readily eluted with either pH ≤3.0 or pH 11 or $1M$ KI or $2M$ MgCl$_2$. Furthermore one can predict that synergistic mixtures (e.g., pH 3.5 and 50% v/v ethylene glycol or pH 9–10 and 50%

Table 7
Typical Eluant Short List

Type	Example
Ligand, target molecule temperature, pH, electric field, distilled H_2O	
Salts	Increasing or decreasing NaCl, Na_2SO_4
Chaotropes	KSCN, KI, $MgCl_2$, $CaCl_2$, $KClO_4$, LIS, urea, guanidine
Alcohols, glycols	n-Propanol, ethylene glycol; pH 2 or 10
Amines	Diaminohexane; dimethylaminopropylamine; triethanolamine, pH 10
Solvents	DMSO, dioxane, formamide, acetonitrile
Detergents	Neutral, anionic, cationic, zwitterionic
Stabilizers	Protein, polymer, detergents, sugars, metal ions

v/v ethylene glycol) would also work well. Figure 6C shows another antibody (AG11) with a high affinity of 4.9×10^{-9} but unlike the previous one, shows ≥60% elution with every eluant of the panel except pH 5.0—it is therefore a candidate for an immunoaffinity chromatography reagent. Both low pH and high pH, as well as chaotropic salts, are all effective eluants on their own. Furthermore, synergistic eluant mixtures (e.g., ethylene glycol or isopropanol) are possible with either weak acid (pH 4.5) or weak alkali (pH 10). Figure 6D shows an antibody (4C6) with a medium to high affinity of 7×10^{-9}, which remarkably shows ≥60% elution with the whole panel of elution reagents. This must be the first choice as an immunopurification reagent; furthermore, one can predict that synergistic mixtures of isopropanol or ethylene glycol will be very effective with either weak acid (pH 5.0) or weak base (pH 9–10). Finally, in 6E we have antibody 1B5, which has a medium to low affinity of 2.2×10^{-7}. In contrast to most of the others described above, this antibody is not readily eluted, except by pH 2.5; however (unusually) it shows significantly better elution in isopropanol than ethylene glycol thus a synergistic mixture of isopropanol and pH 3.0 buffer may be effective, though isopropanol and alkali might be equally effective.

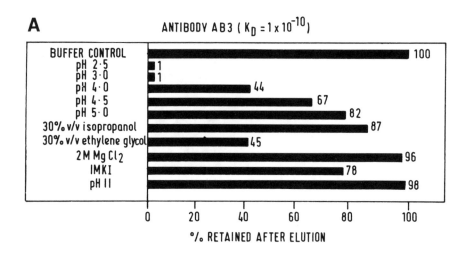

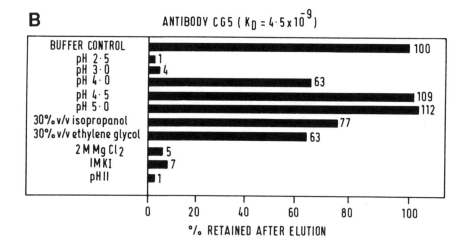

Fig. 6A–E. ELISA assay format 5C used to characterize the "elutability fingerprints" of five different monoclonal antibodies to SBTI. Results are expressed as the percent of color developed after elution and washing relative to the saline elution/wash control. Variation in pH was produced with 0.1M glycine-HCl (for pH 2.5, 3.0, 4.0, 4.5, and 5.0) and 0.1M glycine-NaOH (for pH 11.0). Isopropanol, ethylene glycol, magnesium chloride, and potassium iodide were all dissolved in distilled water. Further experimental details are given in Section 3.3.3.

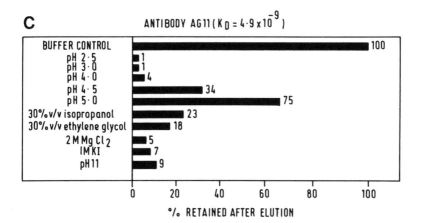

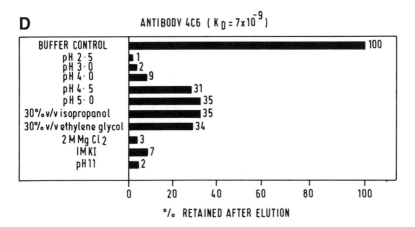

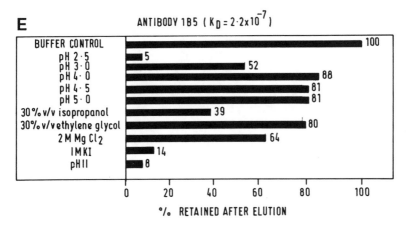

Fig. 6C–E

3.4. Eluant Selection and Epitope Mapping

From the foregoing it can be seen that it is not difficult to find a series of common eluants *(2,38,39)* that efficiently elute target antigen from most antibodies in a panel of monoclonal antibodies; in almost every case acid at a sufficiently low pH will be an effective eluant.

However, just because elution is efficient does not mean that the antibody will survive regeneration for a useful number of cycles, nor does it mean that the target antigen or prototype product are necessarily biologically active or stable. The purpose of eluant selection is thus to optimize the eluant composition in such a way that product activity is preserved as much as possible while still eluting ≥95% of the antigen from the immobilized antibody. A good general strategy *(7)* is to combine two or more eluants that function by different physico-chemical means in a such way that, when combined, their effect is greater than the sum of their separate influences (i.e., synergism). The most common example of this is a mixture of alcohols (ethylene glycol or isopropanol) with acid/alkali *(37)*, which enables one to use pH 5 in ethylene glycol instead of pH 2.5, which might otherwise damage either product or antibody. More interestingly, it is possible to equilibrate the column with buffer of the desired pH (e.g., pH 5.0) and then step-elute the column with 50–100% v/v ethylene glycol in water in such a way that a sharp interface or region of "chaos" exists between the pH buffer and the ethylene glycol. The product is thus only briefly exposed to disruption and is eluted in a sharp peak *(37)*. Another strategy used by Mejan et al. *(40)* if the product (Factor VIII) is unstable in alkali is to use a buffer of higher pH (pH 10 triethanolamine), but to make use of the gel-filtration effect of the porous support, whereby the eluted Factor VIII moves faster than the advancing buffer front and is eluted in a region of pH 9, where it is more stable. In fact volatile, weak alkaline buffers, such as dilute ammonia or triethanolamine *(41,42)* deserve more use. Another strategy (also exemplified with Factor VIII) is to use a stabilizer additive, in this case 1 *M* lysine *(7)*, which itself plays no part in the elution mechanism, but allows the use of higher concentrations of eluant than would otherwise be possible. Thus, urea concentrations up to 4 *M* can be used in the presence of 1 *M* lysine, whereas only 2 *M* urea can be tolerated by Factor VIII in the absence of lysine. Similarly, 1 *M* KI + 1 *M* lysine is well tolerated by Factor VIIIC whereas 1 *M* KI on its own is liable to produce unacceptable losses of biological activity. As a general rule, examination of tabulated data, as in Fig. 6, together with a knowledge of the stability of the target antigen (e.g., with respect to acid, alkali, chaotropes, ureas, organic solvents, and alcohols) will allow selection or rejection of possible eluants, which can then be fine-tuned regarding concentration and/or synergized in mixtures. Furthermore, knowledge of the particular product will also often point the way to further additives that may be used as stabilizers, e.g., amino acids, metal ions, chelators, thiols, polyions, and the like. Failing this, general stabilizers, such as polymers, sugars, or bland proteins, may be appropriate as additives.

Although not important at first sight, the mapping of epitopes on an antigen by means of antibody reactivity can be useful in choosing antibody panels and immunopurification reagents *(43–47)*. It is in the nature of things that most antibodies will be produced to immunodominant epitopes and relatively fewer antibodies will be produced to minor epiopes. Thus, when a large number of antibodies are produced, it is helpful in deciding which ones to discard or use in immunopurification if they are known to react with common or rare epitopes: The latter should be retained at the expense of the former.

A number of general methods have evolved to decide whether two antibodies are redundant, i.e., recognize the same epitope. In essence, a competition assay is set up in which one antibody (the reference antibody) is labeled with ^{125}I or biotin and a second antibody (or panel of antibodies) is tested in excess concentration to see to what extent it can displace the binding of the labeled test antibody or antigen. If no displacement of binding occurs, then the epitopes are unrelated; if complete inhibition of binding occurs, then the epitopes are identical; and if partial inhibition of binding occurs, then the epitopes are adjacent or overlapping. An assay format like 5B or 5C is most suitable, but the "eluant" is replaced by the second (excess) antibody. It should be noted that the results can be plotted in the form of a checkerboard, with each antibody pair in turn tested as labeled or unlabeled excess and each combination therefore tested twice. In theory, whether the labeled or competing antibody is used should give the same result, but in practice this is not always so (ref. *45* and *see* Note 6). The reasons for this are several, but include a change in antibody affinity upon solid phasing or damage resulting from iodination, biotinylation, or the like, such that their reactivity is no longer equivalent to the unlabeled competitor. In addition to such competitive binding assays, one may under favorable circumstances, use active site inhibitors (in the case of antigens that are enzymes) or simple chemical modification reagents *(18,44)*, which may derivatize amino acid functional groups on antigens (*see* Note 5). Apart from their use as epitope maps, the latter may also serve a practical purpose in alerting the user to potential problems if the antigen is liable to chemical modifications leading to loss of critical epitopes.

Because of the extreme specificity of monoclonal antibodies, it is important not to select for immunopurification an antibody that may recognize a labile epitope (i.e., one that is liable to disappear in consequence of proteolysis, reduction/oxidation, glycosylation, deamidation, or the like).

3.5. Tertiary Screening

Hitherto, screening has taken place in surrogate fashion: ELISA/ RIA assays have been set up that to a greater or lesser extent mimic the performance of the desired antibody in an immunopurification format (*see* Note 1). The purpose of tertiary screening is to take purified antibody, immobilize it at a known concentration (e.g., 1 mg/mL gel) by a defined chemistry (e.g., CNBr) to a defined bead chemistry/ porosity/surface area (e.g., Sepharose CL4B) and to determine both the capacity of this adsorbant for antigen (product) binding on a small scale (e.g., 1 mL) and the recovery of biologically active product using the chosen eluant(s) *(7,48)*. It is desirable that this performance should be retained over a number of test adsorption/elution/washing cycles as well as any necessary storage/cleaning/sterilizing cycles. Unfortunately it is not always possible to predict that a promising antibody in primary and/or secondary screening will always function well in tertiary screening, but in the author's experience there is a 90% chance that it will behave as expected.

At this point you should aim to have approx 10 antibodies to test in tertiary screening. The reduction in numbers will have come about because of random attrition (Table 6) as well as rejection because of class (IgM, IgA) and subclass (IgG_3, IgG_{2b} in the cases of murine antibodies) as well as inability to be eluted by any eluant that is inert with respect to the product. Ideally the 10 antibodies for testing should be readily eluted by a mild eluant, yet have medium/high affinity (these two attributes are not mutually exclusive in monoclonal antibodies, *see* Fig. 6B, C, D) and a high functional capacity. A small proportion of antibodies function well in solution, but "die" when immobilized by adsorption to plastic or to CNBr activated Sepharose 4B. Although they can be "rescued" by indirect coupling to protein A plus crosslinking *in situ (49,50)*, it is probably easier to drop them from the tertiary screen unless they have some overriding virtue(s) or are the only remaining candidates (*see* Note 11).

The antibodies should be purified either by sodium sulfate precipitation and/or protein A affinity chromatography as follows:

1. Add 18% w/v anhydrous $Na_2 SO_4$ to the antibody preparation (it may be necessary to concentrate tissue-culture supernatants or dilute ascites) and warm to 37°C to ensure dissolution of the salt. After 5 min at 37°C, centrifuge the precipitate at 5000g for 1 min at 37°C. Discard the supernatant, redissolve the precipitate in distilled water, and repeat the precipitation with added anhydrous Na_2SO_4, 17% w/v. After the final resuspension, if needed, the antibody can be further purified on immobilized protein A. Start with the sample adjusted to pH 8.5–9.0 in 2–3M NaCl to ensure tight binding of the murine IgG_1 subclass and elute with a decreasing salt/pH gradient formed from 0.1M phosphate/0.1M sodium citrate, pH 9.0, and 0.1M phosphate/0.1M citric acid, pH 2.0. The various subclasses of murine IgG elute at characteristic pH values (Section 3.2.3. paragraph 3, refs. *33,34*), which is useful not only to purify the antibodies prior to immobilization, but also to confirm the subclass assignments. The purpose of purification is to ensure that the amount of immobilized antibody protein is known accurately. If this is not done, any comparative evaluations of capacity, yield, leakage, and so on are meaningless. If you have an accurate and reliable assay for antibody biological activity, you may feel confident that you can measure the amount coupled by a depletion assay (i.e., before–after coupling assays); however, bear in mind that a precision of ≤5% is needed to remove the need for purification.

2. It is very important to know the amount of protein coupled to the solid phase to within 5%; furthermore, it makes comparative testing easier if all antibodies from the panel ($n = 10$) are coupled at the same ligand density (e.g., 1.0 mg/mL ± 0.1). One method of estimating the content is to measure the absorbance at 280 nm before and after coupling. The OD × volume units after coupling divided by the OD × volume units before coupling × 100 gives the percentage uncoupled, hence the percentage coupled can be estimated. Unfortunately the method suffers from serious errors caused by dilution by solvent within the beads, aggregation and light scattering of soluble IgG in alkaline solution, and release of UV-absorbing impurities from the solid phase. A direct assay of the amount coupled is likely to be more reliable. The author prefers to calculate the percentage coupled

directly by adding a ^{125}I-tracer to the antibody before coupling, thus counts/mL/100 s in 1 mL of gel before coupling divided into counts/mL/100 s in gel after coupling gives the percentage coupled directly.

If ^{125}I-tracer facilities are not available it is quite feasible to use the Pierce (Rockford, IL) BCA assay on a 100-μL aliquot of gel slurry, since the protein, although present in the solid phase, releases a soluble purple chromogen into the supernatant reagent.

3. Coupling of purified antibody for tertiary screening can be carried out on any activated gel, but for convenience the commercially activated gel CNBr-Sepharose 4B (Pharmacia) is a first choice, since it is widely used and reliable. In certain cases, if the amount of protein to be coupled is high ($\geq$ 10mg/mL) (*see* Note 8) and/or the mol wt of the target ligand is $\geq 1 \times 10^6$, then it may be prudent to use a higher-capacity and/or higher-pore-diameter gel, such as periodate oxidized Sephacryl S-1000 (*see* Chapter 11 and ref. *48*). The main requirement in coupling prior to tertiary screening is that the input level of antibody (e.g., 1 mg/mL) should as closely as possible equal the amount coupled (i.e., 1 mg/mL); thus the main requirement of the coupling method is that it should be $\geq$95% efficient, and CNBr-Sepharose 4B when used at pH 8.6 usually suits this purpose.

4. Swell sufficient CNBr-activated Sepharose 4B to give 1.0 mL of settled gel for each antibody used—dispense this amount of hydrated gel into the bottom of each of 10 tubes (polystyrene 16 × 100 screw-cap tubes). Add 1 mg of the different chosen purified antibodies to each tube in a vol of 1–2mL and adjust each pH to 8.6 ± 0.2. Also add (if possible) 10,000 counts/100 s of ^{125}I-murine IgG antibody tracer. Incubate overnight at 20°C while roller-mixing, and then centrifuge and carefully separate the fluid phase. Count separately the percentage of counts in the gel and fluid phases and calculate the percentage of gel-bound antibody and hence the concentration of immobilized antibody. Block any residual CNBr groups with 0.1*M* ethanolamine, pH 9.0, for 3 h at 20°C.

Wash the gel with pH 4.7 (0.1*M* acetate buffer), then water, then pH 9.5 (0.05*M* borate buffer), then water, and then pH 7.0 (0.1*M* phosphate buffer), and repeat the counting procedure to estimate the percentage of tracer still bound to the solid-phase gel. The purpose of these abrupt pH changes is to remove any noncovalently bound antibody (*see* Note 9) that might leak off later in use and either compromise the binding capacity (by competition) or leak into and contaminate the product. *In extremis*, it may be desirable to subject the solid phase to a series of dummy elution cycles with those eluants chosen from the secondary screening of eluants, e.g., pH 2.5, pH 11.0, 1*M* KI, 5*M* urea, or the like, because when they are first placed in contact with the antibody coupled to solid phase, some leakage is likely to occur.

5. To 0.1 mL of each solid phase add 1 mL of crude product containing 100,000 counts/100 s of labeled tracer antigen (either fluorescent or enzyme labels might be possible in certain circumstances). Roller-mix the solid-phase antibody and product for several hours and remove 0.2 mL aliquots at zero, 10 min, 20 min, 30 min, 1 h, 2 h, 3 h, and so on. Separate bound and free tracer antigen by centrifugation, filtration, washing, and so on, and calculate the maximum percentage of binding of tracer and hence the capacity in units of product per milliliter of gel. This is the static saturation binding capacity. Wash each solid phase three times in saline and then pack into minicolumns (e.g., Biorad PolyPrep™ columns, Richmond, CA). Elute each solid-phase antibody with the appropriate eluant as determined in Section 3.4. (i.e., that eluant that will most effectively elute the product with the least damage to its biological activity). After washing out the product, neutralize (pH 2.5, pH 11.0) or dialyze (1*M* KI, 5*M* urea, and so on) as soon as possible to preserve biological activity. If a ^{125}I-tracer was added to the product, count the amount of tracer eluted and the amount remaining bound to the solid phase (less any counts from labeled antibody coupled) as well as any unbound tracer in the wash-through. Calculate the percentage of units bound that were subsequently recovered in the eluant; also measure by bioassay the amount of biologically active product recovered. It is also useful to measure the total protein recovered in the eluant by the Pierce BCA assay. At this point you can calculate several figures of merit for each column, e.g., capacity as units bound; capacity as yield of units desorbed; (mass of antigen, units of biological activity, percentage of input bound, and so on) and specific biological activity (units/milligram total protein or units/milligram antigen) in the final product. A clear picture should emerge of which antibody/eluant composition has the best overall combination of desirable features.

A prudent experiment is to repeat the above binding/elution cycle a further number of times (e.g., $n = 2$ to $n = 10$) to ascertain that capacity, yield, purity, leakage, and so on are typical figures. As a general rule the parameters obtained in the first cycle of column use are not typical. On subsequent reuse, capacity tends to decrease, percentage yield and purity tend to increase, and antibody leakage tends to decrease. It is worth bearing in mind that some combinations of eluant/gel are not very compatible (e.g., 1*M* KI, 3*M* KSCN with noncrosslinked agarose).

4. Notes

1. A serious criticism of any screening assay is that it does not closely approximate the final form of use, especially when the screen uses a format in which the antigen is immobilized (e.g., Fig. 1A) and the antibody is bound and eluted. Although no formal studies have been published, a variety of experiments in which the same antibody/antigen pairs have been tested in both the normal (immobilized antibody) and reversed (immobilized antigen) format show that in the majority of cases they both predict the same ease and spectrum of eluant feasibility.

2. Screening assays have been described in ELISA format using peroxidase labels. This is purely for convenience, as peroxidase is widely available and cheap. Alkaline phosphatase labels can equally well be used. Although more expensive, they show better sensitivity, linearity, and alkaline stability than peroxidase, and can be made even more sensitive by secondary enzymic cycling. Clearly antigens that contain peroxidase or alkaline phosphatase should not be screened with the same enzyme in the conjugate, to avoid a high background.

3. Though screening here is described primarily for immunopurification, it may be convenient to bear in mind that the same antibodies may also be useful for assays, depletion, or histology, so further characterization might include solid-phase adsorption to plastic, conjugation to enzymes, iodination, and biotinylation, as well as general characterization methods, such as isoelectric focusing and SDS-PAGE (±reduction), which are useful for fingerprinting and purity evaluation.

4. It has been generally assumed that screening is carried out on a panel of antibodies generated "in house." It is of course equally feasible to test a panel that was acquired from external sources. Such antibodies should be treated carefully, as they may be duplicated and/or of uncertain origin, and have little or no data on species, class and/or subclass, antigen and/or epitope purity, or screening method. It makes sense to start the screening of such materials as early as possible in the screening program, even at the primary screening phase, to get some rough idea of titer, affinity, and the like.

5. The chapter does not describe those special situations in which the antigen may have a specific biological activity that can be readily determined by a sensitive and reliable assay. Obvious examples are enzymes and clotting factors. Frequently antibodies that inhibit these

activities may be detected, however, when planning such a screening exercise, it should be remembered that many other positive binding antibodies that are not inhibitory may be missed, so these inhibition bioassays are better suited to use as secondary or characterization, rather than primary, screening assays. Also take into account the requirements for good assay discrimination (S/N ratio) as outlined in Section 3.1.1.

6. Artifacts do occur during screening, and it is as well to be aware of the possibilities, though in most cases they lead to only a small percentage of "lost" antibodies, and this is of little consequence for immunopurification.

During immunization, complexes may form in the host animal if the antigen is an enzyme or inhibitor with the homologous inhibitor or enzyme from the host. The active site(s) are thus lost and antibodies are less likely to arise to such areas on the molecule's surface. One strategy in the case of enzyme antigens is to preblock the antigen's active site with a low-mol-wt covalent inhibitor. This also has the advantage that it reduces the toxicity of enzymes in the host. This same problem can reoccur if the antigen used in the screening assay is an enzyme or inhibitor that can form complexes with the homologous inhibitor or enzyme in the bovine fetal serum added to the HAT medium during tissue culture. Again the solution is to preblock the antigen so that it is biologically inactive.

A variant of this situation can arise when neoantigens are formed by complex formation either in the host and/or in the screening assay plus fetal calf serum. Whether antibodies to such neoantigens are a good or a bad thing depends on the nature of the epitopes that you want to recognize in the immunopurification application.

A number of artifacts are known to arise occasionally in specific screening assays. During immobilization of proteins (antigen and antibody) by direct adsorption onto polystyrene, some antigens may be lost, damaged, or occluded as a result of strong noncovalent binding *(51)*; likewise, so-called plastic neoantigens that do not exist when the molecule is in solution may be generated.

In RIA/IRMA, damage to sensitive epitopes may occur directly as a result of iodination of tyrosine residues in the active site or epitope, or (more likely) as a result of the strong oxidizing conditions used during iodination, which can damage methionine, cysteine, and tryptophan residues. One way to diagnose and/or avoid this is to use the Bolton-Hunter reagent *(52;* Amersham), which combines under milder conditions with the ε-NH$_2$ groups of lysine. Of course this may destroy epitopes that contain essential lysines.

7. There are a number of general problems which it is common sense to avoid. Do not immunize iv with toxic or immunosuppressive antigens: They should be inactivated and/or administered by the ip route. It is unwise to immunize with impure antigen, but if this cannot be avoided, try not to screen with the same impure antigen, otherwise you are likely to pick up positive antibodies to the impurities. Avoid planning a fusion with only one hyperimmune mouse—a back-up No:2 mouse (or other species) is a great time-saver if your best animal dies in consequence of accidents or other causes. Try to avoid screening just for nonimmune murine IgG production in tissue culture. It is far too nonspecific. Do not use assays which have a poor S/N ratio, nor run them without appropriate positive and negative controls on every occasion, as baselines and reagents do drift unpredictably over time.

8. A special case of immunopurification is immunodepletion (53). Here the purpose is removal of antigen activity to very low levels (≤1%) without removing other active materials present. Generally depletion reagents have to have high affinity and/or high loading levels on the solid phase and work with very low concentrations of target antigen (ng–μg/mL levels). An important difference from most immunopurification work is that when regenerating the solid phase only the antibody activity and not the antigen needs to survive the elution step. Thus more aggressive elution reagents can be contemplated. This is an application in which polyclonal antibodies are often superior, since they contain a proportion of antibodies of very high affinity as well as having a wide spectrum of epitope recognition, so the demand for a strong regeneration elution scheme is no drawback.

9. An unavoidable consequence of using immobilized antibodies is that some variable amounts of immunoglobulin leak from the column into the final product. The sum total of leakage is contributed by bead breakdown, chemical coupling cleavage, and antibody degradation as well as desorption of noncovalently adsorbed antibody. Prudent design and operation (especially precycling and extensive washing; *see* Chapter 11) can reduce much of this, but leakage will still contribute 5–10 ng/mL of antibody in most products. These levels can be assessed qualitatively by ^{125}I-tracer incorporation into immobilized antibody or ELISA for quantitation of leaked murine IgG. There is some evidence that even within a subclass, antibodies themselves vary in leakage potential; if this is important, it may well be worth including leakage measurements in the tertiary screening process, as well as avoiding subclasses IgG_{2b} and IgG_3, which are known to be susceptible to proteolysis (35,36).

10. Although "chemical" methods of elution have largely been used in screening here, other methods have been described, which may be appropriate in certain intractable or unusual situations. These include temperature-change elution (up or down) *(54)*; high-low-pressure cycling *(55)* and electrophoretic *(56)* and isotachophoretic desorption *(57)*. Another possible system combining both chemical and electrophoretic approaches is feasible if the sample is placed and eluted at one end of a preparative isoelectric-focusing bed. This has the theoretical advantage that desorbed product will rapidly migrate and focus (concentrate) at its isoelectric point in a stable pH and chemical environment.

11. It is a frequent observation that antibodies that bind well in solution fail to give the expected capacity on coupling to a solid phase. Although chemical inactivation may be the problem in a minority of cases, the usual cause of the problem is steric occlusion, which is common to all immobilized antibodies. The problem can be minimized by coupling the antibody at a low density (0.5–2.0 mg/mL) *(58)* and/ or using a gel support in which the gel pores have a mol-wt exclusion limit much greater than the target molecule. In the author's laboratory good use is made of Sephacryl S-1000 in this respect *(48)*; *see* Chapter 11.

Acknowledgments

I am grateful to Martin Bonde (Department of Biotechnology, Technical University of Denmark) for providing the raw data for Fig. 6 and to Joan Dawes, Brenda Griffin, Valerie Hornsey, Keith James, Liselle Micklem, Ian MacGregor, and Chris Prowse for providing such a fertile environment for our ideas to flourish in.

References

1. Sada, E., Katoh, S., Kiyokawa, A., and Kondo, A. (1988) Characterisation of fractionated polyclonal antibodies for immunoaffinity chromatography. *Biotechnol. Bioeng.* **31,** 635–642.

2. Hill, C. R., Birch, J. R., and Benton, C. (1986) Affinity chromatography using monoclonal antibodies, in *Bioactive Microbial Products III* (Stowell, J. D., Bailey, P. J., and Winstanley, D. J., eds.), Academic, London–New York, pp. 175–190.

3. Sada, E., Katoh, S., Sukai, K., Tohma, M., and Kondo, A. (1986) Adsorption equilibrium in immuno-affinity chromatography with polyclonal and monoclonal antibodies. *Biotechnol. Bioeng.* **28,** 1497–1502.

4. Sheppard, A. J., Hughes, M., and Stephen, J. (1987) Affinity purification of tetanus toxin using polyclonal and monoclonal antibody immunoadsorbents. *J. Appl. Bacteriol.* **62,** 335–348.

5. Miller, K. F., Bolt, D. J., and Goldsby, R. A. (1983) A rapid solution-phase screening technique for hybridoma culture supernatants using radiolabeled antigen and a solid-phase immunoadsorbent. *J. Immunol. Methods.* **59**, 277–280.

6. Kenney, J. S., Hughes, B. W., Masada, M. P., and Allison, A. C. (1989) Influence of adjuvents on the quantity, affinity, isotype and epitope specificity of murine antibodies. *J. Immunol. Methods* **121**, 157–166.

7. Hornsey, V. S., Griffin, B. D., Pepper, D. S., Micklem, L. R., and Prowse, C. V. (1987) Immunoaffinity purification of factor VIII complex. *Thromb. Haemost.* **57**, 102–105

8. Ronald, W. P., Tremaine, J. H., and MacKenzie, D. J. (1986) Assessment of southern bean mosaic virus monoclonal antibodies for affinity chromatography. *Phytopathology* **76**, 491–494

9. Locker, D. and Motta, G. (1983) Detection of antibody secreting hybridomas with diazobenzyloxymethyl paper: An easy, sensitive and versatile assay. *J. Immunol. Methods* **59**, 269–275.

10. Smith, E., Roberts, K., Butcher, G. W., and Galfre, G. (1984) Monoclonal antibody screening: Two methods using antigens immobilized on nitrocellulose. *Anal. Biochem.* **138**, 119–124.

11. Burger, D., Ramus, M.-A., and Schapira, M. (1985) Antibodies to human plasma kallikrein from egg yolks of an immunised hen: Preparation and characterisation. *Thromb. Res.* **40**, 283–288.

12. Hassl, A. and Aspock, H. (1988) Purification of egg yolk immunoglobulins. A two-step procedure using hydrophobic interaction chromatography and gel filtration. *J. Immunol. Methods* **110**, 225–228.

13. Polson, A., von Wechmar, B., and van Regenmortel, M. H. V. (1980) Isolation of viral IgY antibodies from yolks of immunised hens. *Immunol. Commun.* **9**, 475–493.

14. Stuart, C. A., Pietrzyk, R. A., Furlanet, R. W., and Green, A. (1988) High affinity antibody from hens eggs directed against the human insulin receptor and the human IGF-l receptor. *Anal. Biochem.* **173**, 142–150.

15. Weiss, M. and Eisenstein, Z. (1987) A simple method to optimise elution from affinity chromatography. *J. Liquid Chromatogr.* **10**, 2815–2824.

16. Wright, J. F., Micklem, L. R., Scott, A., and James, K. (1983) Testing of monoclonal antibody specificity. *Hybridoma* **2**, 351–353.

17. Griffin, B. D., Micklem, L. R., McCann, M. C., James, K., and Pepper, D. S. (1986) The production and characterisation of a panel of ten murine monoclonal antibodies to human procoagulant factor VIII. *Thromb. Haemost.* **55**, 40–46.

18. Novick, D., Eshhar, Z., and Rubinstein, M. (1982) Monoclonal antibodies to human alpha interferon and their use for affinity chromatography. *J. Immunol.* **129**, 2244–2247.

19. Novick, D., Eshhar, Z., Fischer, D. G., Friedland, J., and Rubinstein, M. (1983) Monoclonal antibodies to human interferon-gamma; production, affinity purification and radiommunoassay. *EMBO J.* **2**, 1527–1530.

20. Graves, H. C. B. (1988) The effect of surface charge on non-specific binding of rabbit immunoglobulin G in solid phase immunoassays. *J. Immunol. Methods* 111, 157–166.

21. Graves, H. C. B. (1988) Noise control in solid phase immunoassays by use of a matrix coat. *J. Immunol. Methods* 111, 167–178.

22. Kenna, J. G., Major, G. N., and Williams, R. S. (1985) Methods for reducing non-specific antibody binding in enzyme linked immunosorbent assays. *J. Immunol. Methods* 85, 409–419.

23. Robertson, P. W., Whybin, L. R., and Cox, J. (1985) Reduction in non-specific binding in enzyme immunoassays using casein hydrolysate in serum diluents. *J. Immunol. Methods* 76, 195–197.

24. Wadsley, J. J. and Watt, R. M. (1987) The effect of pH on the aggregation of biotinylated antibodies and on the signal-to-noise observed in immunoassays utilising biotinylated antibodies. *J. Immunol. Methods* 103 1–7.

25. Wright, J. F. and Hunter, W. M. (1982) A convenient replacement for cyanogen bromide activated solid phases in immunoradiometric assays. *J. Immunol. Methods* 48, 311–325.

26. Hunter, W. M. and Corrie, J. E. T. (1983) The sucrose layering separation: A non-centrifugation system, in *Immunoassays for Clinical Chemistry* (Hunter, W. M. and Corrie, J. E. T., eds.), Churchill-Livingstone, Edinburgh, pp. 170–177.

27. Hall, T. J. and Heckel, C. (1988) Thiocyanate elution estimation of relative antibody affinity. *J. Immunol. Methods* 115, 153–154.

28. MacDonald, R. A., Hosking, C. S., and Jones, C. L. (1988) The measurement of relative antibody affinity by ELISA using thiocyanate elution. *J. Immunol. Methods* 106, 191–194.

29. MacDonald, R. A. and Hosking, C. S. (1988) Thiocyanate elution estimation of relative antibody affinity—reply letter. *J. Immunol. Methods* 115, 155.

30. Sevlever, D., Bruera, M. R., and Gatti, C. A. (1986) Measurement of antigen:antibody binding constants by elution of affinity chromatography columns with continuous concentration gradients of dissociating agents. *Immunol. Inv.* 15, 497–503.

31. Friguet, B., Chaffotte, A. F., Djavadi-Ohaniance, L., and Goldberg, M.E. (1985) Measurements of the true affinity constant in solution of antigen antibody complexes by enzyme linked immunosorbent assay. *J. Immunol. Methods* 77, 305–319.

32. Schots, A., van der Leede, B. J., De Jongh, E., and Egberts, E. (1988) A method for the determination of antibody affinity using a direct ELISA. *J. Immunol. Methods* 109, 225–233.

33. Ey, P. L., Prowse, S. J., and Jenkins, C.R. (1978) Isolation of pure IgG$_1$, IgG$_{2a}$ and IgG$_{2b}$ immunoglobulins from mouse serum using protein A-Sepharose. *Immunochemistry* 15, 429–436.

34. Seppala, I., Sarvas, H., Peterfy, F., and Makela, 0. (1981) The four subclasses of IgG can be isolated from mouse serum by using protein A-Sepharose. *Scand. J. Immunol.* 14, 335–342.

35. Lamoyi, E. and Nisonoff, A. (1983) Preparation of F(ab')$_2$ fragments from mouse IgG of various subclasses. *J. Immunol. Methods* **56**, 235–243.
36. Parham, P. (1983) On the fragmentation of monoclonal IgG$_1$, IgG$_{2a}$ and IgG$_{2b}$ from BALB/c mice. *J. Immunol.* **131**, 2895–2902.
37. Fornstedt, N. (1984) Affinity chromatographic studies on antigen–antibody dissociation. *FEBS Lett.* **177**, 195–199.
38. Robinson, D. R. and Jencks, W. P. (1965) The effect of compounds of the urea guanidinium class on the activity coefficient of acetyltetraglycine ethyl ester and related compounds. *J. Am. Chem. Soc.* **87**, 2462–2470.
39. Robinson, D. R. and Jencks, W.P. (1965) The effect of concentrated salt solutions on the activity coefficient of acetyltetraglycine ethyl ester. *J. Am. Chem. Soc.* **87**, 2470–2479.
40. Mejan, 0., Fert, V., Delezay, M., Delaage, M., Cheballah, R., and Bourgois, A. (1988) Immunopurification of human factor VIII/vWF complex from plasma. *Thromb. Haemost.* **59**, 364–371.
41. Sairam, M. R. and Porath, J. (1976) Isolation of antibodies to protein hormones by bioaffinity chromatography on divinylsulfonyl Sepharose. *Biochem. Biophys. Res. Commun.* **69**, 190–196.
42. Lim, P.-L. (1987) Isolation of specific IgM monoclonal antibodies by affinity chromatography using alkaline buffers. *Mol. Immunol.* **24**, 11–15.
43. Friguet, B., Djavadi-Ohaniance, L., Pages, J., Bussard, A., and Goldberg, M. (1983) A convenient enzyme linked immunosorbent assay for testing whether monoclonal antibodies recognise the same antigenic site. Application to hybridomas specific for the B$_2$-subunit of *Escherichia coli* tryptophan synthase. *J. Immunol. Methods* **60**, 351–358.
44. MacGregor, I. R., Micklem, L. R., James, K., and Pepper, D. S. (1985) Characterisation of epitopes on human tissue plasminogen activator recognised by a group of monoclonal antibodies. *Thromb. Haemost.* **53**, 45–50.
45. Morel, G. A., Yarmush, D. M., Colton, C. K., Benjamin, D. C., and Yarmush, M. L. (1988) Monoclonal antibodies to bovine serum albumin: Affinity and specificity determinations. *Mol. Immunol.* **25**, 7–15.
46. Stahli, C., Miggiano, V., Stocker, J., Staehelin, Th., Haring, P., and Takacs, B. (1983) Distinction of epitopes by monoclonal antibodies. *Methods Enzymol.* **92**, 242–253.
47. Takada, A., Shizume, K., Ozawa, T., Takahashi, S., and Takada, Y. (1986) Characterisation of various antibodies against tissue plasminogen activator using highly sensitive enzyme immunoassay. *Thromb. Res.* **42**, 63–72.
48. Hornsey, V. S., Prowse, C. V., and Pepper, D. S. (1986) Reductive amination for solid phase coupling of protein:A practical alternative to cyanogen bromide. *J. Immunol. Methods* **93**, 83–88.
49. MacSween, J. M. and Eastwood, S. L. (1978) Recovery of immunologically active antigen from Staphylococcal protein A–antibody adsorbent. *J. Immunol. Methods* **23**, 259–267.
50. Schneider, C., Newman, R. A., Sutherland, D. R., Asser, U., and Greaves, M. F. (1982) A one step purification of membrane proteins using a high efficiency immunomatrix. *J. Biol. Chem.* **257**, 10766–10769.

51. Sankolli, G. M., Stott, R. A. W., and Kricka, L. J. (1987) Improvement in the antibody binding characteristics of microtiter wells by pretreatment with antiIgG Fc immunoglobulin. *J. Immunol. Methods* **104,** 191–194.
52. Bolton, A. E. and Hunter, W. M. (1973) The labelling of proteins to high specific radioactivities by conjugation to a ^{125}I-containing acylating agent. Application to the radioimmunoassay. *Biochem. J.* **133,** 529–539.
53. Hornsey, V. S., Waterston, Y. G., and Prowse, C. V. (1988) Artificial factor VIII deficient plasma: Preparation using monoclonal antibodies and its use in one stage coagulation assays. *J. Clin. Pathol.* **41,** 562–567.
54. Ohlson, S., Lundblad, A., and Zopf, D. (1988) Novel approach to affinity chromatography using weak monoclonal antibodies. *Anal. Biochem.* **169,** 204–208.
55. Olson, W. C., Leung, S. K., and Yarmush, M. L. (1989) Recovery of antigens from immunoadsorbents using high pressure. *Biotechnology* **7,** 369–373.
56. Muller, R. F., Palluk, R., and Kempfle, M. A. (1984) The preparative refinement of steroid specific antibodies by affinity chromatography combined with electrophoretic desorption–apparatus and method. *J. Chem. Tech. Biotechnol.* **34,** 263–272.
57. Prusik, Z. and Kasicka, V. (1985) Micropreparative isotachophoretic electrodesorption of monoclonal antibodies from an affinity adsorbent. *J. Chromatogr.* **320,** 81–88.
58. Sada, E., Katoh, S., Kondo, A., and Kiyokawa, A. (1986) Effects of coupling method and ligand concentration on adsorption equilibrium in immunoaffinity chromatography. *J. Chem. Eng. Japan* **19,** 502–506.

CHAPTER 11

Some Alternative Coupling Chemistries for Affinity Chromatography

Duncan S. Pepper

1. Introduction

There are a great many methods published for use in coupling ligands to solid phases to provide affinity purification matrices for macromolecules. Frequently the authors are very enthusiastic about some feature of their work and commercial suppliers too will tell you about their strong points (but not the weak points) in their products.

Many of the people who want to use affinity chromatography are not trained as synthetic organic chemists and don't want to spend time acquiring these skills. An obvious starting point for most people is therefore to use commercial activated gels or precoupled ligands. Unfortunately things can go wrong; as alluded to above, you may be told about the good points but not the bad, also the particular gel particle you need (base polymer, pore diameter, surface area, and particle size) cannot realistically be offered in every possible combination—frequently you will get a choice of only one particle, rarely two, and never more than three. Trying to compare competing claims from manufacturers data is impossible in practice even if you are experienced with these products; either the data is not available or it is presented in such a form (units, and so on) that defy interconversion.

It is the purpose of this chapter to describe two or three approaches that are simple, reliable, cheap, convenient, and can be used in 95%

From: *Methods in Molecular Biology, Vol. 11: Practical Protein Chromatography*
Edited by: A. Kenney and S. Fowell Copyright © 1992 The Humana Press Inc., Totowa, NJ

of cases by people who are not trained chemists. Several principles underly the choice of methods, namely avoidance of organic/anhydrous solvents, ability to easily scale up or scale down quantities, single-step reactions, ease of quantitation of product, activated gels should be stable in aqueous suspension at 4°C for several months, unpleasant or toxic reagents are avoided, and last but not least, the efficiency of coupling of valuable proteins should exceed 90% uptake from concentrations of 1 mg/mL or less.

1.1. A Brief Outline of the Divinyl Sulfone Method

The divinyl sulfone (DVS) method of activation works well with almost any hydroxylic or carboxylic polymer (cellulose, dextran, agarose, Sepharose, Sephacryl, Fractogel, Ultrogel, Trisacryl, and so on). It is fairly rapid ($\leq$ 2 h) and the activated material is stable for up to 12 mo in aqueous suspension at 4°C. The reagent itself is moderately expensive but is used in small amounts, its toxicity is limited—you need only avoid spilling it on skin or inhaling vapor. It is a pale brown oily liquid that slowly dissolves in alkaline buffers or water on shaking. Owing to the impurities, a slight brown color is often imported to the gel. Vinyl sulfone-activated gels couple (Fig. 1) efficiently to proteins and thiols in the pH range of 6.5–10.0 and carbohydrates in the pH range of 9.5–12.5. Because of the alkaline lability of the vinyl sulfone linkage, monovalent ligands (e.g., amino acids) are likely to hydrolyze from the support at pH $\geq$ 8.0; however, proteins (multivalent points of attachment) invariably show less leakage at alkaline pH than is the case with CNBr. The degree of activation of the gel can be quantitated by titration with sodium thiosulfate. The amount of protein coupled to the gel can be determined in several ways, including simple chromogenic assays, most conveniently the BCA assay. There is no single adequately detailed publication on DVS applications but this work is based on short reports from a number of sources (1–7). DVS gels are the nearest thing to a universal activated solid phase, coupling well to both proteins via amino groups and carbohydrates via hydroxyl groups, (see also Note 12).

1.2. A Brief Outline of the Periodate Method

Periodate activation of Sephacryl is the best example of a wide range of methods that depend on the reductive amination of a Schiffs base formed between an aldehyde and an amine (Fig. 2). Usually, the

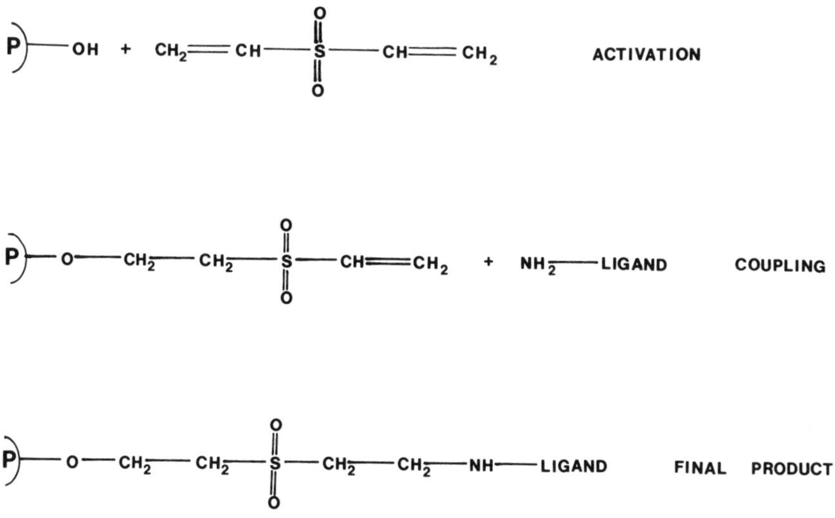

Fig. 1. Chemistry of activation and coupling with divinylsulfone.

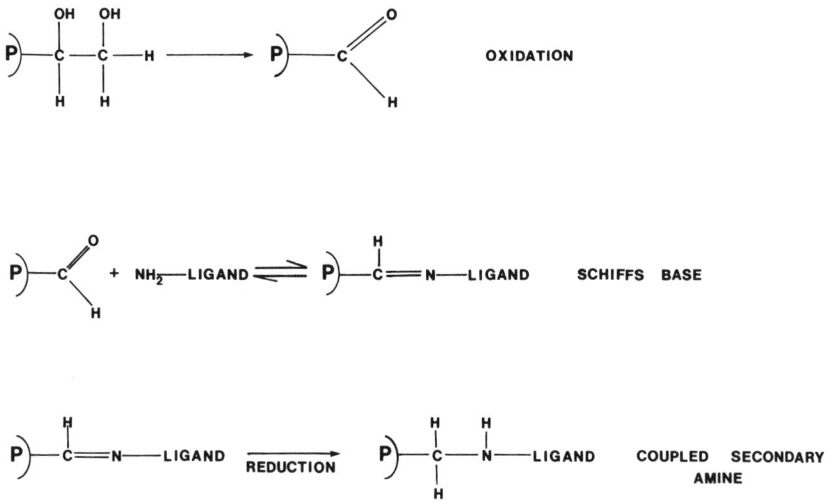

Fig. 2. Periodate activation, Schiffs base formation, and reductive amination.

aldehyde is formed on the solid phase and the amine is provided by a protein, however, it can also work the other way around with an amine (or hydrazide) on the gel polymer and an aldehyde provided by the sugars on a glycoprotein or polysaccharide. In reality, the individual methods differ widely in their convenience and efficiency of coupling. In this author's laboratory, experience has shown that simple periodate oxidation of the commercially available Sephacryl range of gels is most effective. For reasons of steric accessibility, the porosity grades S-500 and S-1000 give the best compromise between capacity and performance. Methods have been published for both analytical *(8)* and preparative *(9)* immunoaffinity matrices. The activated gel can be prepared quickly ($\leq 20'$) and is stable at 4°C in water for at least 1 yr. The presence of reactive aldehyde groups in the gel can be quickly ascertained by the pink color reaction with Schiffs reagent. Proteins can be coupled via amino groups, the pH can be made a variable to control the efficiency of coupling, the degree of inactivation, and also the valency of attachment, and hence leakage. The latter can be further minimized by using a reducing agent either during the protein coupling and blocking steps or during only the latter step (*see* Note 14). Owing to the presence of reducing groups (aldehyde, and the like) the BCA protein assay gives false color with the beads, so alternative protein estimating methods are preferable (*see* Note 6). Periodate-activated Sephacryl can be recommended as the first method to be tried by anyone wanting to couple a protein to a solid phase. It shows good capacity, efficiency of uptake, and uses safe and cheap reagents. Occasionally, enzymes and antibodies show variable degrees of inactivation as a result of high valency coupling (multiple points of attachment). This can often be avoided by coupling at a lower pH *(7–8)* (*see* Notes 9 and 10), but if efficiency of coupling falls below an acceptable level then the DVS method should be tried. The literature contains many diverse methods of reductive amination for immobilization of proteins or saccharides, however, in general, these all suffer from one or more of the problems of low capacity, slow reaction, poor efficiency of uptake, and they cannot be recommended for routine use. Likewise, it is possible to generate aldehydes in other gels than Sephacryl but the same problems of capacity and efficiency will arise, so again it is not recommended to use this method with gels other than Sephacryl. Fortunately, the Sephacryl range of gels show good physical and chemical properties and are available in a very wide range of pore sizes (= MW exclusion limit), which makes them an attractive choice in many applications.

1.3. A Brief Outline of the Epoxide Method

Inevitably, there will be same "problem" ligands that don't couple well to DVS or periodate-activated Sephacryl. Often, these will be devoid of reactive amino or hydroxyl groups or may have a structure or composition that is uncertain (e.g., natural products and synthetic polymers). In the absence of any clear choice of alternative chemistry based on a specific reaction with some known residue(s), it may be worth considering epoxy derivatives. There are a number of possibilities but the best is probably 1,4 butanediol-diglycidyl ether *(10,11)*. As well as problem ligands, one can have also "problem" solid phases, i.e., those that are not very amenable to derivatization; unfortunately, the aggressive conditions of epoxide reactions are not useful where the solid phase is destroyed by pH values above 12–13 and this precludes a major group of particles composed of silica, glass, and so on. Fortuitously, alternative silane-epoxy activation can be used as a route for these special materials but a description of this is beyond the scope of this chapter. Another major useful feature of epoxide activation is as a universal intermediary that can be derivatized to produce other groups that may be more suitable. The details can be sought elsewhere but it is a simple matter to convert epoxy groups to diol (hence, also aldehyde), carboxyl, sulfopropyl, amino, hydrazino, IDA (metal chelate), and DEAE groups, as well as more specific reactions (Fig. 3). Additional advantages of epoxy activation with "BUDGIE" reagent include its low leakage ether bond and flexible extended hydrophilic spacer arm. Against this must be considered the disadvantages of high pH activation, slower reactivity, and the inconvenience and inefficiencies of subsequent step derivatizations if used (*see* Notes 15 and 16).

2. Materials

All three methods used can be operated with the same mechanical equipment and no one item is essential, but it is worth trying to use them all as it makes a far simpler and more reliable system.

A suitable container for both activation of the gel and later coupling is a round borosilicate (Pyrex or Duran) bottle with a wide screw cap mouth (containing no liner), which should also be chemically resistant. Avoid metal and rubber components. A most suitable range is the Schott-Duran series from 50 mL to 10 L. These will accommo-

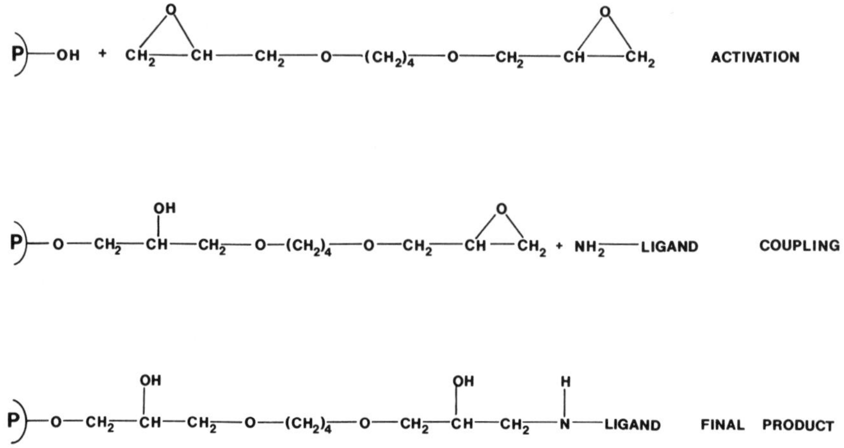

Fig. 3. Chemistry of *bis*-epoxide ("BUDGIE") activation and coupling.

date any vol of gel from 10 mL to 3 L (e.g., Schott P. No 21-801-245 screw-cap laboratory bottle 45 mm ISO cap). If smaller vols of gel are to be used or if more accurate measurement of settled vol is required, then calibrated conical bottomed centrifuge tubes in Pyrex or polypropylene can be used.

Mixing of gels during the various stages is best done on a horizontal roller mixer of suitable size and speed to suit the bottle sizes being used. Variable speed is particularly valuable when working with a two-phase system, such as epoxy ("BUDGIE"), to ensure adequate mixing between the two phases. Magnetic stirrers and propeller-bladed motor stirrers are not recommended as they provide unreliable operation. Traditional gel washing and activation has often been performed with the aid of Buchner funnels, though these are very inconvenient for multiple transfers or large vols. A much better choice is to leave the gel in the bottle and suck the excess fluid out via a sintered glass frit on the end of a glass tube attached to a source of vacuum. A suitable porosity grade of glass sinter is #1 or #2, the former being quicker to use in most cases. The dimensions of the filter tube are not critical but should fit into the working vessel, be long enough, and fit the vacuum tubing. Typical dimensions are 6 mm dia. frit for gel vols 1 mL, 12 mm dia. frit for vols of 1–100 mL and 30 mm frit for larger vols. Shaped frits, such as "pots" and "candles" are also available from many commercial glass-blowers. The length of glass tube can be ca. 300 mm with a diameter of 6 mm. These items are frequently sold as "filter tubes" or "gas distribution tubes" (e.g., Schott 13-mm filter sinter P.No 25-8573-1). Among the many advantages of using these *in situ* filters are: (i) they eliminate the need for Buchner funnel and flask and filter/sinter, (ii) eliminates the need and losses of transfer between reaction vessel and filter, (iii) the filter tube can be used as a mixing rod when the vacuum is turned off, (iv) multiple gels/activation samples can be run in parallel ($n \leq 8$). Quantitative analysis of gel activation/ coupling is also facilitated because one can adjust the gel:fluid ratio to exactly 50:50 in the calibrated bottle by observing the gel:fluid interface while removing excess fluid, also small aliquots (≤ 1 g) moist gel can be retrieved from large vols of stock without the need to suck dry the whole sample by simply removing the filter tube with the vacuum still applied; a small sample of moist gel cake usually remains adherent to the sinter surface.

For situations where vacuum filtration with sinters is not feasible, consider centrifugation in 16 x 100 mm polystyrene tubes, up to eight can easily be processed at one time, and for vols of gel ≤ 1 mL this is a convenient way to couple and wash samples. The fluids can be aspirated by hand with a Pasteur pipet and the whole tube will also fit directly into a multiwell γ-counter for ^{125}I-tracer counting. Other equipment that is desirable but not essential include a water tap suction pump or mechanical diaphragm vacuum pump with liquid trap, centrifuge with eight place rotors, pH meter, with redox Pt electrode, UV/vis spectrophotometer, γ-counter for ^{125}I-tracers and disposable Luer syringes and glass fiber filters. For quantitative analysis of DVS and epoxy activation levels it is also helpful to have a small magnetic stirrer, a 100-μL precision syringe, and disposable plastic pipet tips (100–1000 μL).

2.1. DVS Activation and Coupling

1. 100 mL of a suitable gel, e.g., Sepharose 4B (Pharmacia) or Sephacryl S-1000 (Pharmacia).
2. Divinyl sulfone ($\equiv$ vinyl sulfone), e.g., Sigma, store at 4°C in dark.
3. 1M Sodium carbonate, having a natural pH of 11.3 ± 0.1 and stable indefinitely, 1 L.
4. 1 M Sodium thiosulfate. Dissolve in H_2O to make 100 mL and adjust pH to 6.5–7.5 without a buffer.
5. 0.1N hydrochloric acid, 10 mL of accurately prepared volumetric reagent.
6. Ligand to be coupled:1 mg/mL in water or buffer, e.g., 1M sodium carbonate, pH 9.0.

7. Blocking solution: $0.1M$ ethanolamine pH 9.0, (1 L). Stable indefinitely at +20°C.

2.2. Periodate–Sephacryl Coupling

1. 100 mL of a suitable gel, e.g., Sephacryl S-500 or S-1000 (Pharmacia).
2. Oxidizing solution: 4 mM sodium periodate in 100 mM sodium acetate, pH 4.7, (1 L). This solution is stable in the dark at 4°C for up to 1 yr and should be checked occasionally to have an $A_{280\,nm}^{1\,cm} = 1.10$–$1.20$ and/or redox potential of +700 to +800 mV.
3. Solution of protein to be coupled 1 mg/mL of protein in water or buffer, e.g., sodium carbonate, pH 9.0.
4. Ascorbic acid (dry solid) store at 4°C. Aqueous solutions are unstable at alkaline pH.
5. Blocking solution: Repeat Section 2.1.7. above.
6. Schiffs reagent (decolorized rosaniline-sulfite) (Sigma).

2.3. Bis-epoxide ("BUDGIE") Activation and Coupling

1. 100 mL of a suitable gel, e.g., Sepharose 4B or Sephacryl S-1000 (Pharmacia).
2. 1,4,Butanedioldiglycidylether ("BUDGIE") (Sigma), 1 L, stable at +20°C.
3. 0.6N Sodium hydroxide in water, 1 L. Stable at +20°C provided atmospheric CO_2 is excluded.
4. Ligand to be coupled, preferably at 10 mg/mL in 1M sodium carbonate, pH 9–13.
5. Blocking solution (optional): $0.1M$ beta-mercaptoethanol in 1M sodium carbonate pH 9, freshly prepared.
6. Phenolphthalein indicator solution (optional) (Sigma), 1 mg/mL in ethanol.

2.4. Assessing Degree of Activation and Coupling

1. 1M hydrazine hydrate in water, adjusted to pH 10.0 (Sigma) stable if CO_2 is excluded.
2. 2,4,6 Trinitrobenzenesulfonic acid (picric acid) (Sigma) made up to 1 mg/mL in $0.1M$ sodium carbonate pH9. Prepare freshly and use within 24 h at 4°C or store aliquots frozen at −40°C.
3. 10% v/v Acetic acid in water.
4. Phenolphthalein indicator solution, *see* Section 2.3.6.
5. Bicinchoninic acid protein assay kit (Pierce) or Coomassie dye protein assay kit (Pierce).

Table 1
Comparison of Features
for the Three Activation and Coupling Methods

A

Method of activation	(pH)	Useful with ligands having these groups	at pH	Works with these gels or solid phases
DVS	(11.3)	R-NH$_2$	6–10	agarose, dextran
(Divinyl sulfone)		R-SH	6–9	pHEMA, cellulose
		R-OH	9–11	also "Ultrogels," "Fractogels," and "Sephacryls"
IO$_4$-Sephacryl (Reductive amination)	(4.7)	R-NH$_2$	7–11	"Sephacryls" S-100 to S1000
bis-epoxide ("BUDGIE")	(13.5)	R-NH$_2$	9–11	agarose, dextran
		R-SH	7–9	pHEMA cellulose
		R-OH	11–13	"Fractogels" and
		Other	11–13	"Sephacryls"

B

Method of activation	Expect levels of active groups mol/mL	Expect a maximum protein coupling capacity of mg/mL	Expect a coupling efficiency with 1 mg/mL input of protein of, %
DVS	1–5 (Low) 5–20 (Medium) 20–50 (High)	50	90–95
IO$_4$	5–15	120	90–95
BUDGIE	20–50	10	5–40

3. Methods

Refer to Table 1 for an overall comparison of the features of the three methods and choose one or two in conjunction with those desirable features discussed in the introduction. It may also be helpful to include a commercial activated gel as a "standard" against which to compare your own efforts, the most suitable is CNBrSepharose 4B (Pharmacia).

3.1. Divinylsulfone (DVS) Method (1,6); (see Note 12)

1. Dispense about 200 mL of gel slurry (50:50 v/v) into a 500-mL glass bottle and suck dry *in situ* with a sinter glass filter tube, note the vol (100 mL) and add an equal vol of 1*M* sodium carbonate, pH 11.3. Carefully add 10 mL of divinyl sulfone liquid (wear gloves, work in a fume hood, and use an automatic pipet to avoid skin contact or inhalation of vapor). Replace cap on reaction bottle.
2. Mix reaction bottle well on a horizontal roller mixer, such that all globules of DVS reagent are well dispersed. Within a few seconds of mixing, a pale-pink color appears that gradually turns to brown during the course of the reaction, which should run for 2 h at room temperature, with continuous roller mixing.
3. Stop the reaction by sucking out all liquid via the filter tube. Add excess distilled water (400 mL) to fill the reaction bottle, and while the vacuum is removed from the filter tube use the latter as a stirrer rod to break up the gel clumps and disperse them in the water, re-cap, shake or mix for 1 min on roller mixer (or until all visible clumps have been dispersed) and suck dry *in situ* again and repeat until the pH has fallen below 9.0. This usually takes 3–4 additions of water. After the last filtration, add an equal vol of water, resuspend and titrate the 50:50 slurry to pH 7.0 ± 0.5 by the addition of 1*N* HCl dropwise on a magnetic stirrer. The gel slurry can be stored as such at 4°C for at least one year.
4. The determination of quantitative content of reactive vinyl groups is optional but recommended. Place exactly 1.0 g of moist gel cake (or 2.0 mL of slurry) into a screw cap plastic universal container (e.g., 25 x 100 mm), add 3.0 mL of 1*M* sodium thiosulfate solution (it is important that both of these are pH 7.0 ± 0.5 and contain no buffers) and roller-mix for 20 h at 20°C while tightly capped to exclude CO_2. Add a small magnetic stirring bar immediately after opening the container and measure the pH, which should be around pH 11.0 ± 1.0. Start stirring and slowly add 0.1*N* HCl dropwise from a 100–µL microsyringe. Continue adding HCl until the pH stabilizes at pH 7.0 ± 0.5. Calculate the mol of vinyl groups per mL of gel as follows: Divide the µL of 0.1*N* HCl added by 10 e.g., 130 µL of added 0.1*N* HCl corresponds to 13.0 µmol of vinyl groups per mL of gel.
5. (Optional) Determination of qualitative content of vinyl groups. Proceed as above but simply mix 1 mL of gel slurry (pH 7) with 1 mL of 1*M* sodium thiosulfate (pH 7) and add 1 drop of phenolphthalein indicator. Cap to exclude atmospheric CO_2 and roller-mix for 2–20 h.

The time of appearance of purple color (alkali) and the visible color intensity can be used as a rough guide to the level of activation.

6. Coupling (*see* Notes 9 and 10). Take a vol of gel slurry equal to 2X the numerical value of protein mass in mg to be coupled, e.g., if 10 mg of monoclonal antibody is to be coupled take 20 mL of slurry. The absolute coupling capacity of DVS gels depends on the degree of activation as well as the type of ligand and coupling pH, thus protein can be coupled with efficiencies of ≥ 90% at inputs of up to 10 mg/mL. Whereas sugars can be coupled at much lower efficiency at inputs up to 200 mg/mL at pH 10. Place gel in reaction bottle e.g., 100 mL size and suck dry with filter tube. Add solution of protein in water or buffer (e.g., sodium carbonate, pH 9, *see* Note 9) in a vol of 1–100 mL. Adjust pH to 9.0 with $1M$ NaOH or $1M$ HCl as necessary. Roller-mix overnight at 20°C. Carefully remove unbound fluid protein and add an equal vol or 20 mL of $0.1M$ ethanolamine, pH 9.0, blocking buffer and mix for a further 3 h. Suck dry *in situ* and wash with a 100 mL vol of distilled water by brief mixing and repeat the suction to dryness. Failure to remove most sodium carbonate at this step will cause excessive frothing at the next step. Wash the solid phase in an acid buffer, e.g., sodium acetate, pH 4.7 (or if you know the ligand can stand it, use buffers of pH 3.0 or even 2.0), followed by water (100 mL) and then adjust the pH to neutrality. In general, the more extensive the washing (vols, number of steps, extremes of pH, and pH step changes) the less noncovalently bound ligand will remain on the beads to leak off later.

7. You can estimate the extent of coupling of protein by an appropriate method (*see* Notes 3–7) namely ^{125}I-tracer uptake; UV absorbance difference before/after coupling; BCA protein assay directly on beads, or Coomassie dye uptake indirectly by difference before/after mixing beads and dye. Nonprotein ligands are more difficult to estimate but in general, radioactive tracer or dye binding/uptake methods are most likely to be appropriate, otherwise use functional binding assays as a rough guide to relative ranking of performance.

8. Assuming the activation and coupling were both performed successfully, consider how to evaluate the functional performance (specific biological activity U/mL of gel) of the protein. In general enzymes and antibodies have obvious appropriate specific assays, however, other group-specific or nonspecific ligands will need to be tested in their binding capacity to an appropriate model compound or substance. It should not be assumed that efficient coupling will always lead to a biologically functional ligand.

3.2. Periodate Sephacryl Method (8,9); (see Notes 13,14)

1. Dispense 200 mL of gel slurry (50:50 v/v) of Sephacryl S-1000 in a 500-mL reaction bottle and suck dry *in situ* with a sinter-filter tube. Add 400 mL of oxidizing solution (4 mM NaIO$_4$ in 100 mM Na acetate, pH 4.7) and roller-mix for 15-20 min at 20°C.

2. If the reaction has proceeded properly the $A^{1\ cm}$ of the oxidizer solution will have fallen from a value of 1.1–1.2 AU to 0.1–0.2 AU (be careful to centrifuge or filter off fine particles before measuring UV absorbance). Suck dry the oxidized Sephacryl gel and wash several times in 400 mL vol of H$_2$0.

3. This activated gel should be tested qualitatively for aldehyde groups as follows: dispense 0.1 mL of washed gel beads into a 2 mL clear plastic tube and add 1 mL of Schiffs reagent; observe the time taken (and final extent) for the appearance of a visible pink-purple color in the beads. Color appearing with 5–10 s corresponds to a high degree of activation; within 10–60 s is a medium degree of activation and greater than 60 s is a low degree of activation. The washed, activated aldehyde-Sephacryl can be stored at 4°C in a 50:50 slurry for at least 1 yr.

4. If the gel has been stored activated at 4°C for several months repeat the qualitative color test in Step 3 above to ensure that you have the correct identity and reactivity of gel. Coupling (*see* Notes 9,10 and 14) is carried out with a vol of slurry 2X the numerical mass in mg of protein to be coupled, e.g., for 10 mg of a monoclonal antibody, dispense 20 mL of 50:50 slurry into a 100-mL reaction bottle (the coupling capacity for protein of periodate-oxidized Sephacryl is considerably greater than this, for example, at an input of 50 mg/mL the coupling efficiency is still ≥ 80% at pH 10). Suck dry *in situ* with a sinter tube and add the protein solution to be coupled (*see* Notes) in an appropriate buffer, such as 1M sodium carbonate, pH 9 (or phosphate or citrate). If the vol of protein solution added is small (< gel vol), add further buffer until a 50:50 v/v gel slurry is obtained. Optionally (*see* Note 14), add solid ascorbic acid to give a final concentration of 5–10 mM ascorbic acid (i.e., between 0.9 mg and 1.7 mg of ascorbic acid are added/mL), then mix to dissolve (some foaming occurs if carbonate buffer is used) and recheck and readjust the pH to the desired value. Optionally, the redox value can be checked at this point and should be in the range of -180mV to -200mV, values below -180mV should be rectified by adding more solid ascorbic acid or raising the pH. (If the reaction is proceeding correctly, a pale yel-

low color is apparent within 1–2 h.) Roller-mix the reaction bottle and contents overnight at 20°C.

5. Add an equal vol (= gel slurry vol) of 0.1M ethanolamine, pH 9.0, and make a fresh addition of solid ascorbic acid at the rate of 0.9–1.7 mg/mL to reestablish reducing conditions during the blocking step (ascorbic acid is unstable at pH 9). Roller-mix for 3 h at room temperature, then separate the unbound fluid by *in situ* filtration with a filter tube and retain it if needed to perform a protein assay (Note: ascorbic acid or other reducing agents prevent the use of $A_{280\,nm}^{1\,cm}$ or BCA methods of protein assay therefore either ^{125}I tracer or Coomassie dye binding assays should be used, *see* Notes 3,6.) Carry out extensive washing with water and pH cycling with buffers as detailed in Section 3.1., Step 6, and functional testing as in Section 3.1., Step 8.

3.3. Bis-*Epoxide* ("*BUDGIE*") Activation and Coupling (10); (see *Notes 15,16*)

1. Dispense 200 mL of gel slurry (50:50 v/v) into a 500-mL reaction bottle and suck dry *in situ*. Add an equal vol to the gel (= 100 mL) of 0.6N NaOH and a similar vol (100 mL) of 1,4,butanedioldiglycidyl ether, such that the final concentration of gel is 33% v/v, the *bis*-epoxide is 33% v/v and the sodium hydroxide is 0.2N. The *bis*-epoxide is not very water-soluble and most of it will settle out as refractile heavy globules in the bottom of the gel. Shake the bottle vigorously to disperse it and immediately place on a roller mixer set at a sufficiently fast speed that the globules remain small (≤ 1 mm) and do not settle out as a visible lower oily phase. During the course of the reaction (overnight, 20°C), the oily phase is slowly adsorbed into the gel via reaction or hydrolysis, and the next day, only a small vol of oily brown globules will settle out when mixing is stopped.

2. The gel is sucked dry *in situ* by the usual method of a sinter-filter tube. Note that because of the low solubility of "BUDGIE" reagent in water the strong ethereal smell will persist where it can absorb to plastics, gel, and so on. The use of multiple vols (400 mL) of hot tap water (40–60°C) with mixing for 1–2 min before sucking dry will improve the extraction. Up to ten additions of water may be needed to remove all traces of the reagent. At this point, the pH should have fallen to below 9.0 and the gel should be slurried in an equal vol of water (100 mL) and adjusted on a magnetic stirrer to pH 7.0 ± 0.5 by the addition of 1N HCl dropwise. Avoid the use of any buffers as this can compromise later quantitative titrations of the epoxide group contents, performed as in Section 3.1. 4. or Section 3.1. 5. but only

2 h mixing with the thiosulfate reagent is required prior to titration.

3. For coupling difficult ligands, aim to get the highest feasible concentration of soluble ligand, e.g., ≥ 10 mg/mL and the highest pH at which it is stable (e.g., pH 11–13) or soluble. Dissolve the ligand in 1M sodium carbonate, pH 11, to the maximum feasible concentration and add an equal vol of this solution to a vol of epoxy-activated gel that has been sucked dry *in situ* in the reaction bottle. Mix on a magnetic stirrer and adjust the pH to 11, 12, or 13 (the higher the pH the better the chance of coupling; *see* Notes and Table 1). Because of the strong buffering capacity in this region, it may be necessary to adjust the pH by adding 40% w/w NaOH solution to avoid an excessive increase in vol. Check that at these high pH values the ligand is still soluble. Roller-mix the reaction bottle for 24–48 h (the lower the pH the longer the reaction time) at room temperature. If coupling is insufficient and the ligand will be stable then mixing can take place in an incubator at 37°C for 24–48 h (*10,11*).

4. Separate the reaction mixture by *in situ* filtration. Blocking is not normally necessary because the epoxy group is not very reactive under normal conditions of usage, and prolonged alkaline coupling will also tend to destroy unreacted groups, leaving *cis*-diol structures, which are conveniently hydrophilic. If, however, blocking is desired, this can be achieved with 0.1M ethanolamine at pH 9, 20 h or 0.1M β-mercaptoethanol in 0.1N sodium carbonate, pH 9 (freshly prepared), for 3 h at room temperature. Wash the gel with multiple cycles of water (400 mL) followed by *in situ* filtration and extremes of low and high pH (cf Section 3.1.) according to what you think the ligand will tolerate. In general, a solid-phase coupled ligand will be significantly more stable than the same molecule in solution.

5. Estimate the amount of ligand coupled, by an appropriate method as in Section 3 .1. (*see* Notes 3–7) and also try and estimate the functional performance of the ligand by an appropriate biological binding capacity assay.

4. Notes

4.1. General Notes Applicable To All Methods

1. Cause of failure. The most common cause of failure in using affinity chromatography is failure to couple adequate amounts of functional ligand to the gel. This is best prevented by careful quality control of the various reaction sequences at each step. Always check that the gel is (a) activated adequately (Sections 3.1.4.–5, 3.2.3.) and always try to check (b) the mass of ligand coupled (*see* Notes 3–7) and (c) its functional (binding) capacity.

2. Stable activated stocks. It is very little trouble to prepare a larger vol of any of these gels—consider making 2–10 times more than you need and estimate the content of active groups before you need to use it. All three chemistries are stable in aqueous 50:50 slurry at 4°C in the presence of 3 mM sodium azide for periods in excess of a year (but make sure the pH is 5–7!). It is a simple matter to dispense an accurate vol of settled gel (V) by simply weighing out or volumetrically dispensing (2 V) of slurry. If you are uncertain which gel you will need, consider making a range of all three chemistries on different types of gel, e.g., Sephacryl S-1000, Sepharose 4B, and Fractogel TSK 75, using, for examples, 16 × 100 mm plastic tubes it is quite feasible to prepare eight different coupled ligand/bead/activation chemistries in parallel.

3. ^{125}I-tracer for uptake measurement. The most reliable, accurate, and rapid way of estimating protein (ligand) uptake and content is by adding a ^{125}I-tracer to the ligand before coupling. Aim to add between 10,000 and 100,000 counts/100 s of tracer (i.e., 10^2–10^3 Bq) and aim to count 1–5 mL of settled gel and a similar vol of unbound fluid.

By measuring the total vol of gel and uncoupled fluid, it is a simple matter to calculate the total counts in the system and the percentage of counts bound and free. Corrections may be needed for low counting efficiency with 5-mL vol and also for poor tracer quality when the percentage of trichloracetic acid precipitable tracer is < 95%. This method can be applied to all activation chemistries and gel types and is recommended whenever possible. It can also be used to evaluate content and leakage at a later time, remote from the coupling step.

4. UV decrease for uptake measurement. The most common method used to evaluate the uptake of protein ligands onto a solid phase is to measure the UV absorbance $A_{280\ nm}^{1\ cm}$ and vol before and after coupling and any decrease is ascribed to protein coupled.

Several problems in practice make this an unreliable procedure. Correction is needed for the vol of fluid within the gel pores that will dilute the ligand before coupling. Aggregates formed before or during coupling will overestimate the OD and hence amount of protein in solution. Some gels or activation chemistries release soluble UV absorbing materials into solution thus showing an increase in OD after coupling instead of the expected decrease. The method is also insensitive and inaccurate when the vol are large and the OD is low, as happens with multiple washes postcoupling.

The method can be used as a backup to one of the other methods (*see* elsewhere in these notes) and if agreement is achieved then the UV absorbance/difference method can probably be relied on, but in uncertain cases, it is better to use two independent methods. When reading UV absorbances at 280 nm, always prefilter the sample through a small disposable microfilter directly into the cuvet to eliminate aggregates and gel fragments.

5. Direct colorimetry by BCA for uptake measurement. The commercial bicinchoninic acid (BCA Pierce) kit is quite useful for estimating directly (not by difference) the protein content of beads as the color is released into solution, the beads settle out, and the clear supernatant can be assayed directly.

To 200 µL slurry (100 µL vol settled beads) dispensed via a disposable polypropylene pipet tip (cut back to increase tip internal diameter to = 1 mm) in a 12 × 75 mm polystyrene tube is added 2 mL of BCA reagent, mix occasionally during incubation at 37°C for 30 min. The reagent changes from green to purple and the chromogen can be quantitated by absorbance at 562 nm. The range of the assay is 10–120 µg/0.1 mL beads corresponding to 0.10–1.20 mg of protein/mL of settled beads.

Unfortunately, the reaction is sensitive to reducing reagents that convert Cu^{2+} to Cu^+ and therefore any gel or coupled protein that has previously been exposed to reducing agents (e.g., aldehydes, ascorbic acid, $NaCNBH_3$, $NaBH_4$, etc.) tends to give a false positive color reaction, possibly via remaining R-SH groups. Thus, the BCA assay is recommended for use with the DVS and BUDGIE activation methods but not with the periodate–Sephacryl method. Reagent blanks should always include bland blocked activated gel devoid of protein as well as appropriate soluble protein standards.

6. Coomassie dye depletion for uptake measurement. An alternative dye binding (difference) method can be used in the majority of activation and solid-phase combinations (not recommended with Fractogels owing to nonspecific binding). The Coomassie dye binding kit from Pierce is most convenient, but note that here, it is used in an absorbance difference mode and not in a color-shift mode, thus the readings can be taken at any wavelength that accurately measures the absorbance of the starting dye reagent. Any bound dye reduces this value and when bound to protein in the solid phase, will change from orange/brown to blue, which is readily seen by eye but is not part of the quantitation measurement.

Choose a wavelength between 400 nm and 600 nm that gives an $A^{1\,cm}_{280\,nm}$ value between 0.3 and 1.0 on your spectrophotometer. Add 200 µL gel slurry (100 µL settled gel vol as in Section 4.1.4.) to 2.0 mL of Coomassie reagent and mix in a capped tube on a roller mixer for

15 min. Allow the beads to settle or centrifuge and remeasure the absorbance of the free dye supernatant fluid at the same chosen wavelength. If the optical density has decreased by more than 90%, add a further 2.0 mL of fresh dyestuff and repeat the above steps. If the absorbance fell by less than 10%, add a further 200 µL of (50:50 v/v) gel slurry and repeat the above steps, remember that interstitial fluid in the beads will in any case reduce the $A_{280\,nm}^{1\,cm}$ by 10% in this case.

The data can be expressed in AU of dye bound per 0.1 mL gel (or × 10/mL of gel) or they can be converted to mol of dye bound per mL of gel (MW of Coomassie blue G-250 is 854, so a 0.01% solution has 117 nmol/mL). Most desirably, the reagent can be standardized by using aliquots of beads with a known protein content (determined by another method) with the same type of beads, activation chemistry, and protein coupled. Always run blanks with the original beads (unactivated) as well as beads activated and blocked with a bland (nonprotein) blocking group. Do not try to standardize this dye depletion method with soluble protein standards as the wavelength shift in peak absorbance will severely distort the results.

7. Other dye depletion methods for uptake measurement. Any dye binding interaction can in principle be used as above to quantitate the amount of bound ligand. The author has used 0.005% w/v solutions of (≡ 163 nmole/mL) either toluidine blue or amido blue-black (≡ amido Schwarz, naphthalene black, buffalo black) in distilled water to quantitate immobilized poly acids (heparin, dextran sulfate), and so on, or polybases (Polybrene, protamine, and the like) respectively. Again, the choice of wavelength is wide, sufficient only that the spectrophotometer can accurately measure the absorbance before and after binding, usually any wavelength between 400–500 nm will do, then proceed as in Note 6 above.

8. Changing pore size and support chemistry. If after trying out a particular combination of solid phase bead and activation chemistry you do not get satisfactory functional results and you are able to rule out a technical failure (e.g., both activation and coupling were proven to work) consider trying (i) different gels (Table 1) with the same activation chemistry, (ii) different activation chemistries with the same bead (but see Table 1 for suitable choice of beads), and (iii) try different porosities of bead with the same activation chemistry.

Sephacryl (S-100–S-1000) offers the widest range of commercially available pore size gels, and this factor alone can often have a major impact on the performance of affinity gels particularly when ligand and/or target molecules have MW $\geq 10^5$. It is a good general rule when preparing a "home-made" reagent to also include a commercially proven reagent—a good choice of preactivated material is CNBr-Sepharose 4B from Pharmacia, which is widely used, efficient at coupling and usually biologically functional when the ligand is coupled. Any "new" method of bead preparation, activation, and so on, is worth comparing against CNBr Sepharose as an objective standard of relative performance, which can be imported into any laboratory. If your ligand is a "problem" one, try others as models when evaluating new chemistries.

9. Choosing buffers. Choice of buffers can usually be made from a group of a few generally useful buffers, such as sodium phosphate (pH 6–8) sodium citrate (pH 2–7) and sodium tetraborate (pH 8–10) or a combination of these where a wide coverage of pH range is needed. Generally, a concentration up to $1M$ should be aimed for, which suppresses any tendency for pH shifts attributable to consumption of reagents or ingress of atmospheric CO_2.

Efficient coupling of proteins usually proceeds better at high loadings when the mutual charge repulsion between proteins is suppressed by salt concentrations above $0.2M$. Also, in the case of gels or activation chemistries exhibiting mild hydrophobicity, it is possible to enhance the noncovalent binding of soluble protein onto the activated gel by working at $1M$ salt concentrations, this then leads to subsequent higher covalent coupling efficiency. Provided the solubility limit of the ligand is not exceeded (and this is easily checked by visual appearance before coupling), one can add, for example, $1M$ sodium sulfate to enhance still further this hydrophobic concentrating effect in cases where coupling efficiency is found to be low.

Where specific inactivation of ligands (enzymes, and the like) is a problem, it is usually possible to choose another one of the above buffers that will avoid the problem(s). Carbonate buffers are freely soluble, cheap to use and have strong buffering in the pH ranges 5–7 and 9–11, but they need to be removed before acidification or copious evolution of CO_2 can be a nuisance. Phosphate buffers are useful over the pH ranges of 2–3, 6–8, and 11–13, but solubility can easily be exceeded with $1M$ solutions at certain pH values especially at 4°C, they also tend to precipitate certain metals (Ca^{2+}, and so on) and be incompatible with some enzyme assays. Trisodium citrate is freely soluble and has useful buffering capacity in the range 2–7 but its most useful feature is as a stabilizer of proteins (e.g., immunoglobulins) which might otherwise aggregate when taken outside the physiological pH range for this reason citrate is useful at both pH 2–3 (where it is a buffer) as well as pH 8–13 where it acts solely as a stabilizer and not as a buffer.

If none of these is suitable, then borate buffers (0.05–0.10M sodium tetraborate) can be used in the pH range 8–10, however, 0.1M is a saturating concentration so higher concentrations are not feasible and it can form soluble complexes with certain carbohydrate or glycoprotein residues. Failing this, buffers, such as HEPES, HEPPS, CHES, can also be used in most cases because their nitrogen residues are so deactivated or sterically hindered as to be unreactive in their buffering ranges.

10. Choice of pH for coupling and blocking. Frequently, an unsatisfactory coupling can be improved by changing the pH; raise it to improve coupling efficiency and lower it to reduce inactivation of coupled ligands. In the case of DVS-activated gels that are quite reactive (Note 12 and Table 1), coupling of proteins proceeds well in the pH range of 7–9 (and thiols are equally reactive), however, carbohydrates (monosaccharides and polysaccharides) are less reactive and coupling may need to be done at pH 10–11.

    Proteins with large numbers of available surface ε-NH$_2$ lysine groups will in general show multiple valency attachment, thus coupling at high pH will increase the valency (number of attachment points) and decrease the tendency to leakage whereas lowering the pH will decrease the valency and reduce the tendency to inactivation of biological function but also increase the tendency to leakage. Remember when changing the coupling pH that the blocking should also be changed to the same value, otherwise the valency of attached proteins may increase during blocking at a higher pH than coupling. Periodate-activated Sephacryl shows similar effects to DVS when coupling proteins (or amino compounds) but it cannot be used to couple carbohydrates. The reactivity of certain proteins (pK and availability of ε-NH$_2$ groups) is strongly pH dependent and for instance, protein A and immunoglobulins *(9)* couple much better at pH 10 than pH 7. Epoxide activated gels are generally coupled at much higher pH values (*see* Table 1) because of lower reactivity, e.g., thiols will couple at pH 8–9, proteins (and amino compounds) at pH 9–11 and carbohydrates (sugars and polysaccharides) at pH 11–13 (*see* Notes 15,16).

11. A general qualitative method for the comparative evaluation of the degree of activation for any chemistry or polymer beads. The principle of the method is to saturate the activated group with hydrazine

and after washing to visualize this as a red color with picric acid. By suitable washing and pH control, it is possible to gain a rough idea of the degree of activation of any gel prior to coupling.

To 0.1 mL settled vol of activated gel in a 16 x 100 polystyrene screw-cap tube, add 1 mL of 1M hydrazine, pH 10, and incubate at 37°C for 1 h. Wash well in water (10 mL) 4–5 times using the *in situ* sinter-filter tube and add 1 mL of TNBS ($\equiv$ picric acid) 1 mg/mL in pH 9, 0.1M carbonate or borate buffer. The active gel will quickly turn pink, red, orange, brown, and black, the speed and extent depending on the degree of activation. Because it is difficult to remove all free hydrazine and because the TNBS reagent is very sensitive, free (nonspecific) color is frequently seen in the buffer above the beads if washing was not very efficient.

Suck off the excess fluid via the sinter-filter tube and add 10 mL of 10% v/v acetic acid, mix for 1 min and suck dry (the bead color will tend to lighten to pink, yellow, or orange) resuspend in 10 mL H$_2$O, mix for 1 min and suck dry, note the color and intensity of each sample at this acid (pH 2) point.

Add 10 mL of 0.1M carbonate or borate buffer, pH 9.0, and the bead colors should darken back to their original values but the supernatant fluid should be free of color (if not, repeat acid, water, base, washing steps); note the color and intensity of each sample, they can be kept for reference and as visual standards against which subsequent batches can be checked. Compare only colors that are alike as color response in the eye is not equivalent at different wavelengths.

4.2. Notes Applicable to Specific Activation Chemistries

12. Specific points to note with DVS activation. DVS activation can be used with any hydroxylic-containing polymer, including those mentioned in Table 1. During estimation of the extent of activation, the thiosulfate reagent slowly generates alkali over a period of 20 h and it is difficult to prevent ingress of atmospheric CO$_2$ completely thus partial neutralization may occur and this tends to underestimate the absolute number of vinyl groups.

The activation chemistry is often quite reliable and predictable, so the relative ranking of activation by this method is still valid. It is also possible to modulate the degree of activation by reducing the amount of divinyl sulfone used initially, thus instead of using 10% of gel vol as DVS (giving 40–50 mol/mL gel) one can reduce this to 5%

(giving 20–25 mol/mL) or even 1% DVS (giving 5–6 mol/mL). These lower levels of activation can be useful when coupling proteins, such as monoclonal antibodies or protein A, that have large number of potential surface reactive ε-NH$_2$ groups from lysines. If all those that are available couple, then the protein is firmly fixed in place, may be distorted, and cannot flex to accommodate binding requirements, hence they may be partially inactivated when coupled to highly activated gels.

Because DVS has mild hydrophobic character *(6, 7)*, one can enhance the binding efficiency of proteins (especially at low concentrations and/or low degrees of polymer bead activation) by adding 1M salts, such as sodium sulfate or 5% w/v PEG 6000 (but check first that these do not precipitate your ligand).

More usefully, for immunological work, DVS shows a remarkable preference *(6)* at lower salt concentrations (≤ 0.15M NaCl) to couple preferentially to hydrophobic IgG rather than hydrophilic albumin. This means that one can potentially add raw antiserum or hybridoma tissue culture supernatants to DVS-activated gels in ≤ 0.15M NaCl and the IgG will preferentially couple, whereas most of the other extraneous protein (albumin) will not. This little known advantage of DVS-gels deserves to be more widely used.

13. Specific points to note with periodate-Sephacryl reduction amination. This method does not work well with gels other than Sephacryls (Table 1) but the properties of these gels and the porosity range (MW exclusion limits) make them extremely valuable as universal solid-phase supports.

The qualitative test (Schiffs reagent) used in Section 3.2.3. cannot be made the basis of a quantitative test because the chromogen formed is not soluble and the reaction is not quantitative. An alternative reagent "Purpald" (Sigma P3021) is available that gives a soluble chromogen with solid phase aldehydes and this has been made the basis of a quantitative assay *(9)*. The periodate oxidation method is sufficiently reliable that qualitative checking with Schiffs reagent is usually all that is required, but for those that are interested in developing this method further, the "Purpald" assay is carried out as follows:

Freshly prepare a stock solution of 1 mg "Purpald" (4-amino-3-hydrazino-5-mercapto-1, 2, 4 triazole) in 1N NaOH and add 1 mL of this to 0.1 mL settled gel vol (200 μL 50:50 v/v gel slurry) in a 12 × 75-mm polystyrene tube and shake the tubes in an open rack on a horizontal vibratory table, such that atmospheric oxygen is available to the reaction mixture. After 30 min at 20°C, a purple color is formed in the supernatant fluid and the beads are allowed to settle or centrifuged down and the supernatant fluid measured in a spectrophotometer at 552 nm.

Suitable standards are unknown as the structure formed within the oxidized Sephacryl is not known, so a water-soluble simple aldehyde, such as propionaldehyde, can be used as an equivalent standard. This assay is sensitive to aldehydes over the range of 1–10 mol/0.1 mL settled gel.

14. The choice of reducing agent in reductive amination. The choice of reducing agent in reductive amination is one that is important but rarely is discussed adequately. Sodium borohydride ($NaBH_4$) was originally used but it has several disadvantages: it is unstable below pH 8.0, it powerfully reduces both free aldehydes as well as Schiffs bases (*see* Fig. 2) thus removing the active groups on the gel, so it is only suitable for postcoupling stabilization of Schiffs bases and "blocking" (by removal) of unused free aldehyde group, so it should not under any circumstances be added during the coupling reaction.

Further problems with $NaBH_4$ include its ability to reduce disulfide bonds in proteins (e.g., immunoglobulins) to free-SH groups and the tendency to generate hydrogen bubbles inside the gel beads when they are washed to neutrality. Because of these problems, a more suitable reducing agent, sodium cyanoborohydride ($NaCNBH_3$) was used and this is stable down to pH 3 and does not reduce free aldehyde groups below pH 7, thus it can be added during the coupling reaction to reduce the unstable reversible Schiffs base (Fig. 2) to a stable secondary amine, the net effect of this being to drive the reaction forward and improve coupling efficiency.

Unfortunately, many reductive aminations only proceed effectively above pH 7 (*see* Fig. 4 in ref. *9*) and in this region, $NaCNBH_3$, becomes a sufficiently strong reducing reagent to reduce free aldehyde groups, and again this detracts from coupling efficiency. A more suitable reducing agent for use in the pH 7–11 range is ascorbic acid (*9*), which reduces only the Schiffs bases and not the free aldehyde groups, one can use 5–10 mM ascorbic acid during both the coupling (e.g., protein) and blocking (e.g., ethanolamine) stages.

Disadvantages of using ascorbic acid include its strong absorbance in the UV (preventing $A_{280 \, nm}^{1 \, cm}$ measurement of coupled/free protein) and its instability in alkaline solutions, especially above pH 10, requiring freshly prepared solutions and additions. Alternatively, a combined reducing and blocking reagent (0.1M, pH 9 hydroxylamine) can be used that does not absorb at 280 nm. The purpose of reducing the Schiffs base is twofold: It drives the reaction forward, increasing coupling efficiency and prevents reversible dissociation and leakage of ligand and blocking groups (which would regenerate free aldehyde groups). Where leakage of ligand or coupling efficiency are not important the reducing step can be eliminated. Where proteins are coupled via multiple points of attachment then reversible dissociation of Schiffs bases is unlikely to provoke leakage but for simple ligands such as monovalent amines (and blocking with ethanolamine), it is wiser to include a reducing agent.

4.3. Notes Specific to the bis-epoxide ("BUDGIE") Activation

15. Avoid sodium borohydride. The original method *(10)* used sodium borohydride to prevent alkaline "peeling" reactions in the agarose gel, however, omission of this reagent does not seem to produce any obvious ill effects, possibly because the agarose gel is crosslinked as well by the *bis*-epoxide reagent, and the other gels used (Table 1) are resistant to alkaline hydrolysis. There are a number of good reasons for omitting $NaBH_4$ from the activation reaction; it generates hydrogen gas continuously and the buildup of pressure is likely to induce leakage of sodium hydroxide onto the roller-mixer and incubator or burst the reaction bottle. If the hydrogen buildup takes place inside a 37°C incubator there is also the risk of an explosion. Lastly, during washing to neutrality the $NaBH_4$ tends to decompose, liberating more hydrogen gas inside the beads which tend to float, stick to glassware, and do not wash or sediment properly, which makes subsequent processing steps difficult. As mentioned in Note 5 above, the use of a convenient BCA protein assay is also prevented if reducing agents or residues are present in the gel beads/coupled ligand, so $NaBH_4$ should be avoided for this reason too.

16. Groups that couple easily to epoxide. The epoxide method works well at neutral pH or weakly alkaline pH where groups, such as –SH, –NH–NH_2, or aromatic amines *(11)* are present. If you can arrange to introduce these into your ligand prior to coupling, then this potentially valuable method could find much wider application.

References

1. Porath, J., Laas, T., and Jansson, J.–C. (1975) Agar derivatives for chromatography, electrophoresis and gel-bound enzymes. III Rigid agarose gels, cross linked with divinyl sulfone (DVS). *J. Chromatogr.* **103**, 49–62.

2. Sairam, M. R., Clarke, W. C., Chung, D., Porath, J., and Li, C. H. (1974) Purification of antibodies to protein hormones by affinity chramatography on divinylsulfonyl Sepharose. *Biochem. Biophys. Res. Commun.* **61**, 355–359.

3. Fornstedt, N. and Porath, J. (1975) Characterization studies on a new lectin found in seeds of *Vicia ervilia. FEBS Letts.* **57**, 187–191.

4. Sairam, M. R. and Porath, J. (1976) Isolation of antibodies to protein hormones by bioaffinity chromatography on divinylsulfonyl Sepharose. *Biochem. Biophys. Res. Commun.* **69**, 190–196.

5. Allen, J. J. and Johnson, E. A. Z. (1977) A simple procedure for the isolation of L-fucose-binding lectins from *Ulex europeus* and *Lotus tetragonolobus. Carbohydrate Res.* **58**, 253–265.

6. Lihme, A., Schafer-Nielsen, C., Larsen, K. P., Muller, K. G., and Bog-Hansen, T.C. (1986) Divinylsulfone activated agarose-formation of stable and non-leaking affinity matrices by immobilisation of immunoglobulins and other proteins. *J. Chromatogr.* **376**, 299–305.

7. Jorgensen T. (1987) High performance liquid chromatography on hydroxyethyl methacrylate. *Biochem. H.S.* **368**, p.752.

8. Wright, J. F. and Hunter, W. M. (1982) A convenient replacement for cyanogen bromide-activated solid phase in immuno-radiometric assays. *J. Immunol. Methods* **48**, 311–325.

9. Hornsey, V. S., Prowse, C. V., and Pepper, D. S. (1986) Reductive amination for solid-phase coupling of protein. A practical alternative to cyanogen bromide. *J. Immunol. Methods* **93**, 83–88.

10. Sundberg, L. and Porath, J. (1974) Preparation of adsorbants for biospecific affinity chromatography. 1. Attachment of group-containing ligands to insoluble polymers by means of bifunctional oxiranes. *J. Chromatogr.* **90**, 87–98.

11. Gelsema, W. J., De Ligny, C. L., Roozen, A.M.P., and Wilms, G. P. (1981) Optimal conditions for the coupling of aromatic amines to epoxy activated Sepharose 6B. *J. Chromatogr.* **209**, 363–368.

CHAPTER 12

Exploiting Weak Affinities

Sten Ohlson

1. Introduction

Affinity chromatography (AC) *(1)* is a technique that is based on the specific interactions between two molecular species. The molecules involved are usually of biochemical nature, such as antibody–antigen, enzyme–inhibitor, and receptor–hormone.

Traditional AC is performed in three discrete steps:

1. The substance to be isolated (ligate, Lt) is adsorbed in the presence of various noninteracting solutes onto the chromatographic column containing the immobilized complementary molecule (ligand, Ln);
2. The contaminants are washed off the column; and
3. The ligate is then eluted in a purified form by various means, e.g., changes in pH and ionic strength. AC has mainly been used as a purification procedure where, e.g., proteins have been isolated to homogeneity in one step with purification factors of 10–100-fold.

The fundamental process of AC is the reversible biospecific formation of a complex between ligand and ligate. The strength of this complex—the affinity—is often expressed in terms of binding constants, e.g., the dissociation constant (K_{diss}):

$$K_{diss} = \frac{(Ln)\,(Lt)}{(LnLt)} \qquad \text{at equilibrium} \qquad (1)$$

where $(Ln), (Lt)$, and $(LnLt)$ are the concentrations of ligand, ligate, and complex between ligand–ligate, respectively.

From: *Methods in Molecular Biology, Vol. 11: Practical Protein Chromatography*
Edited by: A. Kenney and S. Fowell Copyright © 1992 The Humana Press Inc., Totowa, NJ

AC has traditionally been used as an adsorption/desorption technique, where K_{diss} is in the range of $10^{-4}-10^{-15}M$. For antibody–antigen complexes, K_{diss} is usually $10^{-6}-10^{-12}M$. In rare cases, the affinities are extremely high, such as the $K_{diss}= 10^{-15}M$ for the binding of biotin to avidin.

However, for many years, weak (low) affinity ($K_{diss} >$ approx $10^{-4}M$) has been considered to be of a nonspecific nature and, therefore, is of no practical use in AC. This can be questioned, especially when considering what is occurring in biological systems, where individual weak binding events between molecular species are rather common. In some of these cases, cooperativity plays an important role, where a multitude of weak interactions triggers the biological activity.

It is only recently *(2,3)* that weak-affinity chromatography (WAC) has shown its potential as a separation technology, where similar carbohydrate antigens were separated by bound weak-affinity monoclonal antibodies. With WAC under optimized conditions, separations can be performed where chromatography is more dynamic, and fast isocratic or mild elution procedures with high performance (narrow peaks) can be realized without jeopardizing the specificity introduced by the weak-affinity ligand. This means that it is now possible with affinity chromatography to retard differently, closely related molecular species in crude mixtures with analysis times of minutes.

One of the major obstacles to the use of WAC has been the lack of a suitable support. Traditional soft gel supports (50–100 μm) are not good alternatives owing to limitations in performance and mechanical strength. However, the introduction and development of high-performance liquid chromatography (HPLC) *(4)* based on small (approx 10–30 μm) and rigid particles (silica, synthetic polymers, or crosslinked polysaccharides) have enabled the rapid progress of most modes of liquid chromatography, including gel-permeation, ion-exchange, and reversed-phase chromatography. In AC, HPLC was first applied approx 10 yr ago *(5)* to create high-performance liquid affinity chromatography (HPLAC).

HPLAC provided efficient separations, i.e., rapid isolations with high resolution and sensitive detection. The interested reader is advised to go to Chapter 8 by Clonis on HPLAC in this volume. The two other critical factors apart from performing HPLAC in developing WAC are (1) selection of a weakly interacting ligand or finding chromatography conditions where weak affinity is prevailing, and (2) achievement of a high amount of active ligands on the support.

The technology of WAC is still in its launching phase with only a few applications available. It is envisaged that WAC can be of use in various disciplines, such as biochemistry, clinical chemistry, and biotechnology. WAC can be designed for difficult separations where quantitative data can be gathered about molecular interactions. Since weak affinities are expressed rather frequently in the communication between molecules (free or immobilized on cells), we will have a new tool in the laboratory to study natural binding events.

An important question should be addressed: What are the limits of WAC? Because we are studying progressively weaker interactions, we will eventually face a borderline situation where specificity is lost in terms of its meaning as a specific interaction with a certain target. As a result, weak binding with any molecular target will be seen. In other modes of liquid chromatography, such as ion-exchange or reversed-phase chromatography, calculations indicate that chromatography often is performed under weak binding conditions. Significant retention is achieved in this case by a high number of repeated interactions or cooperativity effects (the ligate can bind several ligands at the same time).

It is conceivable that affinity chromatography can be conducted under a whole range of very weak affinities ($K_{diss} \gg 10^{-4} M$), where many ligand candidates can be selected for useful separations. In the short term, WAC may find useful applications in "analytical" affinity chromatography for quantitative monitoring of structurally related biomolecules, such as steroid hormones, pesticides, and drug metabolites. The experiences gained from analytical weak-affinity separations will eventually lead to increased interest in preparative chromatography, since isocratic or "mild" elution procedures can enhance dramatically the recovery of fragile biomolecules.

The weak-affinity approach to chromatography as presented in this chapter may set a stage for a revolution in affinity chromatography. The development of weak affinity chromatography will to a large extent be dependent on the availability of appropriate ligands selected for weak affinities and how to arrange them in a proper column format. The recent establishment of new technologies for production of specially designed fragments of macromolecules, such as heavy-chain fragments of monoclonal antibodies, will enable the researcher to study the different aspects of this new technology. The most well-documented example of weak-affinity chromatography—separation of similar antigens with weak-affinity monoclonal antibodies—will be presented here in detail.

2. Materials

2.1. Preparation of WAC Column

1. Produce hybridomas, as described by Nowinski et al. *(6)*, from Balb/ cJ mice immunized with a glucose-containing tetrasaccharide hapten, Glcα 1-6-Glcα 1-4Glcα 1-4Glc (named [Glc]$_4$) *(7)* immobilized to keyhole limpet hemocyanin. Clone positive cells by limiting dilution-secreting immunoglobulins that interact with polyvalent ([Glc]$_4$-bovine serum albumin [{Glc}$_4$-BSA]). Select cell lines with weak affinities ($K_{diss} > 10^{-4}M$) as measured from inhibitions studies using monovalent oligosaccharide inhibitors (*see* Chapter 10). One clone termed 39.5 is grown as ascites in Balb/cJ mice *(8)*.
2. Purify 39.5 monoclonal antibody (mouse IgG2b) by precipitation with ammonium sulfate, and then subject it to protein A-chromatography (protein A-Sepharose) according to the protocols by Gersten and Marchalonis *(9)*. The 39.5 is purified to approx 95% homogeneity by sodium dodecyl sulfate-polyacrylamide gel electrophoresis (SDS-PAGE) and with a recovery of >90% of the antigen binding activity for (Glc)$_4$-BSA as determined by Enzyme Linked Immunosorbent Assay (ELISA).
3. Immobilize the purified 39.5 monoclonal antibodies *in situ* onto a SelectiSpher 10 tresyl-activated microparticulate silica column (10 μm, 300 Å, 100 × 5 mm, HyClone AB, Sweden). The column is in a Teflon™- coated stainless-steel format to minimize any undesirable interactions with metal components.

2.2. Use of WAC Column

1. Use a microprocessor-controlled HPLC-instrument from Shimadzu, Japan (LC 4A) equipped with a variable wavelength UV -detector or other suitable HPLC system.
2. Solutes (ligates): Use only chemicals of analytical grade. Test para-nitrophenyl (PNP) derivatives of various carbohydrates including α-maltose, β-maltose, α-glucose, β-glucose, α-mannose, Dissolve the carbohydrates either in the mobile phase (*see below*) or in a simulated crude extract containing 4% fetal bovine serum (HyClone Inc., USA) and 0.1% sodium azide. Make fresh solutions every day.
3. Mobile phase: 0.02M sodium phosphate buffer, 0.2M sodium chloride pH 7.5 with 0.1% sodium azide. Filter the mobile-phase solution (0.45 μm) and degas before use by vacuum filtration. The buffer solution is stable for weeks when stored at 4–6°C. Sodium azide is a toxic chemical and should be handled with care.

4. Coupling solution: 0.2*M* sodium phosphate buffer, 0.5*M* sodium chloride pH 7.5 plus 0.1% sodium azide. Deactivation solution: 0.2*M* Tris-HCl, buffer pH 8.0. These solutions are stable for at least 1 wk when stored at 4-6° C.

3. Methods

3.1. Preparation of WAC Column

1. Couple the 39.5 antibody directly onto the activated tresyl HPLC column already attached to the HPLC instrument running at a flow rate of 1 mL/min. Pump the 39.5 (60 mL, 2 mg/mL) onto the column in the coupling solution with added panose (0.1 mg/mL). The panose, which has affinity for 39.5, is added to shield the antigen binding site possibly from coupling to the active tresyl groups of the support. Follow the coupling of antibodies on-line by measuring the UV at 280 nm at the outlet of the column.

   The coupling is very fast and efficient. Usually only one cycle is needed for maximum binding of antibodies. After complete coupling, wash the HPLC-column with 60 mL of deactivation solution at 1 mL/min to remove any remaining tresyl groups on the support. Wash away the bound panose by elution with 100 mL 0.1*M* citrate-HCl, pH 3.0, at 1 mL/min. Carry out the coupling procedure at room temperature (20–23°C).

2. Produce, as a suitable reference, a blank column without any 39.5 in an identical manner as described above. By this coupling procedure, 86 mg of 39.5 is bound (43 mg/mL column vol), and 55% of the antibody still expresses its antigen-specific binding activity. Determine the antigen-binding capacity by frontal-affinity chromatography *(10)* using tritiated $(Glc)_4$ as the ligate.

3.2. Use of WAC Column

1. Perform all chromatography under thermostatted conditions, where the affinity column and a 3-m stainless-steel tubing (between pump and injector) are immersed in a controlled temperature bath capable of regulating temperatures 5–50°C. The tubing is included to ensure proper temperature at the inlet of the column.

   Set flow rate to 1 mL/min. Monitor the derivatized saccharides at 300 nm where the PNP group shows high absorptivity. Carry out sample application (0.1 µg) with a syringe into a Rheodyne injector (USA) of 20-uL loop size. The loop size determines the amount of injection and shows an accuracy of ±1%.

2. The best way to illustrate the results from chromatographic runs is to
study the actual chromatogram. As seen from Fig. 1, the separation
of PNP-α and β-maltose is shown under isocratic conditions at 45°C.
The saccharides α-maltose and β-maltose are specifically retarded by
the 39.5 column, whereas others, such as α- and β-glucose, α-mannose, α-lactose, and α-cellobiose, cannot interact with the affinity column. This indicates that the PNP-group *per se* does not bind to the
antibody support and that the specific retardation comes from the
interaction with the saccharide structure. In addition, the secchar-
ides cannot bind to the reference column, which further supports
the specific attraction addressed by the antibody.

As seen from Fig. 1, the contaminants in the crude mixture are
not retarded and appear in the void volume (the retention volume of
the solutes that do not interact with the solid phase). A minor front
peak is seen, which probably reflects the presence of excluded material (large molecules that cannot penetrate the pores of the silica particle). The derivatives of α- and β-maltose are completely resolved under
isocratic conditions within 15 min. The separation is highly reproducible and shows no decline after repeated injections of at least 100
samples. Quantification of eluted substances can easily be performed
by using the HPLC integrator, and with an appropriate autosampler,
automatic analysis of 50–100 samples can be executed unattended
every day. When not in use, flush the column with mobile phase and
store in the same in the cold room. To avoid extensive corrosion of
any stainless-steel components of the HPLC equipment, wash all flow
lines thoroughly with water, and store them in water.

4. Notes

1. The retention in affinity chromatography is generally governed by
several parameters, including active amount of bound ligand, the
affinity, usually expressed as the K_{diss} the amount of ligate, and support characteristics, such as porosity. A simple equation can be
derived that is valid when separations are performed under linear
adsorption isotherm conditions (i.e., the retention of a ligate is independent of its injected amount:

$$k' = \text{const.}*\frac{(\text{Ln,active})}{K_{diss}} \qquad (2)$$

k' is the capacity factor (k' = $t_r - t_o/t_o$; t_r is the measured retention
time, and t_o is the retention time of a noninteracting similar ligate)
[Ln, active] denotes the active concentration of ligand that interacts

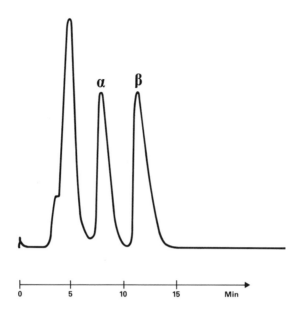

Fig. 1. Separation of PNP α- and β-maltose in a crude extract using WAC. Experimental procedures are described in materials and methods.

with the ligate under the selected chromatography conditions. The const. in (2) is a function of the column features (porosity and exclusion volume) and is usually in the range of 0.5–1.0.

When examining eq. 2, it can be seen that, in order to perform WAC with a reasonable retention ($k' > 1$), high levels of active, bound ligand must be achieved (in the millimolar range). It is also important to conduct chromatography under maximum retention, i.e., the amount of injected ligate should not diminish retention. Under certain conditions, this can only be realized by injecting very minor amounts of ligate.

Based on the above discussion, a rough estimate of the affinity range for the separation in Fig. 1 can be estimated, since the concentration of active ligand is known (0.1–1 mM) and the retention is approximately independent of ligate concentration. It should be pointed out that the tetrasaccharide antigen can only bind with one antigen binding site of the antibody at the same time. By calculation as described above, the saccharides in Fig. 1 are chromatographed in the affinity range $K_{diss} = 10^{-3} - 10^{-4}M$. The practical limit of WAC is controlled by the amount of ligand that can be bound in an active state. For instance, sterical hindrances set the maximum amount of ligand that will interact with the ligates. (*See* Section 3.2.2.).

2. Another important fact to consider in WAC is the selection of support. Highly efficient HPLC-stationary phases with small particle sizes (5–20 μm is recommended) have to be used in order not to jeopardize the high resolution obtained by the minute differences in weak affinity between the ligand and various ligates. The HPLC-supports minimize nonkinetic effects, which is usually the major cause of broadening of eluting peaks. Table 1 gives a summary of commercially available activated gels that are possible candidates for binding ligands in applications for WAC.

   Different coupling chemistries can here be tested to find the best choice for WAC. It is our experience that aldehyde- and tresyl-activated supports are good selections, and can be used to couple ligands in high amounts that carry amino and thiol residues.

   In addition, selection of a proper porosity range of the stationary phase is essential to establish the maximum surface area available for binding of ligand and its subsequent interaction with ligates. The performance of the separation in terms of resolution can also be increased by extending column length (k' stays the same). However, there are practical barriers to having very long columns, because the pressure drop will be significant and analysis times can be impractically long. (*See* Section 2.1.3.)

3. The choice of ligand is of tremendous importance to realize the potential of WAC. The key elements to have in mind when selecting an appropriate ligand for WAC include experimental procedures to find ligands with weak affinity, technology to produce the ligand in high amounts cost efficiently, handles on the ligand for immobilization that do not discriminate the active binding sites for ligate, and high stability when facing various elution conditions. Small (low-mol-wt) ligands are preferred to create high molar loadings if they can be bound with preserved high activity (e.g., using spacer molecules).

   With the advent of hybridoma technology for producing monoclonal antibodies or fragments thereof and various methods for producing recombinant proteins/peptides, we now have a powerful means to provide high-density supports with ligands selected for weak affinities. In the application described above (Fig. 1), monoclonal antibodies were selected for weak affinity and were manufactured in high amounts by ascites production (each mouse gave 3–5 mL of 5-10 mg/mL). The 39.5 clone produced monoclonal antibodies that were highly temperature-sensitive, and therefore, affinity could be regulated by changes in temperature. One disadvantage with using antibodies is that they are rather large molecules, where major por-

Table 1
Suppliers of Activated HPLAC Matrices

Type	Trade name	Matrix	Particle size, μm	Main applications	Supplier[a]
Aldehyde	SelectiSpher Aldehyde	Silica	10	Coupling of amino groups	PB
	Bakerbond Glutar-Aldehyde	Silica	40	Coupling of amino groups	BAK
Chloroformate	HyPAC	Silica	10,30	Coupling of amino groups	SE, ST, C
	Carbonite N	Silica	5–10	Coupling of amino groups	BA
Epoxy	Ultraffinity-EP	Silica	10	Coupling of amino groups	B
	Eupergit	Methacrylate	30	Coupling of amino groups	R
	Duraspher AS	Silica	7	Coupling of amino groups	A
	Bakerbond Epoxy	Silica	40	Coupling of amino groups	BAK
Hydroxysuccini-mide	Affi prep 10	Polymer	40–60	Coupling of amino groups	BR
Tresyl	SelectiSpher Tresyl	Silica	10	Coupling of amino, thiol groups	PB, P
	Shim-pack AFC-TR	Silica	10	Coupling of amino, thiol groups	S
	TSK gel tresyl-5PW	S-DVB[b]	10	Coupling of amino, thiol groups	TS
	YMC-Pack AFS Tresyl	Silica	10	Coupling of amino, thiol groups	Y

[a]Key to suppliers: PB–Perstorp Bioolytica AB (Sweden); SE–Serva (West Germany); ST–Sterogen (USA); C–Chromatochem (USA); BA–Barspec (Israel); B–Beckman (USA); R–Rhom-Pharma (West Germany); A–Alltech (USA); BR–BioRad (USA); P–Pierce (USA); S–Shimadzu (Japan); TS–Toso Haas (Japan); Y–YMC (Japan); BAK–J. T. Baker (USA). [b]DVB–Divinylbenzene

tions of the molecule are not involved in the binding event. The F_c-tail in the IgG antibody does not play any role in the recognition *per se* but has other missions, including attracting F–binding cells, such as certain macrophages. One way to increase the loading of ligand is to use fragments of antibodies, such as F_{ab}, heavy or light chains, or even the antigen binding site. The emerging technology of using recombinant DNA technology for synthesis of single-chain antibodies with moderate mol wt *(11)* might offer value for the future exploitation of WAC. In addition, procedures to incorporate molecular handles at specific positions in the ligand molecule will provide an efficient means for immobilization.

An alternative approach is to use smaller ligands, such as selected peptides *(12)*, that can mimic the antibodies' antigen-recognition site. The technology involves the construction of a representative set of analogs that, when conjugated to a solid support, combine the benefits of HPLAC with the specificity of weak-to-moderate affinity antibodies. In addition, traditional high-affinity ligands can be modified into the weak-affinity format by structural modifications. Molecular imprinting *(13)* is capable of providing a range of substrate- and enantio-selective ligands that may be potential candidates for WAC .

4. During the last 10 yr, we have witnessed an explosive growth of HPLC systems with special emphasis on the design of new detector principles, data-handling equipment, and sampling systems. For example, detectors for carbohydrates (pulsed amperometric devices) are now available that can monitor nanogram amounts of carbohydrates without any need for derivatization. Furthermore, there are now efficient scintillation detectors that can monitor the eluate for radioactivity continuously.

Packing of columns can now be performed with different slurry packing techniques, where the support is packed under high pressure (approx 200–300 atm). The activated tresyl gel was packed in the mobile phase by use of a Haskel pump (USA), giving initial flow rates of 50–100 mL/min. Larger particles (>20 µm) can be dry-packed (*see* Section 2.1.3.).

5. The eluting conditions for WAC can be varied in many ways, including changes in pH, ionic strength, and temperature. It is also possible to add various organic solvents to the mobile phase to change the elution pattern. By a careful selection of elution characteristics, the optimal conditions for running WAC can be found. One of the key features of WAC, namely the option to run isocratic chromatography, will make experiment protocols easy to manage. For instance, the problems with ghost peaks appearing when changing elution

profiles in analytical affinity chromatography will be solved. The "mild" elution procedures in WAC will probably improve the yield of fragile biomolecules. It is expected that, if gradient elutions are to be used, WAC can be performed with very minor changes of the mobile phase, such as increasing ionic strength by a fraction.

In the above application, temperature was a powerful way to modify the elution profile from very weak affinities at 50°C ($K_{diss} = 10^{-4}-10^{-3}M$) to rather high affinities at 4°C ($K_{diss} = 10^{-6}-10^{-5}M$). It is not recommended to go above 55–60°C, facing the risk of irreversible denaturation of the immobilized antibodies with major loss of retention capacity.

6. As seen from Fig 1, the 39.5-HPLAC column has the ability to discriminate between similar forms of derivatives of saccharide. In fact, the column can quite unexpectedly separate the α- and β-anomers of the maltose with a high degree of performance. In other experiments *(3)*, we have been able to follow the mutarotation of maltose, which is the major component in the crystalline state.

7. The interaction between biomolecules, including proteins, is of fundamental importance in the understanding of biological behavior. WAC might provide an easy-to-use tool to address more accurately such issues as affinity (in both relative and absolute terms) and localization of molecular targets for interaction. For example, retention volumes and the spreading of the ligate peak in the chromatogram can give valuable information about affinity and kinetic speed constants for the binding of ligates to its target: the ligand *(14,15)*. The technology of WAC can also be helpful in drug design, here potential activator and inhibitor molecules can easily be screened for their ability to interact with various receptor targets.

References

1. Scouten, W. H. (1981) Affinity Chromatography in *Chemical Analysis* Vol. 59 (Elving, P. J., and Winefordner, J. D., eds.) John Wiley, New York.
2. Ohlson, S., Lundblad, A., and Zopf, D. (1988) Novel approach to affinity chromatography using weak monoclonal antibodies. *Anal. Biochem.* **169,** 204–208.
3. Wang, W. T., Kumlien, J., Ohlson, S., Lundblad, A., and Zopf, D. (1989) Analysis of a glucose-containing tetrasaccharide by high-performance liquid affinity chromatography. *Anal. Biochem.* **182,** 48–53.
4. Meyer, V. (1988) *Practical High Performance Liquid Chromatography* (John Wiley, New York).
5. Ohlson, S., Hansson, L., Larsson, P. O., and Mosbach, K. (1978) High performance liquid affinity chromatography (HPLAC) and its application to the separation of enzymes and antigens. *FEBS Lett.* **93,** 5–9.

6. Nowinski, R. C., Lostrom, M. E., Tam, M. R., Stone, M. R., and Burnette, W. N. (1979) The isolation of hybrid cell lines producing monoclonal antibodies against the pl5(E) protein of ecotropic murine leukemia viruses. *Virology* **93,** 111–126.

7. Hallgren, P., Hansson, G., Henriksson, K. G., Hager, A., Lundblad, A., and Svensson, S. (1974) Increased excretion of glucose-containing tetrasaccharide in the urine of a patient with glycogen storage disease type II (Pompes Disease). *Eur. J. Clin. Invest.* **4,** 429–433.

8. Lundblad, A., Schroer, K., and Zopf, D. (1984) Radioimmunoassay of a glucose—containing tetrasaccharide using a monoclonal antibody. *J. Immunol. Methods* **68,** 217–226.

9. Gersten, D. M. and Marchalonis, J. J. (1978) A rapid, novel method for the solid-phase derivatization of IgG antibodies for immune-affinity chromatography. *J. Immunol. Methods* **24,** 305–309.

10. Kasai, K. I., Oda, Y., Nishikata, M., and Ishii, S. I. (1986) Frontal affinity chromatography: Theory for its application to studies on specific interactions of biomolecules. *J. Chromatogr.* **376,** 33–47.

11. Bird, R. E., Hardman, K. D., Jacobson, J. W., Johnson, S., Kaufman, B. M., Less, S. M., Lee, T, Pope S. H., Riordan, G. S., and Whitlow, M. (1988) Single-chain antigen-binding proteins. *Science* **242,** 423–426.

12. Kauvar, L. M., Cheung, P. Y. K., Gomer, R. H. and Fleischer, A.A. (1990) Paralog Chromatography. *Biotechniques* **8,** 204–209.

13. Ekberg, B. and Mosbach, K. (1989) Molecular imprinting: A technique for producing specific separation materials. Tibtech., **7,** 92–96.

14. Chaiken, I. (1979) Quantitative uses of affinity chromatography. *Anal. Biochem.* **97,** 1–10.

15. Wade, J. L., Bergold, A. L., and Carr, P. W. (1987) Theoretical description of non-linear chromatography, with applications to physicochemical measurements in affinity chromatography and implications for preparative-scale separations. *Anal. Chem.* **59,** 1286–1295.

CHAPTER 13

Biospecific Affinity Elution

Robert Kerry Scopes

1. Introduction

Affinity chromatography's main achievement has been to make use of a biologically significant interaction between a macromolecule and its ligand to separate the desired component from a complex mixture. Having used this interaction to pull out the required component, it is necessary to reverse the process so that it is liberated and the adsorbent regenerated. Although this can usually be done by a nonspecific method, any contaminants that have also been attracted to the adsorbent are likely to be eluted along with the desired component. Use of biospecific affinity elution, in which the interaction is displaced by inclusion of free ligand in the eluting buffer, can minimize co-elution of unwanted contaminants.

The term "affinity chromatography" has been extended from its original concept to include many interactions between adsorbent and macromolecule that are not physiological. Thus adsorbents, such as dyes, metal chelates, and other nonbiological selective materials, have been called affinity adsorbents. This is not an incorrect use of the word, since any interaction can be called an affinity. To describe the original meaning of affinity chromatography it is now necessary to include the word "biospecific," implying an interaction between a natural binding site and the *natural* ligand (whether it be enzyme–substrate, enzyme–inhibitor, or other protein–ligand interaction). Similarly, the

From: *Methods in Molecular Biology, Vol. 11: Practical Protein Chromatography*
Edited by: A. Kenney and S. Fowell Copyright © 1992 The Humana Press Inc., Totowa, NJ

term "biospecific affinity elution" can be used to describe the use of natural ligand in eluting the desired component from a column, rather than the less descriptive term "affinity elution."

By using a natural ligand in an eluting buffer that succeeds in removing the desired component (without significantly affecting other properties of the eluant, such as ionic strength or pH, which might cause nonspecific elution), we can assume that the ligand has bound to the adsorbed macromolecule, and in doing so has caused its attachment to the column to be weakened. But this does not in itself define *how* the macromolecule was interacting with the column. It is not necessary for the adsorbent to be a biopecific one for this type of affinity elution to succeed. Indeed, many of the most successful examples of biospecific affinity elution have used ion exchangers *(1)* or dye ligands (e.g., *2,3*) as adsorbent.

Thus biospecific affinity elution can in theory apply to any type of adsorbent and any type of molecule bound to that adsorbent, provided that the eluting ligand has an affinity for the bound molecule, and it is this interaction which results in elution. For practical purposes we need not be concerned about why affinity elution works, though some understanding of likely mechanisms helps in designing methodology.

Ion-exchange chromatography is widely used, and the two modes of nonspecific elution are to increase the ionic strength and, less often, to change the pH. Biospecific affinity elution from ion exchangers has been very successful in a number of cases, provided that certain criteria are met. First (and this applies to all affinity methods) the macromolecule, which we shall assume is a protein, must have a recognizable specific binding site for a ligand, which we shall assume to be a small molecule. For instance, all enzymes by definition have substrate binding sites, and often subsidiary activator or inhibitor sites. Many soluble proteins (e.g., blood proteins) naturally interact with small molecules. On the other hand, structural or storage proteins may not have a physiologically significant binding site, and so would not be amenable to *biospecific* affinity techniques. Second, the protein must bind to an ion exchanger, and be eluted, at a pH that allows interaction with the ligand. Third, on ion exchangers, the elution has in nearly all cases been a result of adding a *charged* ligand; the binding of this ligand *reduces* the overall charge of the macromolecule, hence weakening its interaction with the exchanger. Since most biological molecules are negatively charged (if at all), this means that the macromolecule should have a net positive charge, so the ion exchanger must be a cation exchanger. This at once reduces the range of proteins that can be successfully purified using affinity elution from ion exchangers to those with an isoelectric point sufficiently high to allow them to bind to a cation exchanger at a usable pH. Whereas animal proteins frequently have high isoelectric points, plant and bacterial proteins are generally more acidic.

Given the requirement that the protein has a physiologically significant binding site, affinity elution from other adsorbents does not have the same requirements as described above, since isoelectric point is not necessarily a stringency for binding to the adsorbent. Both true affinity adsorbents with immobilized biological ligands, and pseudoaffinity adsorbents that happen to interact with the protein through the natural binding site are ideal candidates for a carefully designed biospecific affinity elution procedure. It is also found that when the protein molecule changes configuration upon binding its ligand, this in itself (regardless of the specificity of the protein–adsorbent interaction) may cause elution.

Because each protein and enzyme has particular properties that distinguish it from all others, it is not possible to give a detailed protocol of methods for affinity elution, only general principles. It is not even predictable whether affinity elution will be successful in any particular case, so the best advice is to try it according to the procedures described below.

2. Materials

2.1. Buffers for Use in Affinity Chromatography

One of the key features in any protein or enzyme work is the choice of a buffer system. Features to be considered are:

1. Suitable pH range (stability of protein charge on protein, and whether the protein has a physiological interaction with ligand at that pH).
2. Ionic strength (stability/solubility of protein; efficient interaction with adsorbent *and* with free ligand at chosen ionic strength).
3. Compatibility (should not interfere with binding of ligand).
4. Presence of stabilizing compounds (metal ions; glycerol; complexing agents).
5. Toxicity (to technicians); cost (for large-scale work).

The nature of the buffer also depends on the type of adsorbent. Some materials, especially ion exchangers, may require buffers of low ionic strength with optimum buffering power. Other adsorbents, such as true affinity adsorbents, may operate better at a physiological ionic strength, of about 0.15 or even higher, to minimize nonspecific binding; the buffer may then make up only a part of the ionic strength, and a salt is included. (*See* Note 1.)

2.2. Buffers for Cation Exchangers

Adsorption of proteins to cation exchangers, such as carboxy-methyl cellulose generally requires a pH <7, as few proteins have significant positive charges above this value. Buffers that are uncharged in the acidic form should be used. (*See* Notes 1 and 2.)

Stock solutions should be made up as follows:

1. MES buffer (pH 6.0–6.8): Add sufficient KOH so that the final solution will be $0.1 M$ in potassium ions. Make the volume to about 90% of the final volume, then dissolve MES (2-[*N*-Morpholino] ethanesulfonic acid) into the solution until the pH is the required value. Make up to volume; and add a few drops of 20% sodium azide for long-term preservation. Use at 10X dilution.
2. MOPS buffer (pH 6.8–7.8): exactly as for MES above, using MOPS (3-[*N*-Morpholino]propane sulfonic acid).
3. Tricine buffer (pH 7.5–8.5): exactly as for MES above, using Tricine *N*-Tris[hydoxymethyl]methylglycine).
4. Acetate buffer (pH 4.0–5.5): exactly as for MES above, using acetic acid.
5. Buffer for pH range 5.5–6.0: If it is necessary to operate in this range, picolinic acid is a possible buffer. This has a major disadvantage in having a strong UV absorption at 280 nm, preventing monitoring of column eluants by UV. MES can be used also, but rather large amounts of this expensive buffer are needed in this pH range.

There are several other zwitterionic buffers covering the pH range 7–9, which are equally good.

2.3. Buffers for Dye–Ligand Chromatography and for Affinity Adsorbents

Adsorption to dyes has some of the characteristics of cation exchange; the dyes are generally negatively charged. However, it may be desirable to minimize cationic exchange behavior, to take advantage of the pseudoaffinity behavior that is less dependent on ionic strength. The same applies to true affinity adsorbents, especially when the natural ligand is charged. Dye–ligand adsorption is relatively little affected over the ionic strength range 0.05–0.2, in contrast to a pure cation exchanger. On the other hand, adsorption increases significantly as pH falls. For an initial screening of dye adsorbents to maximize binding, a pH 6.0 MES buffer, containing Mg^{2+} ions *(4)* and a little salt to make the ionic strength up to 0.05 is useful:

1. 10X stock solution: $0.1 M$ KOH, adjusted to pH 6.0 with MES (*see above*), containing 20 mM $MgCl_2$ and $0.3 M$ KCl or NaCl.

   This buffer can also be made to pH 6.5 if desired.
2. For large-scale work MES and some other zwitterionic buffers can be quite expensive. For circumstances in which a low ionic strength is not essential, we have developed a buffer system for use over the pH range 4–8 that contains only cheap reagents, and has a nearly constant ionic strength regardless of the pH to which it is adjusted *(3)*. The buffering components are acetate (pH 4.0–5.5), succinate (pH 5.5–6.5), and phosphate (pH 6.5–8.0):

   10X stock solution: $0.3 M$ K_2HPO_4 plus $0.2 M$ NaCl. After tenfold dilution, the pH is adjusted to the desired value with a mixture containing 2M acetic acid and 0.5M succinic acid. Mg2+ ions may be added, together with any other component regarded as useful (note that many metal ions will precipitate with the phosphate), after the pH has been reduced. The ionic strength will be about 0.1.

2.4. Stock Solutions of Biochemicals Used in Affinity Elution

Affinity elution uses biochemicals, such as ATP, NAD^+, substrates and substrate analogs, some of which are unstable in solution, and most of which are susceptible to bacterial action if stored over a long period unfrozen. Most can be made into a concentrated stock solution, which can either be frozen or stored in the refrigerator in the presence of a preservative, such as azide (ca. 5 mM). Solutions which are acid (e.g., NAD^+, disodium ATP) should not be treated with azide, as it evolves toxic hydrazoic acid gas and leaves the solution unprotected (on the other hand, acid solutions are less likely to become contaminated with microorganisms). It is often a good idea to neutralize all such solutions, since addition of an acid substrate to a weak buffer will cause an unplanned pH change. When neutralizing, use a weak base (e.g., Tris) to avoid overshooting to a high pH, at which the substance may be unstable.

NAD⁺ is unstable in alkali, whereas NADH is unstable in acid. NADH should not be made up in water, but always dissolved in a weakly basic solution, such as 20 mM Tris base. In this case, *do not neutralize,* as NADH solutions decompose in a day or two at pH 7, but will keep quite well for a couple of weeks in Tris base containing azide.

The amounts and concentrations of stock solutions made up will depend on circumstances, including cost, but it is a good idea to standardize on a concentration, e.g., 100 mM, for all such reagents.

3. Methods

3.1. Types of Adsorbents

As indicated above, the effectiveness of affinity elution does not require any particular character of the adsorbent; it may be anything from the most sophisticated high-performance chromatographic packing to a column of crushed rock. The requirement is that inclusion of the ligand causes elution. The following have been used:

1. True affinity adsorbents, in which the ligand is a substrate, inhibitor, or other natural compound that interacts with the protein.
2. "Pseudoaffinity adsorbents," in which the ligand is not natural, but mimics the substrate, or at least binds at the substrate site, so that affinity elution can be considered a displacement from the immobilized ligand.
3. Ion exchangers, with the usual case being elution of a high-isolectric-point protein from a cation exchanger using a negatively charged ligand.
4. Hydrophobic adsorbents. In this case elution may be a result of conformational changes on binding of substrate, or a direct displacement if the substrate is itself hydrophobic.
5. Other adsorbents, e.g., hydroxyapatite, metal chelates, inorganics in general, and specialized "one-off" adsorbents developed for a particular isolation.

3.2. Adsorption Conditions

1. Sample volume and concentration. For proteins that adsorb "totally," and do not slowly migrate down the column during application, the volume of sample is not important provided that the column is not overloaded. Thus 100 mL of 1 mg/mL protein should give the same ultimate result as 10 mL of 10 mg/mL. On the other hand, it will take longer to apply the larger sample, which would be an important consideration with a small, narrow column. (*See* Note 3.)

2. Buffer concentration and pH: Buffer concentration will depend on many things; in the case of ion exchange, it is important to minimize the ionic-strength contribution from the buffer (see above), but there must be enough buffer to minimize undesirable pH changes on the column. Enough buffer is generally *at least* 4–5 mM of buffering molecules *on either side* of the pH being used. Thus 10 mM of a buffer at, or within 0.3 U of its pK_a should be considered a minimum concentration; similarly up to 100 mM of a buffer is needed 1 pH unit from its pK_a (*see* Note 4).

3. Size and shape of columns: True chromatography requires a long column, with solutes adsorbed on only a small fraction of the adsorbent at the top of the column. This maximizes the plate number in such a way as to get good resolution of similar components. Affinity chromatography is more of an "on–off" process, in which much of the adsorbent may be used in the initial binding step, and true chromatographic behavior during elution is not required and is sometimes not even desirable. Thus, short, squat columns are most suitable, since volumetric flow rates are high, and so processing can be fast. A column with a height-to-diameter ratio of no more than 2 is best, and columns with ratios < 1 are common in large-scale work. The column size need be no more than 2–3X the minimum needed before "breakthrough" from overloading occurs.

4. Flow rates: The optimum flow rate for chromatography has been investigated extensively (5,6) and depends mainly upon the size of the solutes (slower for slow-diffusing molecules, such as proteins) and on the size of the bead particles in the column (the smaller the bead, the faster the rate). It also depends on temperature, since the whole operation is a mass-transfer process that relates to diffusion coefficients, which increase with temperature.

   For routine "soft" chromatographic materials, such as agarose beads with diameters in the range 50–150 μm, linear flow rates of about 10–15 cm/h at 4°C, and 20–30 cm/h at 20–25°C, have been found optimum. The linear flow rate is the flow rate mL/h divided by the cross-sectional area of the column (cm^2). For high-performance materials, with bead sizes of 10 μm, the value increases to about 200 cm/h. Whereas rates faster than these may be quite suitable in many cases, slower rates will not generally improve the performance of the adsorption step. During elution, however, it may be desirable to decrease the flow rate by a factor of 2 to ensure that interaction of the biospecific elutant can take place efficiently.

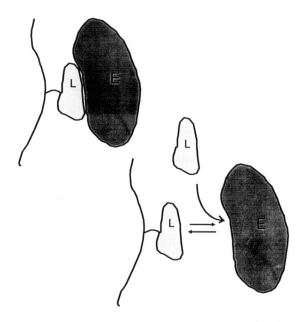

Fig. 1. Mode of displacement of an enzyme (E) by ligand (L). The enzyme must first dissociate from the immobilized ligand before interaction with the free ligand can take place. The free enzyme–ligand complex is unable to rebind to the adsorbent.

3.3. Preelution Buffer Washes

Having adsorbed the desired protein on a suitable column, it is now necessary, before elution is attempted, to carry out a preelution wash in a buffer that begins to desorb the protein. This is because, for the ligand to bind to the protein, it is normally necessary for the protein to be free in solution, and not adsorbed (Fig. 1). Consequently there must be a significant proportion of protein partitioned into solution for affinity elution to succeed. The same argument applies, though for slightly different reasons, for ion exchangers as it does for more specific adsorbents (Fig. 2).

Thus the preelution buffer for a cation exchanger will be at a higher pH or ionic strength than the application buffer. For a more specific adsorbent, weakening of the interaction can occur as a result of increases in salt concentration. Affinity elution is even successful from hydrophobic adsorbents, for which the weakening is achieved by decreasing the salt concentration *(7)*.

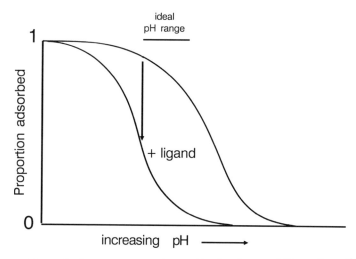

Fig. 2. Effect of the presence of negatively charged ligand on the binding of a protein to a cation exchanger. At the lowest pH, the binding is too strong for the presence of ligand to make much difference. Ideal conditions are found only in a narrow range of pH, as indicated.

3.4. Elution Procedure

This is the key step, in which the enzyme is hopefully eluted in a highly purified form. The main consideration is the optimum amount and concentration of the biospecific ligand, which for convenience we shall call the substrate. Not only science, but also economics, is involved, as some substrates are very expensive to use on a large scale. Affinity elution will work better the higher the concentration of substrate that is used, but care must be taken to allow for nonspecific effects (*see* Notes 6 and 7) as well as the cost factor. Also, to maximize recovery from the column, a long pulse of substrate may be needed. Having decided how much can be afforded, it is a matter of deciding whether to put through a short pulse of concentrated substrate or to rely on a larger volume of more dilute substrate. The latter has disadvantages in that (a) the quantity of any nonspecific protein that is trickling off the column regardless will be greater, and (b) the desired enzyme will be eluted in a large volume, so any subsequent concentration step will take longer. The short concentrated pulse has the disadvantage of risking more nonspecific elution (notwithstanding any dummy wash), and it may leave some of the enzyme behind once the pulse has passed by.

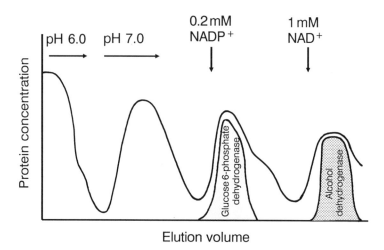

Fig. 3. Purification of two dehydrogenases by affinity elution from the dye adsorbent Procion Scarlet MX-G-Sepharose CL-4B. Extract of the bacterium *Zymomonas mobilis* was applied at pH 6. After a preelution wash at pH 7, glucose 6-phosphate dehydrogenase was eluted with 0.2 mM NADP⁺ (which does not bind to alcohol dehydrogenase); then alcohol dehydrogenase was eluted with 1 mM NAD⁺.

The actual value of substrate concentration needed depends on the dissociation constant (in that buffer) of the enzyme-substrate complex. For efficient elution of a monomeric enzyme, a minimum value of $10 \times K_d$ is found to be needed; in this case the enzyme should start to be eluted close to the substrate front if the elution-buffer composition is correct. But for dimers and larger oligomers with several binding sites it can be shown that a smaller level of substrate will suffice (8). Normally substrate levels used are in the range 0.2–2 mM for successful affinity elution. (*See* Fig. 3).

3.5. Concentration of Eluted Protein

The eluted protein will in most cases be very dilute, and before anything else is done will need to be concentrated. There are three general ways of concentrating protein solutions: removal of water, adsorption, and precipitation.

Removal of water may be done by lyophilization (freeze-dying), but this is not generally recommended except for long-term storage. More often it is done by ultrafiltration, in which the water and low-mol-wt solutes are forced through a semipermeable membrane. A volume of 200 mL can be concentrated to about 10 mL in 1 h through a 30-cm^2 membrane (mol wt cut off 10,000), provided that the final protein concentration is no more than 10 mg/mL.

Adsorption may be very useful provided that there is a convenient adsorbent that will totally bind the protein in the buffer conditions of elution. Obviously this will not normally be the adsorbent just used for the affinity step. We have conveniently adsorbed low-isoelectric-point enzymes affinity-eluted from dye columns, using DEAE-cellulose batchwise, where 20 g wet wt is sufficient for 500 mL of 0.5 mg/mL eluted enzyme; the DEAE-cellulose is then poured into a small column and eluted with 0.5 *M* NaCl. (*See* Note 8.)

Precipitation followed by centrifugation is convenient: the precipitant is normally ammonium sulfate. However, incomplete precipitation and denaturation are common if the initial protein concentration is less than about 2 mg/mL. Thus ammonium sulfate treatment should only be carried out when the eluted protein concentration is high, or after a preliminary ultrafiltration or adsorption step. Keeping proteins as ammonium sulfate precipitates is one of the best forms of long-term storage.

4. Notes

1. On ion exchangers the counter ion should not be a buffer ion, since this accumulates in the microenvironment, where proteins are binding, which results in large pH changes. To maximize buffering power for a given ionic strength, one of the buffering species (the acid form for cation exchangers) should be uncharged.
2. Proteins with an isoelectric point high enough to bind to cation exchangers may not do so in the presence of nucleic acids. The positively charged proteins may bind to soluble nucleic acids, rather than the adsorbent. Treatment with neutralized protamine sulfate to precipitate out nucleic acids will alleviate this problem.
3. At a concentration much over 15 mg/mL, the importance of the protein as a contributor to ionic strength is significant, and osmotic effects can cause gel shrinkage. In most cases a protein concentration of 2–10 mg/mL is best.
4. For ion exchangers the ionic strength may be critical, so the composition of the buffer should be quoted so that the ionic strength is clearly defined. For instance, "20 m*M* Tris, pH 7.8" is not clearly defined, especially if the temperature at which the pH was measured is not stated. On the other hand, "10 m*M* HCl adjusted to pH 7.8 with Tris" defines the ionic strength as 0.01, even if the temperature is not quoted.

5. Despite a general assumption that enzyme work should be carried out in the cold, most enzymes, and certainly most nonenzyme proteins are stable enough for a purification procedure to be carried out at ambient temperature. This has the advantage that all processes can be speeded up about twofold, as diffusion coefficients are about 2X greater. On the other hand, more care is needed to avoid bacterial or fungal contamination.

6. In certain circumstances it may be desirable to include in the preelution buffer what can be termed a "dummy substrate." This is for cases in which the introduction of the true substrate results in a significant alteration of the buffer's properties, in particular, ionic strength. Elution with 1 mM ATP results in an increase in ionic strength of > 0.005, which itself may cause nonspecific elution of unwanted proteins. In this case a dummy substrate of 1 mM EDTA can be used, before the true affinity-elution step with ATP.

7. There are many examples in which concentrations as high as 50 mM (e.g., of ATP) have been necessary because the workers had not appreciated the requirements of the preelution conditions. Often these expensive experiments were succeeding partly because of nonspecific ionic-strength effects.

8. When using dye columns, treatment after elution using DEAE-cellulose has an advantage in that any dye that may have leaked from the column is tightly bound and is not eluted even by 1M NaCl. Even if the desired enzyme does not bind to DEAE-cellulose, any color is quickly removed.

9. Protein adsorbents, whether they be ion exchangers, affinity adsorbents, dyes, or some other material, become contaminated with proteins and other substances that do not wash off with the usual buffers. Each material may have extreme conditions that should not be exceeded during a clean-up process; e.g., a maximum or minimum pH, concentration of organic solvent, and so on. This will depend on the nature of the adsorbent, especially the matrix material. Polysaccharides, such as cellulose or agarose, are generally quite stable to strong alkali, so, *provided that the ligand is also stable,* sodium hydroxide is suitable. Most biological macromolecules are soluble in alkali, so a washing scheme including NaOH is desirable. High concentrations of urea (5–8M) are also used to remove recalcitrant proteins; urea–NaOH combinations may be used. A suitable mixture is 5M urea in 0.1M NaOH. After a prolonged storage period it may be useful to wash this before commencing equilibration with buffer. If

the material is unstable to alkali, a detergent such as sodium dodecyl sulfate, may be considered, and acid or organic solvent washes may be used in some circumstances.

Much preferable to constantly replacing an adsorbent because it cannot be cleaned up successfully is not letting it become contaminated in the first place. If a disposable "pre-column" can be used, which takes out most of the undesirable material without adsorbing the product, this is advisable. However, it is not always easy to design a system that does not take out your protein as well. One method is to batch-treat with just enough adsorbent to avoid taking out your protein, filter, and then run on the column. The most strongly adsorbing and potentially disruptive proteins and other compounds are removed, and the column remains relatively clean.

Storage of adsorbents must be in conditions that prevent bacterial or fungal contamination. This is especially important with matrices that are biological polymers. Use a buffer that does not contain organic compounds (e.g., phosphate, *not* acetate); add a bacteriocide, such as sodium azide, chlorobutanol, or Hibitane®; and store in a cold room.

References

1. Scopes, R. K. (1977) Purification of glycolytic enzymes by affinity elution chromatography. *Biochem. J.* **161**, 253–263.
2. Rajgopal, C. S. and Vijayalakshmi, M. A. (1982) Purification of luciferase by affinity elution chromatography on Blue Dextran columns. *J. Chromatogr.* **243**, 164-168.
3. Pawluk, A., Scopes, R. K. and Griffiths-Smith, K. (1986) Isolation and properties of the glycolytic enzymes from *Zymomonas mobilis*. The five enzymes from glyceraldehyde phosphate dehydrogenase through to pyruvate kinase. *Biochem. J.* **238**, 275–281.
4. Hughes, P. and Sherwood R. F. (1987) Metal ion-promoted dye-ligand chromatography, in *Reactive Dyes in Protein and Enzyme Technology* (Clonis, Y. D., Atkinson, A., Bruton, C. J. and Lowe, C. R., eds.), Macmillan, U.K., 86–102.
5. van Deemter, J. J., Zuderweg, F. J., and Klinkenberg, A. (1956) Longitudinal diffusion and resistance to mass transfer as a cause of nonideality in chromatography. *Chem. Eng. Sci.* **5**, 271–279.
6. Janson, J.-C. and Hedman, P. (1982) Large-scale chromatography of proteins. *Adv. Biochem. Eng.* **25**, 43–99.
7. Scopes, R. K. and Porath, J. (1989) Differential salt-promoted chromatography for protein purification. *Bioseparation* **1**, 3–7.
8. Scopes, R. K. (1987) *Protein Purification, Principles and Practices* 2nd Ed. Springer Verlag, New York, pp. 162–166.

CHAPTER 14

Size-Exclusion High-Performance Liquid Chromatography of Proteins

David A. Harris

1. Introduction

In chromatography, a solute partitions between a mobile and a stationary phase. Uniquely in size-exclusion chromatography, both phases have identical physicochemical properties; differential partition is effected by restricting the *access* of the solute to the stationary phase. At its simplest, the stationary phase is held inside pores with a distribution of sizes, and some pores are too small to permit the entry of larger molecules—hence *size exclusion* (SE) (or the alternative *gel permeation*) chromatography.

In "pure" SE chromatography, interactions among solute, solvent, and support should all be equal. Thus, solute will not move preferentially from one phase to the other, but simply diffuse through all the space accessible to it. Similarly, one solvent molecule will suffer no change in thermodynamic activity when a nearby solvent molecule is replaced by a group on the support. Meeting these conditions is a tall order for the polymer chemist, and so some solute/support interactions do inevitably occur in SE columns (see next section). Normally, solvent composition (ionic strength, pH, and so on) are chosen to minimize such interactions *(below)*; however, if the primary purpose of size exclusion high-performance liquid chromatography (SE-HPLC) is to *isolate* separated proteins, even the weak interactions of commercial supports may be exploited in a mixed size/charge separation *(1)*.

From: *Methods in Molecular Biology, Vol. 11: Practical Protein Chromatography*
Edited by: A. Kenney and S. Fowell Copyright © 1992 The Humana Press Inc., Totowa, NJ

In considering applications of SE-HPLC, a brief consideration
its quantitative aspects is useful. A protein, introduced into a colun
containing pores of different sizes, may potentially enter the volume
outside the support beads (V_o) and that contained inside the pores
(V_i). However, the macromolecules are prevented by their size from
entering some pores, so only a fraction of V_i is available to them. This
fraction is given by $K_d V_i$ where K_d is a form of partition coefficient. The
protein thus emerges from the column at an elution volume of:

$$V_e = V_o + K_d V_i \tag{1}$$

As can be seen, K_d can take values only between 0–1, i.e., V_e lies
between V_o and $(V_o + V_i)$, the total volume of solvent in the column (V_t).

This has several relevant consequences.

1. Because V_e is thus restricted, the resolving power of the method will
   be less than chromatographic methods based on adsorption where
   V_e is not limited by the physical size of the column, and gradient elution is possible.
2. Because partition occurs by diffusion, the sample must be applied in
   a small volume *(v)* of concentrated solution *(v = 0.01 V_t)*. Both these
   points restrict the preparative use of SE-HPLC.
3. For optimal resolution, the protein must have time to equilibrate into
   the pores, by diffusion. This requires small particle sizes (to minimize diffusion distances) and relatively slow flow rates—the larger
   the protein, the slower the required flow.

HPLC technology is well versed in providing supports with small
particle sizes (5–10 µm) and accurate pumps to produce adequate
driving pressures (but *see below*); the reproducibility, resolution, and
relative speed of this approach have fitted SE-HPLC particularly for
analytical procedures, such as mol-wt determination, assessment of
protein purity, and the study of protein–ligand interactions as discussed below.

2. Materials

2.1. Hardware

A typical HPLC setup is shown in diagrammatic form in Fig. 1, and
can be assembled, or purchased as a package, from a number of manufacturers. The following points specific to SE-HPLC (of proteins) should
be noted.

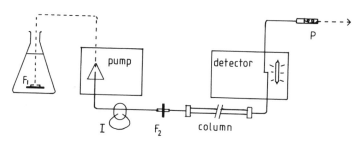

Fig. 1. Schematic of a setup for analytical SE-HPLC. In addition to pump, column, and detector *(see text)*, the buffer is filtered at the inlet from the reservoir (F_1), and prior to the column (after the injection port) via a frit or guard column (F_2). After the detector, a pressure regulator, P, (e.g., Anachem U462) in the flow line minimizes outgassing. High-pressure tubing (solid line) is 316 stainless steel (0.25 mm id); low-pressure tubing (dotted line) is Teflon™. The injection valve, I, (e.g., Rheodyne 7125) brings the (low-pressure) sample into the high-pressure flow line.

2.1.1. Pump

SE-HPLC typically operates at low flow rates (<1 mL/min) compared to standard (reverse-phase) HPLC. A suitable pump must be controllable accurately to 0.01 mL/min (e.g., Beckman 116, Waters 590); this may not be possible on a general-purpose instrument.

2.1.2. Detector

Detection at 280 nm (down to 0.1 mg/ml) or 210–220 nm (down to 5 μg/mL) is convenient; ideally, a variable wavelength photometric detector should be available (e.g., Beckman 166, Waters 481).

2.1.3. Liquid path

Conventional HPLC uses stainless-steel tubing (ss 316, 0.25 mm id). Labile proteins may be denatured by metal ions leached from this tubing, and assemblies that minimize metal/solution contact are available (e.g., System Prep inert purification system, Pharmacia, LKB, Uppsala, Sweden). However, few proteins show such high sensitivity, particularly if EDTA (0.5–2 mM) is included in the buffers as a protective agent.

2.2. Column

The low-pressure technique equivalent to SE-HPLC is gel filtration, which has conventionally used beaded dextran, agarose, or polyacrylamide for separation. However, such gels cannot resist the pressures (10–40 bar) required for the higher resolution HPLC methods. Three approaches have been used in developing supports for SE-HPLC.

1. To use the convenient surface properties of existing gel filtration supports (e.g., agarose) and modify mechanical stability by crosslinking (e.g., the Superose range, Pharmacia, Uppsala, Sweden).
2. To use the convenient mechanical properties of existing HPLC supports (porous silicas) and modify the surface chemically to minimize solute/support interactions (e.g., TSK SW series [Toya Soda Co., Tokyo, Japan], Zorbax range [DuPont, Wilmington, DE]).
3. To design a novel, porous, hydrophilic polymer (e.g., TSK PW range).

A comprehensive list of supports available for SE-HPLC is given in ref. *(2)*. For many purposes, however, the columns TSK 3000SW (mol-wt range 2000–500,000) and TSK 6000PW (mol-wt range 10^4–10^8) cover a suitable range and exhibit minimal solute/support interaction. The SW series has a greater resolving power (V_i/V_o larger), but is less stable. Both are obtained as prepacked, steel columns from a variety of suppliers (Waters Protein PAK 300 SW; Pharmacia/LKB Ultropac TSK 3000 SW; Altex/Beckman [Berkeley, CA] Spherogel 3000SW; Anachem [Luton, UK] Anagel TSK G3000 SW; Bio-Rad [Richmond, CA] Bio-Sil SEC 250).

2.3. Solvents

High-grade deionized water (preferably passed through a carbon bed to remove organic materials) should be used to prepare buffers. HPLC-grade water may be purchased, but a Millipore or Elga laboratory systems yield water of sufficient quality. Buffers can be prepared from high-purity (AnalaR) reagents in the normal way, but should be checked for absorbance at the monitoring wavelength (especially if this is 220 nm) before use.

Clearly, any buffer must be compatible with stability of the protein studied. In addition, the solution should have:

1. An ionic strength of 0.15 M or greater (to minimize ionic interactions between support and proteins). This ionic strength can be due to the buffer itself (0.1 M or above) or achieved by addition of NaCl.
2. A suitable pH. This must be below pH 7.5 for silica-based supports (e.g., TSK SW), since these slowly dissolve at higher pH values. (The PW series may be used up to pH 12.)
3. Up to 1 mM EDTA and 0.4 mM dithiothreitol may be added to stabilize labile proteins.

Table 1
Buffer Solutions Suitable for Determination
of Molecular Weight on TSK Columns[a]

Buffer	Attributes	Reference
0.1M Sodium phosphate	[b]	*1,3,4*
0.02M Sodium phosphate, 0.15M NaCl	[b]	2
0.01M Na$_2$HPO$_4$, 0.0018M KH$_2$PO$_4$, 0.17M NaCl, 0.003M KCl[+] (phosphate buffered saline)	[b,c]	5
0.01M Sodium phosphate, 0.3M NaCl, 10% DMSO	[b]	9
0.2M Triethylammonium formate	[d]	8
0.03M HEPES, 0.01M imidazole, 0.2M sodium acetate, 0.5 mg/mL dodecyl octaethylene glycol monoether (C$_{12}$E$_8$)	[f]	7
0.1M Sodium phosphate, 0.1% sodium dodecyl sulfate	[b,e]	7
0.01M Sodium phosphate, 6M guanidinium hydrochloride	[e]	2
6M Urea (deionized), pH 3.15 (formic acid)[+]	[e]	6
0.1% Trifluoroacetic acid/36% acetonitrile[*,+]	[d,e]	8

[a]This list is not intended to be exhaustive; a wide variety of buffers can be used as long as ionic strength and the stability of the column material are considered (*see text* and instructions on column use from individual suppliers). All buffers are pH 7.0, unless otherwise indicated[+]. [*]Indicates compatibility with PW range of columns only. [b]Low absorbance at 220nm. [c]Physiologically compatible. [d]Volatile (can be removed by freeze-drying). [e]Denaturing. [f]Suitable for membrane proteins.

An illustrative list of suitable buffers for use with TSK columns is given in Table 1. Sodium (or potassium) phosphates have been widely employed in SE-HPLC, since they have low UV absorbance down to 200 nm and pK values in the range 6–7.5. Modifications to introduce other potentially desirable properties (denaturing ability, volatility, physiological compatibility) are included in Table 1.

2.4. Standard Protein Solution

This should contain four or five proteins of mol wt spanning that of the unknown protein. A convenient mixture is 20 μg/mL of each of immunoglobin G, bovine albumin, ovalbumin, and soybean trypsin inhibitor in the running buffer. A more complete list of suitable standards is given in Table 2.

Table 2
Calibration Proteins Suitable for SE-HPLC *(see refs. 1,3,7,8)*

Protein	Mol wt, $\times 10^{-3}$
Bacitracin	1.4
Insulin (B-chain)	3.4
Aprotinin	6.5
Parvalbumin	12
Myoglobin	17.8
Trypsin inhibitor (soybean)	22.1
β-lactoglobulin	35
Ovalbumin	43
Serum albumin (bovine)	67 (monomer), 134 (dimer)
Alkaline phosphatase	86
Immunoglobulin G	158
Catalase	232
Urease (jack bean)	483
Thyroglobulin	670
Spectrin	460 (dimer), 920 (tetramer)
Hemocyanin	1700
MS2 phage	3600
Tomato bushy stunt virus	8700

3. Method

1. Connect the pump (via a suitable filter) to a reservoir containing pure water. Prime the pump according to the manufacturer's instructions, and operate at 0.1 mL/min until liquid emerges from the distal end of the injection port (Fig. 1).
2. Connect a "guard column" (a short column containing the separation support) TSK SW (7.5 mm × 75 mm) or a 2-μm porous frit (Waters 84560/Anachem A315) to the injection port. (This protects the expensive separation column from fouling.) Operate the pump until liquid emerges from the guard column. (*See* Note 1.)
3. Connect this to the separation column (TSK 3000 SW, 7.5 mm × 300 mm) according to the manufacturer's instructions. *(Note that only one direction of flow is permitted.)* Again, operate the pump until liquid emerges. Then connect this assembly to the detector. The outflow is led to waste through a pressure regulator to minimize outgassing (Fig. 1).
4. Increase the flow rate to 0.5 mL/min (*see* Note 2) and flush the system with 50 mL water to remove any organic solvent in which the column is stored. If this step is omitted, precipitation of buffer components may occur.

5. Filter 500 mL of buffer (0.1 *M* sodium phosphate, pH 7.0) by vacuum filtration through a 0.45-μm hydrophilic filter (Millipore Durapore, Gelman FP Vericel or Nylaflo) at room temperature to remove suspended material. This also serves to degas the solution sufficiently for most purposes.

6. Pump 5–10 column vol (100–200 mL) of buffer through the column. It is often convenient to pump at 0.2 mL/min overnight, *provided that a switch is made to freshly filtered (degassed) buffer on the day of use.*

7. On day of use, switch on the detector, and monitor the eluate at 220 nm (280 nm for higher protein concentration). Adjust the buffer flow rate to 0.5 mL/min. The operating pressure should be below 500 psi (35 bar, 3.5 MPa) and should be noted to monitor column performance over repeated runs. When A_{220} reaches a low, constant value (<0.2 AU), zero the detector. Adjust the detector range to 0.2 AU full-scale deflection, and zero the recorder such that zero detector output lies at 20% on the recorder scale (to allow for negative drift).

8. Dummy injection. Load 200 μL of buffer into the injection loop using a blunt-ended syringe (Hamilton 700 series). Inject this onto the column by bringing the injection loop into the solvent line, and simultaneously start the recorder to follow A_{220} for 30 min. The recorder should show a small deflection on injection, but no further movement from the baseline. (*See* Note 3.)

9. Standardization. Centrifuge 250 μL standard protein solution in a microfuge (e.g., Heraeus Biofuge A) for 5 min at 20,000*g*, to remove particulate matter. Sample 200 μL from the supernatant, and inject as above, following A_{220}. Plot log of mol wt vs elution time. A typical calibration curve is shown in Fig. 2.

10. Prepare and inject samples, containing >10 μg/mL protein, as in the previous step, and read the mol wt from the calibration curve. (*See* Note 4.)

11. Closedown. Replace the buffer by pumping through approx 100 mL of pure water. (Removal of salts is essential to prevent corrosion and/or crystallization, leading to mechanical damage to the system.) Care must be taken to flush through all parts of the system; if two pumps are used, both must be made to pump water, and water must pass through both the high- and low-pressure sides of the injection valve. The flow rate can then be reduced to zero.

12. For prolonged storage, the system should be filled with methanol:water (1:9) to prevent bacterial growth. Azide is not recommended as an antibacterial agent, since its salts are explosive. The column can then be removed, stoppered, and stored if required.

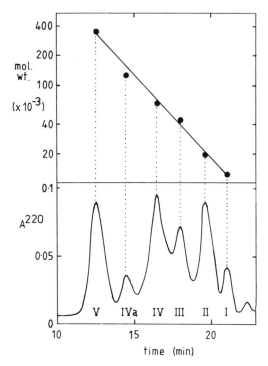

Fig. 2. Construction of a calibration curve for SE-HPLC. A Waters Protein PAK 300SW (7.5 mm x 300 mm) was equilibrated with 100 mM phosphate (pH 6.5) and operated at 0.5mL/min. At t = 0, 100 µL of the following protein mixture (in the same buffer) were injected. I. Cytochrome c (1 µg) mol wt 12,300; II. Trypsin inhibitor (soybean) mol wt 22,100; III. Ovalbumin mol wt 43,000; IV. Serum albumin (bovine) mol wt 68,000; V. Mitochondrial F$_1$-ATPase mol wt 360,000; Five micrograms were injected, except where indicated, and absorbance monitored at 220 nm. The elution times are plotted vs log mol wt in the upper box. Note that (a) bovine serum albumin gives a small peak representing a dimer of the protein (IVa), and (b) peak O represents low-mol-wt contaminants eluting at V_t. (Data taken from ref. *4*.)

4. Notes

1. When setting up, check that liquid is emerging from the eluate waste tube. If not: (a) if the operating pressure is low, check all the joints for leakage (conveniently by wrapping them in absorbent tissue), check that the pumps are correctly primed, and check that the injection port is not voided to waste; (b) if the operating pressure is raised, there is a blockage in the system. Remove items in the flow path (Fig. 1)

individually until the blockage is located, and replace/clean the of-fending item. If the blockage is in the guard system, the frit may be sonicated in $3M$ nitric acid, or the guard column perfused, in the reverse direction, with $0.4M$ NaCl or 20% DMSO in water.

2. All alterations to the flow rate should be made gradually (e.g., 0.1–0.5 mL/min over 30 min) to prevent mechanical damage to the column packing.

3. If peaks are observed after the dummy injection, there is contamination in the syringe or, more commonly, the injection loop. Flush well with the running buffer before using again.

4. Inject amounts of protein above 1 μg; even if the sensitivity is high enough to detect smaller amounts of protein, losses owing to adsorption (e.g., on frits) become unacceptable below this level.

5. If no peaks are observed with the standard proteins, check that:
   a. The detector is switched on, and the (correct) lamp is operating;
   b. The A_{220} of the solution, before zeroing the detector, is low;
   c. The detector cell is full of liquid, and its windows are clean and
   d. The guard column or separation column is not clogged with material. To do this, the injection loop may be connected directly to the detector and the standards injected again.

6. If no peaks are observed with the injected sample (but peaks are obtained with the standards), the protein may not be soluble in the running buffer (check for a pellet after centrifuging the sample), which must then be changed. Alternatively, the protein may have been adsorbed onto the column. In this case, the column should be washed before further use. Suitable washing media are $0.1M$ phosphate (pH 7.5) containing 0.1% sodium dodecyl sulfate, $0.4M$ NaCl, or 20% DMSO in water. (Other protocols may be suggested by the manufacturer.) If no suitable eluent can be found for the protein under investigation, it may be necessary to abandon this technique for an alternative method.

7. If the observed trace is "noisy": (a) if the noise is abolished when the flow is stopped, the cause is outgassing of the buffer. Use a buffer that has been freshly vacuum filtered, and check that the pressure regulator (Fig. 1) is present and operative; (b) if the noise occurs in the absence of flow, the cause is electrical. Check that the detector response time is at a suitable level (0.5–1 s), and that nearby equipment is not interfering. If there is no obvious cause, the detector lamp may be failing.

8. Keep records of standard separations from different days to monitor the efficiency of the column. If resolution falls, inspect and clean the

guard and/or separation columns as in Step 5 above. If the above precautions are followed, columns should last for a year or more without a significant deterioration in performance.

9. A plot of log mol wt vs elution time (Fig. 2), or V_e, is convenient for estimating mol wt over a fair range of standards. Alternatively, log mol wt vs K_d (as defined above) may be used *(3)*. In this case, it is necessary to measure the column parameters, V_0 and V_t, in addition to V_e. These can be measured from the elution times of Dextran Blue 2000 and sodium azide, respectively.

10. K_d depends not only on mol wt, but also on molecular shape. The value obtained above reflects an "apparent mol wt," which assumes that the protein tested is of similar shape to the standards (and, of course, that the protein does not interact with the column support). The standards in Table 2 are chosen as globular proteins; examples of proteins giving anomalous mol wt in this system are fibrinogen and protein A (both rod shaped: elute ahead of expected mol wt) and cytochrome c (positively charged: retarded because of residual negative charges on silica of column). For a more detailed discussion of the relationship among K_d mol wt, and radius of gyration of a protein, *see* ref. *3*.

11. Aggregated or associated proteins elute in advance of their individual components. SE-HPLC has been used to follow association (and its concentration dependence) between the mitochondrial ATPase and its naturally occurring inhibitor protein *(4)*, as well as between antibodies and protein A *(5)*.

12. Hydrophobic proteins can be separated using SE-HPLC as with conventional gel filtration, provided solubilizers are present in the running buffer. Separations have been achieved in $6M$ urea/formic acid (pH 3.15) *(6)*; 0.1% SDS, 0.1 mg/mL $C_{12}E_8$ and other biocompatible detergents *(7)*; and in 40% acetonitrile/0.1% trifluoracetic acid *(7)* (see Table 1). Mol-wt determination in detergents, however, is difficult *(7)* because of the large size of the detergent/protein complexes.

5. Appendix: Alternative Applications

5.1. SE-HPLC in Assessment of Purity and Yield

In SE-HPLC, a protein is characterized by its elution time, and the amount of protein in a peak is related to peak area (peak height). Thus, the technique can be used to assess purity of a protein preparation (a pure protein gives a single peak), to follow purity during puri-

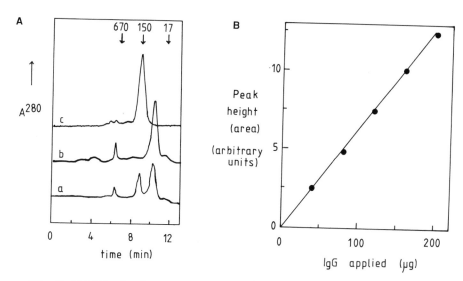

Fig. 3. **(A)** Monitoring monoclonal antibody production by SE-HPLC. A Bio-Sil SEC 250 column (7.5 mm x 300 mm) (Bio-Rad) was equilibrated with 0.3M NaCl, 10 mM sodium phosphate, pH 6.8, containing 10% DMSO at a flow rate of 1 mL/min. The following samples were injected and A$_{280}$ monitored: (a) 20 µL unfractionated mouse ascite fluid (1–2 mg/mL protein) (b) 20 µL of (a) after passage through a column bearing immobilized protein A (Affi-gel protein A; Bio-Rad. (c) 20µL protein eluted from affinity column (b) with 0.1M sodium citrate, 0.15M NaCl (pH 5.0).Calibration points for mol wt of 670,000, 150,000, and 17,000 are indicated. **(B)** Calibration of SE-HPLC for determination of yield. A column was set up as in Fig. 3A. Bovine IgG (40–200 µg) was injected from a 10 mg/mL stock solution and peak height (or area) monitored from A$_{280}$. (Data taken from ref. 9.)

fication and, simultaneously, to follow yield. A single sample can be tested in 20 min and requires < 5 µg of protein. Comparable techniques, such as electrophoresis, take longer and do not readily provide information on yield. Thus, SE-HPLC is convenient for monitoring purity after successive steps during purification of a protein. As an example, the assessment of a monoclonal antibody purification procedure is given in Fig. 3 (after ref. 9).

5.2. Studies on Conformational Changes in Proteins

For proteins of the same mol wt, K_d is related to molecular shape; compact, globular proteins enter smaller pores than more open mol-

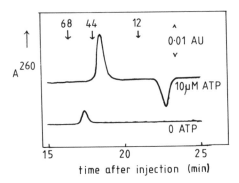

Fig. 4. ATP binding to the catalytic (β) subunit of the chloroplast ATPase (F_1). A Waters Protein PAK 300SW (7.5 mm x 300 mm) column was equilibrated in 100 mM sodium phosphate (pH 7.0) (lower curve) or 100 mM sodium phosphate (pH 7.0) containing 10 μL ATP (upper curve). 5 μg of purified β subunit, equilibrated in 100 μL of the respective buffer, were injected, and A_{260} monitored. In the upper curve, the background absorbance owing to ATP was subtracted electronically by zeroing the instrument (*see* Method section). Calibration points for mol wt of 68,000, 44,000, and 12,000 are indicated. Note the trough at V_t in the upper curve, corresponding to the depletion of ATP from the solution by binding to protein. This ATP is eluted in the foremost peak. Note also that β + ATP elutes behind the β subunit itself, owing to a conformational change in the protein on ATP binding (*see* Note 11).

ecules. SE-HPLC can be used to distinguish holo- and apo-forms of cytochrome c_{550} from *Paracoccus denitrificans* (Page, M. D. and Ferguson, S. J., personal communication), and (if ligands are included in the running buffer) to detect conformational changes owing to ligand binding (*see* Fig. 4) *(11)*.

5.3. Measurement of Ligand/Protein Association

Hummel and Dreyer (*see* ref. *10*) described a procedure for estimating association constants based on the coelution of bound ligand and protein from a gel filtration column. This was slow and expensive in protein using conventional chromatography; however, the rapidity and sensitivity of SE-HPLC make it more feasible.

In a typical experiment, the detector is set to measure *ligand* absorbance, and the column equilibrated with the lowest concentration of ligand tested. Protein, *equilibrated in the same solution,* is injected onto the column, and absorbance monitored. The bound ligand elutes as a positive peak, with the protein. This is followed by a negative peak, marking the solution depleted of the ligand by binding (Fig. 4). The area under either peak corresponds to the ligand bound at the ambient ligand concentration: The area of the trough is normally taken since it does not require correction for protein absorbance. The column is then equilibrated with the next ligand concentration, and the procedure repeated, generating a curve of bound vs free ligand concentration, and hence, the binding constant.

5.4. SE-HPLC in Protein Preparation

Variations in K_d (reflecting molecular size) can provide a basis for protein purification, just as can variations in charge, hydrophobicity, and so on, as described elsewhere in this volume. SE-HPLC is a gentle process, since proteins do not interact with (potentially denaturing) highly charged or hydrophobic interfaces, and it is thus well suited for the isolation of labile proteins.

Generally, however, laboratory scale SE-HPLC is not ideal for large-scale preparation of proteins. Restrictions on elution volume *(above)* limit resolving power (only about 10 proteins can be typically resolved), and restrictions on sample volume mean that the protein must be concentrated prior to separation. A maximum of 3–5 mg of protein can be applied to the 7.5 mm x 300 mm column described above, as compared to 50–100 mg for a comparable ion exchange column and, although larger columns are available (up to 210 mm x 600 mm), they are expensive for routine use.

Nonetheless, the above protocol can be simply modified for preparative purposes. The pressure regulator at the end of the eluate line is removed (to minimize dead volume), and the eluate led to a fraction collector via fine bore (0.1 mm id) Teflon™ tubing. In this case, degassing of solutions before use becomes more critical, and bubbling solutions with helium (and maintaining under a helium atmosphere) may be advantageous.

For amounts of protein larger than 5 mg, the column and injection loop may be increased in size (up to 2 mL, containing 50–100 mg protein, can be injected onto a 210 mm x 600 mm column). In this case, the flow rate should be increased in proportion to the cross-sectional area of the column.

References

1. Kopaciewicz, W. and Regnier, F. E. (1982) Non-ideal SE-chromatography of proteins: Effects of pH at low ionic strengths. *Anal. Biochem.* **126,** 8–16.
2. Johns, D. (1987) Columns for HPLC separations of macromolecules, in *HPLC of Macromolecules—A Practical Approach* (Oliver, R. W. A., ed.) IRL Press, Oxford, UK, pp. 1–7.
3. Potschka, M. (1987) Universal calibration of gel permeation chromatography. *Anal. Biochem.* **162,** 47–64.
4. Harris, D. A., Husain, I., Jackson, P. J., Lunsdorf, H., Schafer, G. and Tiedge, H. (1986) Interactions between the soluble F_1-ATPase and its naturally occurring inhibitor protein. Studies using hydrophilic HPLC and immunoelectron microscopy. *Eur. J. Biochem* **157,** 181–186.
5. Das, C., Mainwaring, R., and Langone, J. J. (1985) Separation of complexes containing protein A and IgG or Fc fragments by HPLC. *Anal. Biochem.* **145,** 27–36.
6. Edelstein, C. and Scanu, A. M. (1986) HPLC of apolipoproteins. *Meth. Enzymol.* **128,** 339–353.
7. le Maire, M., Aggerbeck, L. P., Montheilhet, C., Andersen, P., and Moller, J. V. (1986) The use of HPLC for determination of size and molecular weight of proteins; a caution and list of membrane proteins suitable as standards. *Anal. Biochem.* **154,** 525–535.
8. Swergold, G. D. and Rubin, C. S. (1983) High performance gel permeation chromatography of polypeptides in a volatile solvent; rapid resolution of proteins and peptides on a column of TSK-G3000 PW. *Anal. Biochem.* **131,** 295–300.
9. Juarez-Salinas, H., Bigbee, W. L., Lamotte, G. B., and Ott, G. S. (1986) New procedures for the analysis and purification of IgG murine MAbs. *Int. Biotech. Lab.* **April,** 20–27.
10. Ackers, G. K. (1973) Studies on protein-ligand binding by gel permeation techniques. *Meth. Enzymol.* **XXVII,** 444–449.
11. Nadanaciva, S. and Harris, D. A. (1990) *Current Research in Photosynthesis.* (Baltscheffsky, M., ed.) Kluwer, Amsterdam, pp. 41–44.

CHAPTER 15

Chromatofocusing

Chee Ming Li and T. William Hutchens

1. Introduction

Chromatofocusing is a protein-separation technique that was introduced by Sluyterman and his colleagues between 1977 and 1981 *(1–5)*. Chromatofocusing combines the advantage of high-capacity ion-exchange procedures with the high resolution of isoelectric focusing into a single chromatographic focusing procedure. During chromatofocusing, a weak ion-exchange column of suitable buffering capacity is equilibrated with a buffer that defines the upper pH of the separation pH gradient to follow. A second "focusing" buffer is then applied to elute bound proteins, roughly in order of their isoelectric (pI) points. The pH of the focusing buffer is adjusted to a pH that defines the lower limit of the pH gradient. The pH gradient is formed internally during isocratic elution with a single focusing buffer; no external gradient forming device is required. The pH gradient is formed as the eluting buffer (i.e., focusing buffer) titrates the buffering groups on the ion exchanger. Peak widths in the range of 0.05 pH unit and samples containing several hundred milligrams of protein can be processed in a single step. Chromatofocusing is therefore a powerful analytical probe of protein surface charge, as well as an effective preparative technique for protein isolation. The application of chromatofocusing to silica-based stationary phases for use in a high-performance mode *(6)* has extended the utility of this technique.

From: *Methods in Molecular Biology, Vol. 11: Practical Protein Chromatography*
Edited by: A. Kenney and S. Fowell Copyright © 1992 The Humana Press, Inc., Totowa, NJ

This chapter provides a simple description of the experimental considerations necessary to make both open-column and high-performance chromatofocusing a useful investigative tool and a successful separation method even for those with little or no previous chromatographic experience. The theoretical considerations and details of chromatofocusing have been provided *(2–5)* and reviewed *(7)* elsewhere.

2. Materials

2.1. Stationary Phase

The separation of proteins by chromatofocusing requires weak anion exchangers on stationary phases compatible with the separation of large biopolymers (e.g., large pore diameters). The most appropriate stationary phases have a high protein-binding capacity and an even buffering capacity over the pH range of interest; the gel must be stable over this particular pH range. Another fundamental requirement of the stationary phase is an absence of nonspecific interactions for the proteins (or the protein-bound ligands) of interest (e.g., *see* refs. *6* and *8*). Several column types that have been used successfully for both high-performance analytical-scale and preparative-scale chromatofocusing are listed in Table 1. The desired volume of the chromatofocusing column depends on the type and quantity of the sample protein (*see* Notes 2 and 4).

2.2. Mobile Phases for Column Equilibration and Generation of Descending Internal pH Gradients

There are only two mobile phases necessary to develop a linear pH gradient during chromatofocusing: the column-equilibration buffer and sample-elution or focusing buffer. The key to success is the focusing buffer. The focusing buffer is actually a set of buffer constituents that provide an even buffering capacity over the pH range in which separation is to take place. The ionic strength of the focusing buffer should be kept low to avoid a salt-displacement (ion-exchange) effect. Polymeric ampholytes designed for chromatofocusing (e.g., Polybuffers 96 and 74 from Pharmacia-LKB Biotechnology Inc., Piscataway, NJ) are commercially available for chromatographic focusing over relatively narrow pH ranges (i.e., 2–3 pH units). Separations requiring

Table 1
Stationary Phases for Chromatofocusing

Column or stationary phase	Immobilized ligand or bonded phase	Stationary-phase composition	Average pore diameter	pH stability
Bakerbond WP-PEI (JT Baker, Inc., Phillipsburg, NJ)	PEI	Silica	250-300 Å	pH 2–12
SynChroPak AX-300	PEI	Silica	300 Å	<pH 8.5
SynChroPak AX-500	PEI	Silica	500 Å	<pH 8.5
SynChroPak AX-1000 (SynChrom, Inc., Lafayette, IN)	PEI	Silica	1000 Å	<pH 8.5
Polybuffer Exchanger 94 (Pharmacia LKB Biotechnology Inc., Piscataway, NJ)	3° and 4° amines	Crosslinked agarose 6B	porous	pH 2–12
Polybuffer Exchanger 118 (Pharmacia)	3° and 4° amines	Crosslinked agarose 6B	porous	pH 2–12
Mono P (Pharmacia)	3° and 4° amines	Crosslinked hydrophilic polymer	porous	pH 2–12

broader pH ranges can also be achieved with mixtures of these polyampholytes in various relative concentrations (Fig. 1) *(6,8)*. The polymeric ampholytes more commonly used for isoelectric focusing can also be used effectively for chromatofocusing (Fig. 2). Alternatively, the design and use of nonpolymeric focusing buffers composed of low-molecular-mass constituents has been described (e.g., *9–12*). These reagents, however, are not commercially available. Depending on the buffering capacity of the stationary phase chosen, these so-called simple focusing buffer systems can be used to generate pH gradients over a wide range of pH values *(11,12)*.

Examples of column-equilibration buffers and elution (focusing) buffers for developing pH gradients over several different pH ranges is provided in Table 2.

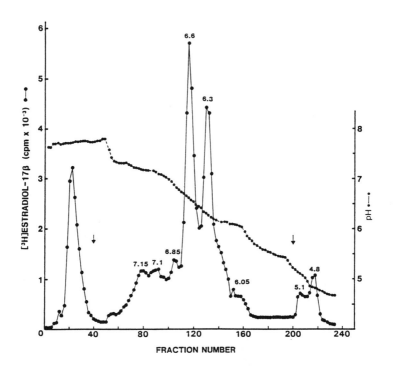

Fig. 1. Open-column chromatofocusing of human uterine estrogen receptor proteins on a column (0.7 cm id × 9 cm) of PBE 94. Initiation of the primary pH gradient (arrow at fraction 40) was by elution with Polybuffers 96 and 74 (30/70) diluted 1/15 at pH 4.0. Initiation of the secondary pH gradient (arrow at fraction 200) was by elution with Polybuffer 74 diluted 1/15 at pH 3.0. Fractions of 1 mL each were collected at 0.5 mL/min.

2.2.1. Column Equilibration Buffer

The pH of the column-equilibration buffer is normally adjusted 0.2–0.5 pH units above the desired starting pH to compensate for fluctuations in pH at the beginning of the gradient (*see* Note 1).

Column equilibration buffer: 25 mM Tris-HCl, 1 mM dithiothreitol, pH 8.2. Dissolve 3.03 g of Tris in approx 980 mL distilled water; adjust the pH to 8.2 (at 4°C) with HCl. Add 0.154 g of dithiothreitol to the solution and dilute to 1.0 L.

2.2.2. Preparation of Focusing Buffers

1. Focusing buffer 1: Polybuffers 96/74 (30/70) diluted 1/10; pH 4.2. Add 30 mL of Polybuffer 96 and 70 mL of Polybuffer 74 to 880 mL of distilled water, and adjust the pH to 4.2 with 6N HCl. Dilute to 1.0 L.

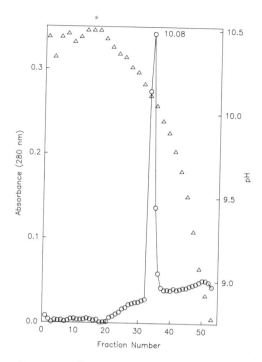

Fig. 2. High-performance chromatofocusing of purified human lactoferrin on a Pharmacia Mono P column (0.7 cm id × 10 cm). The column was equilibrated in 25 mM triethylamine (pH 11.0 at 25°C). The pH gradient was initiated by elution with Pharmalyte 3-10 diluted 1/100 with water and adjusted to pH 9.0. Fractions of 0.5 mL each were collected at 1.0 mL/min.

2. Focusing buffer 2: Polybuffer 74 diluted to 1/10; pH 3.0. Dilute 100 mL of Polybuffer 74 to 880 mL with distilled water, and adjust the pH to 3.0 with 6N HCl. Dilute to 1.0 L.

3. Methods

For temperature-labile proteins and enzymes, adjust all buffers to their final pH at 0–4°C (ice bath). Perform the separation in a cold room. Most high-performance liquid chromatography (HPLC) systems are now compatible with use in a cold room (temperature jackets for the columns are also available). This is not necessary for proteins that are stable at room temperature. Filter all buffers through a Millipore 0.45-μm membrane and degas before use (dissolved CO_2 may cause a plateau in the pH 5–6 region).

Table 2
Mobile Phases for Chromatofocusing[a]

Desired pH gradient	Column-equilibration buffer	Column eluent (focusing buffer)	Dilution factor
11–9	20 mM triethylamine; adjust to pH 11.5 (HCl)	pH 9–11 ampholyte or pH 3–10 ampholyte at pH 9 (HCl)	1/50–1/100
10–7	20 mM triethylamine; adjust to pH 10.5 (HCl)	Pharmalyte 8–10.5 at pH 7.0 (HCl)	1/40
9–6	20 mM ethanolamine; adjust to pH 9.4 (HCl)	Polybuffer 96 at pH 6.0 (CH_3COOH)	1/10
8–5	25 mM Tris; adjust to pH 8.2 (HCl)	Polybuffer 96 (30%) Polybuffer 74 (70%) at pH 5.0 (HCl)	1/10
8–3.0	25 mM Tris; adjust to pH 8.2 (HCl)	Polybuffer 96 (30%); and Polybuffer 74 (70%) at pH 4.2 (HCl)	1/10
		Polybuffer 74 at pH 3.0 (HCl)	1/10
7–4	25 mM imidazole; adjust to pH 7.2 (HCl)	Polybuffer 74 at pH 4.0 (HCl)	1/10
5–3.5	25 mM piperizine; adjust to pH 5.5 (HCl)	Polybuffer 74 at pH 3.2	1/10

[a]The final concentrations (dilutions) of focusing buffers listed in this table are based upon the general use of either Pharmacia Polybuffer Exchanger 94 or Mono P. Custom applications on different anion-exchange resins should begin with these dilution factors, with the expectation that some modifications may be required. The pH gradients indicated will be formed after approx 10–15 column vol of eluent (focusing buffer) have passed. Always prepare extra focusing buffer in case greater volumes are required to complete the pH gradient.

The following example is for the generation of a pH gradient from 8.2 to 3.0 in a 10-mL column of Pharmacia Polybuffer Exchanger 94 (PBE 94). This protocol also works well for high-performance chromatofocusing on certain HPLC anion-exchange columns: try a 25-cm SynChroPak® AX-500 (or AX-300 column) or a 25-cm Bakerbond WAX polyethyleneimine (PEI) column.

3.1. Preparation of the Stationary Phase (PBE 94) for Chromatofocusing

1. Begin the equilibration procedure by suspending the PBE 94 gel particles in equilibration buffer. After allowing the gel to settle, remove the supernatant buffer (*see* Note 2).
2. Repeat this procedure several times until the pH and conductivity of the gel slurry reaches the pH and conductivity of the column-equilibration buffer.
3. Prepare the PBE 94 gel as a 50% (v/v) gel suspension in equilibration buffer. To do this, allow the gel slurry to settle to a constant volume in a graduated container.
4. Decant the supernatant and add exactly 1 vol of equilibration buffer. In this manner, a known column-bed volume is easily prepared. The PBE 94 or other prepacked stationary phases (e.g., high-performance anion-exchange columns) may also be equilibrated to the pH and conductivity of the column-equilibration buffer by pumping the buffer directly through the packed column (be sure that all organic solvents are removed first).

3.2. Packing the Column with Equilibrated PBE 94

1. Degas the 50% PBE gel slurry, and gently mix the slurry to form an even suspension.
2. To prepare a 10-mL column, pour 20-mL of the 50% gel suspension into the vertical open column (1.5 cm id; outlet closed). Let the gel settle for 2–3 min before connecting the column to a peristaltic pump (open the column outlet).
3. Pack the column by pumping equilibration buffer through at a flow rate of 1.2–1.4 mL/min.
4. After the column is packed to a constant bed volume, close the column outlet, disconnect the column from the pump, gently fill the column with equilibration buffer, and gently insert a frit on top of the gel. Old prepacked PD-10 G-25 desalting columns from Pharmacia make nice minicolumns for chromatofocusing; save both frits (e.g., *8*).
5. Reconnect the column to the pump and run for another 5–10 min at a flow rate of 1 mL/min. Confirm the pH of the column eluent.

3.3. Sample Application

1. Equilibrate the sample into the column-equilibration buffer either by simple gel filtration (Sephadex G-25) or by dialysis. The applied

sample volume should not exceed much more than half of the gel bed volume (i.e., <5 mL in this case).

2. Close the column outlet, disconnect the column inlet line from the pump, remove excess equilibration buffer from above the column, load the sample into the column by gravity (open the column outlet), and collect the column eluent.

3. Rinse the side of the column with 1 mL of equilibration buffer and let it run into the column.

4. Add approx 1.5 mL of equilibration buffer to the column and begin pumping column-equilibration buffer (a minimum of 3–4 column volumes) to remove unbound or loosely bound protein.

3.4. Sample Elution

No special gradient apparatus is required, since the gradient is formed internally by isocratic elution. In this example we illustrate the tandem use of two focusing buffers to generate a wide-range pH gradient.

1. To initiate pH-gradient formation, begin pumping focusing buffer 1 into the column.

2. Collect 1-min (1-mL) fractions. The pH of the gradient can be monitored during the run using a pH meter equipped with a flow-through electrode. Alternatively, the pH of individual fractions can be determined directly, either during or after completion of the gradient. It is advisable to monitor the eluent pH during the first several experiments with any given elution protocol, or until the pH-gradient characteristics are firmly established (*see* Note 3). Reproducible results require accurate pH measurements. Calibrate the pH electrode and measure the pH of collected fractions at the same temperature at which the fractions were collected. Allow some time for the pH electrode to become properly equilibrated (especially for the Polybuffers).

3. Monitor eluted protein peaks with a flow-through UV monitor at 280 nm. Diluted Polybuffers absorb little at 280 nm. Other monitoring techniques can also be used. In our studies, bioactivity or the elution positions of radiolabeled proteins were monitored by scintillation counting *(6,8,13)*.

4. End the separation experiment by elution with high salt (e.g., 1–2M NaCl) and/or low pH (pH 3.0) to remove tightly bound protein.

3.5. Sample Recovery

The separation of proteins from polymeric focusing-buffer constituents is not always necessary. When desired, this can be achieved by size-exclusion chromatography, precipitation with ammonium sulfate, hydrophobic-interaction chromatography, or affinity chromatography. Difficulties in protein–polyampholyte separations have been reported *(11,14)*. Therefore, use caution in defining adequate removal of focusing-buffer constituents from the proteins of interest.

3.6. Column Regeneration

Remove any remaining proteins strongly adsorbed to the column by washing the column with 1–2M NaCl, 8M urea (or 6M guanidine-HCl), or ethylene glycol (e.g., 50%). Lipids can be removed with organic solvents, but only after the salts have been washed away. Wash the column with water; then reequilibrate the column with column-equilibration buffer. Monitor both pH and conductivity.

4. Notes

1. Proteins have a net charge of zero at their isoelectric point (pI). This does not indicate that they have no charge. The initial column-equilibration pH should typically be above the pI of the protein of interest (i.e., protein to be adsorbed) to ensure that it has a net negative surface charge. However, this is not always necessary. Proteins can have regions or domains of negative charge at pH values below their pI. (*See* ref. *7* for a more thorough discussion of elution pH vs pI.)

2. The best way to determine the amount of gel (or column type) and the concentration of elution buffer is empirically. A gradient volume equal to 10–15 times the gel bed volume is appropriate. The sample volume should initially be limited to <25% of the column bed volume.

3. In order to optimize resolution in chromatofocusing, you must consider the applied sample volume and composition, column dimensions and bed volume, and the concentration of elution (focusing) buffer, as well as the slope and range of the pH gradient. In general, the slope and volume of the pH gradient is determined by the concentration and pH of the elution (focusing) buffer.

   It is often best to use a shallow pH gradient. This can be achieved over a wide-range pH gradient by using focusing buffers at a low concentration (large volume required to titrate the stationary phase) or by increasing the column length. You may also choose to select a very narrow pH range by titrating the focusing buffer to a pH close to that of the initial column equilibration buffer.

   Be careful: Very dilute focusing buffers will result in very shallow, but often uneven, pH gradients. If this happens, use a more concentrated focusing buffer. A successful separation is often a compromise

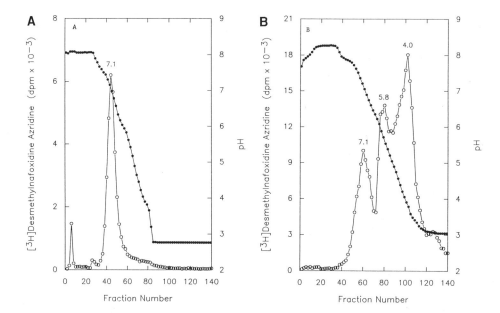

Fig. 3. Surface charge properties of purified, DNA-binding estrogen receptor forms revealed by chromatofocusing in the presence and absence of 3*M* urea. In the presence of 3*M* urea (**A**), the chromatofocusing elution pH (●) of the affinity-purified receptor dimer covalently labeled with [³H]desmethylnafoxidine aziridine (O) revealed a major peak at pH 7.1. In the absence of urea (**B**), the chromatofocusing elution profile was quite different.

between a stable but narrow pH gradient. Keep your goal in mind. Overall resolution or the achievement of a specific separation goal may sometimes be achieved without a "good"-looking (i.e., linear) pH gradient (e.g., ref. *6*).

4. The use of various stationary phases, especially high-performance weak anion-exchange resins, is encouraged, even though the particular column developer or manufacturer may not have evaluated their columns for such purposes. It is often easier to develop a particular pH gradient on a new stationary phase than to evaluate its buffering capacity and theoretical potential.

5. The use of uncharged mobile-phase-modifying reagents (e.g., up to 6*M* urea, as well as glycerol, ethylene glycol, and detergents), is often quite helpful to achieve resolutions not always possible in their absence (Fig. 3). A good example of this is presented in ref. *15*.

6. EDTA may not be included in either the column-equilibration buffers or focusing buffers. The presence of EDTA causes significant alterations in the pH gradient.

Acknowledgment

This work was supported, in part, by the US Department of Agriculture, Agricultural Research Service Agreement No. 58-7MNI-6-100. The contents of this publication do not necessarily reflect the views or policies of the US Department of Agriculture, nor does mention of trade names, commercial products, or organizations imply endorsement by the US Government.

References

1. Sluyterman, L. A. AE. and Wijdeness, J. (1977) Chromatofocusing: Isoelectric focusing on ion exchangers in the absence of an externally applied potential, in *Proc. Int. Symp. Electrofocusing and Isotachophoresis*, (Radola, B. J. and Graesslin, D., eds.), Walter de Gruyter, Berlin, pp. 463–466.
2. Sluyterman, L. A. AE. and Elgersma, O. (1978) Chromatofocusing: Isoelectric focusing on ion exchange columns. I. General principles. *J. Chromatogr.* **150**, 17–30.
3. Sluyterman, L. A. AE. and Wijdeness, J. (1978) Chromatofocusing: Isoelectric focusing on ion exchange columns. II. Experimental verification. *J. Chromatogr.* **150**, 31–44.
4. Sluyterman, L. A. AE. and Wijdeness, J. (1981) Chromatofocusing. III. The properties of a DEAE-agarose anion exchanger and its suitability for protein separations. *J. Chromatogr.* **206**, 429–440.
5. Sluyterman, L. A. AE. and Wijdeness, J. (1981) Chromatofocusing. IV. Properties of an agarose polyethyleneimine ion exchanger and its suitability for protein separations. *J. Chromatogr.* **206**, 441–447.
6. Hutchens, T. W., Wiehle, R. D., Shahabi, N. A., and Wittliff, J. L. (1983) Rapid analysis of estrogen receptor heterogeneity by chromatofocusing with HPLC. *J. Chromatogr.* **266**, 115–128.
7. Hutchens, T. W. (1989) Chromatofocusing, in *Protein Purification* (Janson, J. and Ryden, L., eds.), VCH, New York, pp. 149–174.
8. Hutchens, T. W., Gibbons, W. E., and Besch, P. K. (1984) High-performance chromatofocusing and size-exclusion chromatography: Separation of human uterine estrogen binding proteins. *J. Chromatogr.* **297**, 283–299.
9. Hutchens, T. W., Li, C. M., and Besch P. K. (1986) A nonpolymeric focusing buffer of defined chemical composition developed for chromatofocusing. *Protides Biol. Fluids* **3**, 765–768.
10. Hutchens, T. W. (1986) Requirements of the mobile and stationary phases during development of chromatographic focusing techniques. *Protides Biol. Fluids* **34**, 749–752.
11. Hutchens, T. W., Li, C. M., and Besch, P. K. (1986) Development of focusing buffer systems for generation of wide-range pH gradients during high-performance chromatofocusing. *J. Chromatogr.* **359**, 157–168.
12. Hutchens, T. W., Li, C. M., and Besch, P. K. (1986) Performance evaluation

of a focusing buffer developed for chromatofocusing on high-performance anion-exchange columns. *J. Chromatogr.* **359,** 169–179.

13. Hutchens, T. W., Dunaway, H. E., and Besch, P. K. (1985) High-performance chromatofocusing of steroid receptor proteins in the presence and absence of steroid. An investigation of steroid-dependent alterations in surface charge heterogeneity. *J. Chromatogr.* **327,** 247–259.

14. Rodkey, L. S. and Hirata, A. (1986) Studies of ampholyte–protein interactions. *Protides Biol. Fluids* **34,** 745–748.

15. Hutchens, T. W., McNaught, R. W., Yip, T-T., Li, C. M., Suzuki, T., and Besch, P. K. (1990) Unique molecular properties of a urea- and salt-stable DNA-binding estrogen receptor dimer covalently-labeled with the antiestrogen [³H]desmethylnafoxidine aziridine. A comparison with the estrogen-receptor complex. *Mol. Endocr.* **4,** 255–267.

CHAPTER 16

Ion-Exchange Chromatography of Proteins

Andrew C. Kenney

1. Introduction

Ion-exchange chromatography is the most widely used technique in protein chromatography (1). This is because it is nearly always possible to develop successful ion-exchange separations for proteins, and the materials required are relatively inexpensive.

All proteins carry charge as a result of the ionization of amino acid side-chain functional groups situated on the protein surface. At approx neutral pH, surface arginine, lysine and histidine residues are likely to carry positive charges, whereas the aspartate and glutamate residues will carry negative charges. The nature and extent of the overall protein charge can be manipulated by means of changes to solvent pH. In most cases, it is the changes to the extent of histidine ionization that leads to the change in the net charge on the protein, because most surface histidine residues will have pK values between 6–7. The diversity of protein sequences and tertiary structures ensures that very few proteins behave identically in ion-exchange systems (2).

An ion-exchange separation involves selecting a column packing that carries functional groups that have either positive or negative charges (anion-exchangers and cation-exchangers, respectively—see Table 1). Under normal conditions, these charges are balanced by counterions. Under appropriate conditions of pH and ionic strength, a mixture of proteins is applied to the column, such that some or all of

From: *Methods in Molecular Biology, Vol. 11: Practical Protein Chromatography*
Edited by: A. Kenney and S. Fowell Copyright © 1992 The Humana Press Inc., Totowa, NJ

Table 1
Some Common Ion Exchanger Functional Groups

Q	$-CH_2-\overset{+}{N}(CH_3)_3$	Quaternary methylammonium
S	$-CH_2-SO_3^-$	Methylsulfonate
DEAE	$-O-CH_2CH_2-\overset{\overset{\displaystyle CH_2CH_3}{+\,\big|}}{N}-H \atop \big| \atop CH_2CH_3$	Diethylaminoethyl
SP	$-O-CH_2CH_2CH_2SO_3^-$	Propylsulfonate
CM	$-O-CH_2COO-$	Carboxymethyl
QA	$-CONH-CH_2CH_2-\overset{\overset{\displaystyle CH_2CH_3}{+\,\big|}}{N}-CH_2CH_3 \atop \big| \atop CH_2CH_3$	Quaternary amino
PEI	$-CH_2CH_2=\overset{+}{N}H$	Ethylimino
CBX	$-CH_2CH_2COO^-$	Carboxyethyl

the proteins become bound by interactions between the surface charge of the protein and the charge carried by the ion exchanger. As a result of this interaction, the counterions are displaced from the ion exchanger.

By gradually increasing the ionic strength of the solvent passing through the column, the bound proteins may be released in a sequence determined by their strength of interaction with the ion-exchanger. The increasing salt concentration releases proteins essentially by competition between the salt ions and the charges carried by the protein. Hence, the most strongly bound proteins are eluted last. The salt ions replace the protein charges and become the new counterions.

A bewildering variety of ion-exchangers suitable for protein separations are now available. Originally, the choice was limited to cellulose or dextran-based materials with diethylaminoethyl (DEAE) or carboxymethyl (CM) substituents. This choice has been widened by the introduction of ion exchangers based on agarose, silica, and synthetic polymer beads (all of which are more convenient to use than the original cellulose and dextran materials), and by the introduction of strong ion-exchange groups that have broadened the operational pH range available.

Table 2 lists some of the more commonly available exchangers. Traditional low-pressure protein separations are possible on all of the materials listed with bead sizes >35µm. Exchangers with bead sizes that are smaller than this are intended for use with high performance liquid chromatography (HPLC) systems. The cellulose-based materials are the most economic, but the quality of separation obtainable and the ease of use often render them inferior to some of the more recently introduced materials based on agarose or synthetic organic polymer beads. Exchangers based on crosslinked dextran can swell and contract with changes in buffer ionic strength. This can have a catastrophic effect on the separation, and these materials are best avoided except for batch adsorption/elution.

Preliminary experiments leading to the development of an ion-exchange separation can most easily be performed batchwise. Samples of the protein mixture of interest (dialyzed or desalted into a suitable buffer) can be contacted with small quantities of either anion or cation exchanger in test tubes (*see* Note 1). After allowing 60 min for adsorption to take place, the ion exchanger may be settled or centrifuged, and samples of the supernatant analyzed for both the target and total protein. If multiple experiments are performed across a range of pH values (5.0–9.0) with both anion and cation exchangers, suitable conditions for ion-exchange chromatography can be quickly established. The subsequent chromatography is now illustrated by means of an example, in which the practicalities of an ion-exchange protein separation are demonstrated by the separation of the three main proteins present in egg white: ovalbumin, ovotransferrin, and lysozyme.

2. Materials

1. Fresh egg white (10 mL).
2. 5 mM Sodium phosphate buffer, pH 6.0.
3. 5 mM Sodium phosphate buffer, pH 6.0, containing 0.33M sodium chloride.
4. 5 mM Sodium phosphate buffer, pH 6.0, containing 1M sodium chloride.
5. 0.1M Sodium hydroxide solution. Handle with care!
6. 10 mL S-Sepharose Fast Flow (Pharmacia LKB, Uppsala, Sweden).
7. 0.2 m Disposable membrane filter (Anotec Separations, Banbury, UK).
8. LC Equipment: Peristaltic pump (0–5 mL/min):
   a. UV spectrophotometer (280 nm)

Table 2
Some Commercially Available Ion Exchangers

Name	Type	Supplier	Average bead size, μm	Base matrix
Q Sepharose Fast Flow	Strong anion	Pharmacia LKB	90	Agarose
S Sepharose Fast Flow	Strong cation	Pharmacia LKB	90	Agarose
DEAE Sepharose Fast Flow	Weak anion	Pharmacia LKB	90	Agarose
CM Sepharose Fast Flow	Weak cation	Pharmacia LKB	90	Agarose
DEAE Sepharose CL–6B	Weak anion	Pharmacia LKB	90	Agarose
CM Sepharose CL–6B	Weak cation	Pharmacia LKB	90	Agarose
DEAE Sephacel	Weak anion	Pharmacia LKB	100	Cellulose
Mono Q	Strong anion	Pharmacia LKB	10	Synthetic polymer
Mono S	Strong cation	Pharmacia LKB	10	Synthetic polymer
Q Sepharose HP	Strong anion	Pharmacia LKB	34	Agarose
S Sepharose HP	Strong cation	Pharmacia LKB	34	Agarose
DEAE Trisacryl M	Weak anion	IBF Biotechnics	60	Synthetic polymer
CM Trisacryl M	Weak cation	IBF Biotechnics	60	Synthetic polymer
SP-Trisacryl M	Strong cation	IBF Biotechnics	60	Synthetic polymer
QA Trisacryl M	Strong anion	IBF Biotechnics	60	Synthetic polymer
DEAE Spherodex M	Weak anion	IBF Biotechnics	70,200[a]	Coated silica

CM Spherodex M	Weak cation	IBF Biotechnics	70,200	Coated silica
Bakerbond WP PEI	Weak anion	J.T. Baker	5,15,40	Coated silica
Bakerbond WP CBX	Weak cation	J.T. Baker	5,15,40	Coated silica
TSK DEAE 5PW	Weak anion	TosoHaas	10,13	Coated silica
TSK CM 5PW	Weak cation	TosoHaas	10,13	Synthetic polymer
TSK SP 5PW	Strong cation	TosoHaas	10,13	Synthetic polymer
DE52	Weak anion	Whatman	30	Cellulose
QA52	Strong anion	Whatman	30	Cellulose
CM52	Weak cation	Whatman	30	Cellulose
SE52	Strong cation	Whatman	30	Cellulose
Fractogel TSK DEAE-650	Weak anion	E. Merck	35,70	Synthetic polymer
Fractogel TSK CM-650	Weak cation	E. Merck	35,70	Synthetic polymer
Fractogel TSK SP-650	Strong cation	E. Merck	35,70	Synthetic polymer
PL-SAX	Strong anion	Polymer Laboratories	8,10	Synthetic polymer
Zorbax Bioseries SAX	Strong anion	DuPont Biotechnology	6	Coated silica
Zorbax Bioseries WAX	Weak anion	DuPont Biotechnology	6	Coated silica
Zorbax Bioseries SCX	Strong cation	DuPont Biotechnology	6	Coated silica
Zorbax Bioseries WCX	Weak cation	DuPont Biotechnology	6	Coated silica

[a] *See* Note 9.

b. Gradient former (*see* Note 2)

c. Fraction collector (optional)

d. Tubing, connectors, and so on

e. Chart recorder

f. Conductivity monitor (if available).

9. Suitable column. Approximately 1.0×15 cm bed needed (Pharmacia LKB C10/30) and packing reservoir.

10. Low-speed centrifuge.

3. Method

1. Prepare the egg white by adding 10–50 mL of the 5 mM phosphate buffer and mixing. Chill the solution at 4°C for 30 min.

2. Remove any precipitate by centrifugation of the egg white solution at $3000g$ for 30 min. Decant away the supernatant, taking care to avoid disturbing the pellet.

3. Store the solution at 4°C until required. Immediately before use, filter 10 mL through a 0.2-μm membrane filter (*see* Note 3).

4. Pack the column with the S-Sepharose Fast Flow (*see* Note 1) according to the manufacturer's instructions. Space does not permit a complete description of column packing techniques, but some guidelines can be found in Note 4. Install the column downstream of the pump and upstream of the UV spectrophotometer and conductivity monitor (if available). Follow the manufacturer's instructions in setting up the instrumentation.

5. Make sure that the ion exchanger is equilibrated by pumping the 5 mM phosphate until the outlet pH and conductivity have similar values to those of the buffer (*see* Note 5). For a 1.0×15 cm, column set the flow rate to 1 mL/min (*see* Note 6).

6. Load 10 mL of the diluted egg white solution onto the column at 1 mL/min. Follow this by washing the column with the 5 mM phosphate buffer at the same flow rate. Continue this until the absorbance at 280 nm returns to the baseline value (*see* Fig. 1 for a typical chromatogram). Collect the unbound proteins for later analysis.

7. Set up the sodium chloride elution gradient (*see* Note 2). The example in Fig. 1 depicts a gradient from 0– $0.33M$ sodium chloride (in 5 mM phosphate) at a gradient rate of 1%/min. This means that the gradient will take 100 min at 1 mL/min. The total vol will therefore be 100 mL (about 6 column vol), and so you will need 50 mL of each buffer. Pump the gradient onto the column at 1 mL/min.

8. When the gradient is exhausted, continue the elution with the 5 mM sodium phosphate, pH 6.0, containing 1M sodium chloride (*see* Note

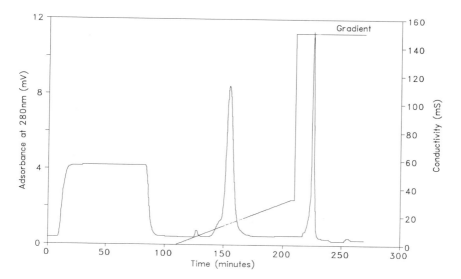

Fig. 1. Separation of egg white proteins on a strong cation exchanger.

7). The final chromatogram should resemble that depicted in Fig. 1. Three main peaks should have been obtained. The first contains the unbound proteins, the second is eluted by the gradient, and the third is eluted by the final step up to 1*M* sodium chloride.

9. Regenerate the column by pumping the sodium hydroxide (*see* Note 8) in the opposite direction to the direction of flow during loading. This will help to sweep away any particulate material that may have accumulated on the surface of the bed or its support net.

10. Analyze the protein fractions obtained. The major protein in each of the three peaks should be ovalbumin, ovotransferrin, and lysozyme, with mol wt of 43,000, 76,000, and 14,000, respectively.

4. Notes

1. Choose an appropriate ion exchanger by considering the following:
   a. pH stability of the target protein;
   b. Stability of the target protein at low ionic strength;
   c. Isoelectric point of the target protein.
   If the target protein is stable over a wide range of pH and ionic strengths, and has an isoelectric point that is well separated from those of the bulk contaminants, then selecting the appropriate ion exchanger will be easy. For acidic proteins (low isoelectric point), select an anion exchanger, and for a basic protein (high isoelectric point), select a cation exchanger. In the egg white example, lysozyme

is eluted last from the cation exchanger because its isoelectric point is the highest of the protein mixture. Often, the real situation is less than ideal, and selection of the appropriate exchanger will have to be based on experiment. In general, having both a strong anion exchanger and a strong cation exchanger available will give the most flexibility.

2. Gradient formers are commercially available (BioRad, Richmond, CA), but can easily be fabricated by using two beakers of appropriate vol and a glass or plastic U-tube. Place the beakers next to each other on a magnetic stirrer. Place a stirrer bar in one of the beakers, and put the lower ionic strength buffer in the same beaker. Add a similar volume of the higher ionic strength buffer to the other unstirred beaker. Fill the U–tube with the low-ionic-strength buffer, and place the tube with one arm of the U in each beaker, taking care not to break the syphon by introducing air bubbles. Pump from the low-ionic-strength side onto the column with gentle stirring. An approximately linear conductivity gradient should be produced.

   Liquid chromatography and HPLC instrumentation is usually provided with automatic binary (two buffers) gradient generation capability.

3. Sometimes prefiltration at 5 μm will be necessary, especially after prolonged storage. The purpose of the filtration is to extend the column lifetime. Where this is unimportant, filtration may be omitted.

4. Mount the column upright with the lower bed support in place. Pump up the column to expel air from the lower bed support. Mix the gel as a 50% slurry in the 5-mM phosphate buffer. Pour the slurry into the column in one go. If there is insufficient space, you will need a packing reservoir or a longer column. Fit the upper bed support, and begin pumping down the column according to the instructions supplied with the ion exchanger. When the settled bed height has stabilized, stop pumping and adjust the upper bed support, so that it comes to rest on the bed surface. In doing this, make sure that any air bubbles are expelled from the upper tubing as the bed support is pushed down. Make sure that the packing flow rate is higher than the flow rate that you subsequently intend to use.

5. Choice of buffer is of paramount importance in ion-exchange chromatography. Always choose a buffer ion with a pK within 0.5 pH U of the desired pH. This is important, because loading protein onto the column has to take place at low ionic strength, so buffering capacity is limited. It is also important to choose a buffer ion that bears the same charge as the exchanger (or is neutral); otherwise, the buffer ion itself will become bound to the column, and variations in its con-

Table 3
Some Useful Buffer Substances for Ion-Exchange Chromatography

Name		pK
Acetate		4.8
MES	2-(*N*-morpholino)ethanesulfonic acid	6.2
ADA	N-(2-acetamido)iminodiacetic acid	6.6
Phosphate		7.2
MOPS	3-(*N*-morpholino)propanesulfonic acid	7.2
NEM	N-ethylmorpholine	7.7
Glycylglycine		8.3
2-amino-2-ethyl-1,3 propanediol		8.8
Ethanolamine		9.5

See Ref. *(3)* for a complete list of suitable buffer substances.

centration as proteins bind and elute will cause fluctuations in pH within the column. Some useful buffer substances and their pK values are detailed in Table 3.

6. Faster flow rates may be used to equilibrate and regenerate the column. Refer to the manufacturer's instructions when selecting the highest operational flow rate, and remember not to exceed the packing flow rate. Slower flow rates must be used when loading and eluting the column in order both to ensure that efficient protein binding takes place and that peak resolution and dispersion are minimized.

7. The gradient and step elution suggested here has been developed from an initial broad gradient experiment (0–1*M* sodium chloride over 20 column vol). After determining the elution condition for each peak of interest, the gradient may often be replaced with a series of step changes in salt concentration, or a mixture of gradient and step elution as depicted here. This approach will nearly always result in a more rapid separation with little or no loss in separation performance.

8. If the elution sequence does not raise the salt concentration to approx 1*M*, then you should flush the column with 1*M* sodium chloride to remove any strongly bound proteins before regeneration with alkali. Sodium hydroxide is an effective way to prolong many ion exchanger lifetimes. However, it is important to check the pH stability of the ion exchanger in use. Never use high-pH solutions with silica- or glass-based exchangers: They may be dissolved. The sodium hydroxide will help to dissolve and release any precipitated or nonspecifically bound protein. It will also provide a degree of sanitization by killing

bacteria and spores. This killing is far more efficient at room temperature than in the cold room. Only store ion exchangers for prolonged periods in sodium hydroxide if the manufacturer recommends it.

9. Some manufacturers offer a range of bead sizes. This is to allow easy scale-up of the separation. The method can initially be developed on a small scale using HPLC and the smaller bead sizes. The large bead sizes can then be used for preparative and process separations, where it may be more convenient to operate at lower pressure and with cheaper ion-exchange materials. Users should note that moving to large bead sizes generally results in poorer resolution because of longer diffusion paths and so forth. Often the method development and optimization is best carried out on the same bead size as is intended for preparative use.

References

1. Bonnerjea, J., Oh, S., Hoare, M., and Dunnill, P. (1986) Protein purification: The right step at the right time. *Biotechnology* 4, 954–958.
2. Skopes, R. K (1987) Protein purification principles and practice. Springer-Verlag, New York.
3. Dawson, R. M. C., Elliott, D. C., Elliott, W. H., and Jones, K. M., eds. (1986) *Data for biochemical research.* Oxford University Press, Oxford, UK.

CHAPTER 17

Displacement Chromatography of Proteins

Steven M. Cramer

1. Introduction

Displacement chromatography is rapidly emerging as a powerful preparative bioseparation technique because of the high throughput and purity associated with the process *(1)*. The operation of preparative elution systems at elevated concentrations has been shown to result in significant tailing of the peaks with the concomitant loss of separation efficiency *(2)*. In contrast, displacement chromatography offers distinct advantages in preparative chromatography as compared to the conventional elution mode *(1,3,4)*. The process takes advantage of the nonlinearity of the isotherms, such that a larger feed can be separated on a given column with the purified components recovered at significantly higher concentrations. Furthermore, the tailing observed in elution chromatography is greatly reduced in displacement chromatography owing to self-sharpening boundaries formed in the process. Whereas in elution chromatography the feed components are diluted during the separation, the feed components are often concentrated during displacement chromatography *(3–7)*. These advantages are particulary significant for the isolation of biopolymers from dilute solutions, such as those encountered in biotechnology processes.

In displacement chromatography, a front of displacer solution traveling behind the feed drives the separation of the feed components into adjacent pure zones that move with the same velocity as the

From: *Methods in Molecular Biology, Vol 11: Practical Protein Chromatography*
Edited by: A. Kenney and S. Fowell Copyright © 1992 The Humana Press Inc., Totowa, NJ

displacer front. The displacer is selected such that it has a higher affinity for the stationary phase than any of the feed components. Displacement chromatography can be readily carried out using existing chromatographic systems with minor modifications to enable the sequential perfusion of the column with the carrier, feed, displacer, and regenerant solutions, as shown in Fig. 1 *(8)*. The column is first equilibrated with a carrier in which the components to be separated have a relatively high affinity for the stationary phase. A feed mixture is then pumped into the column followed by a displacer solution. During the introduction of the feed, the components saturate the stationary phase at the top of the column and frontal chromatography occurs. As the displacer front traverses the column, the feed components are displaced and separated into adjacent bands as they compete for the adsorption sites on the stationary phase. The order of the zones corresponds to the increasing affinity of the components for the stationary phase. The concentration of each component in the final displacement train is determined solely by its adsorption isotherm and the concentration and isotherm of the displacer as shown in Fig. 2. Thus, displacement systems can often result in significant concentration of the feed components during the separation process. The chord joining the origin to the point on the displacer isotherm corresponding to the inlet concentration of the displacer is termed the operating line. Displacement of the feed components is possible only when the operating line intersects the individual isotherms of the feed components. The width of each zone is proportional to the amount of the component present in the feed. Upon the emergence of displacer, the column is regenerated by removal of the displacer with an appropriate solvent followed by reequilibration with the carrier.

Although the physicochemical basis of the displacement mode of chromatography was established by Tiselius in 1943 *(9)*, early attempts to develop the displacement mode into a practical separation process were largely unsuccessful owing to the lack of efficient chromatographic systems *(1)*. However, progress in HPLC with the availability of sorbents exhibiting rapid kinetics and mass transfer have stimulated the recent research on displacement chromatography *(1,3,4)*.

We have demonstrated that displacement chromatography can be successfully employed for the simultaneous concentration and purification of peptides, antibiotics, and proteins *(5)*. A mathematical model

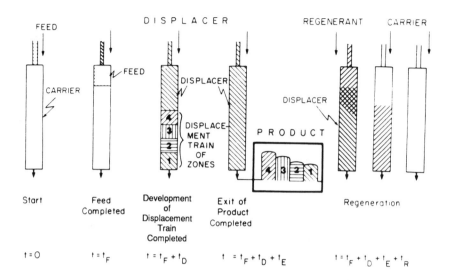

Fig. 1. Schematic of the operating steps in displacement chromatography. The process is carried out by subsequent step changes in the mobile phase composition at the column inlet to the carrier, feed, displacer, and regenerant solutions. (Reproduced with permission from ref. *[8].*)

for the simulation of nonideal displacement chromatography has also been developed to facilitate the optimization of these separations *(10).* We have extended the scope of biopolymer displacement chromatography to more complex feed mixtures, elevated flow rates, and illustrated how protein isotherms can be manipulated to effect the efficient purification of biopolymers by displacement chromatography *(11).* The scale-up of biopolymer displacement to large particle and column diameter systems has also been examined *(12).* This chapter will focus on some of the practical considerations of displacement chromatography.

Although this chapter is by no means exhaustive, it serves to illustrate some of the practical considerations in developing displacement chromatographic methods for protein purification. It is hoped that this brief chapter will bring displacement chromatography to the attention of many users of chromatography, and will entice them to try their own hand at this powerful technique for simultaneous condition and purification of biopolymers.

1.1. Methods Development in Displacement Chromatography

At present, the major obstacles to the widespread implementation of displacement chromatography for the simultaneous concentration and purification of biopolymers is the availability of appropriate nontoxic displacers for a variety of adsorbent systems, and the perceived difficulty of methods development. In this section, a rather simplistic general approach to displacement methods development. This approach can be employed for a variety of adsorptive systems. The reader is referred to more comprehensive treatments of displacement presented elsewhere *(1,3,4,13)* for more detailed information. Following this general treatment of displacement methods development, specific experimental details for protein displacement on cation-exchange materials will be presented.

1. Prior to any displacement experiments, linear elution chromatography should be employed to select an appropriate stationary-phase material with sufficient selectivity for the feed components of interest.
2. Measure the adsorption isotherms of the feed components (*see* Note 1). If pure material of the feed compounds of interest is available, use microbore frontal chromatography to measure the single component adsorption isotherms in a selected carrier mobile-phase solution. For ion-exchange systems, the carrier is typically a low salt aqueous solution buffered at an appropriate pH. Single component protein isotherms can be determined by frontal chromatography according to the technique of Jacobson et al. *(14)* by using a 50×1 mm microbore column packed with the selected stationary-phase material. An injection valve system containing two 2-mL loops is used for successive introduction of increasing concentrations of the given feed component. This technique involves the measurement of breakthrough volumes for these step increases in feed concentration along with the use of a simple mass balance equation to calculate the stationary-phase concentrations in equilibrium with each mobile phase concentration. The mass balance equation can be written as:

$$q(c_b) = q(c_a) + [(C_b - C_a)(V_F - V_D)]/V_{SP} \qquad (1)$$

where $q(c_a)$ = concentration of the adsorbed solute in equilibrium with the mobile-phase concentration c_a; V_F = breakthrough volume of the front, V_D = dead volume; V_{SP} = volume of stationary phase in the column. The reader is referred to the paper by Jacobson et al. *(14)*

for more detailed information on isotherm determination by microbore frontal chromatography. (*see* Note 1.)

3. If the isotherms of the components of interest are "concave downwards" (a necessary condition for displacement chromatography) and do not cross, then move on to step 5. For relatively simple mixtures, this will usually be the case.

4. If the isotherms of the components of interest either cross or are not concave downwards, make appropriate changes in the mobile-phase composition (e.g., pH, salt, organic modifier, and so on) and reexamine the adsorption isotherms of the components of interest. Continue modifying the mobile-phase composition until the resulting isotherms are "concave downwards" and do not cross.

5. Use microbore frontal chromatography to examine the "regenerability" of potential displacers. This can be done by measuring the breakthrough volume of a known protein in frontal chromatography before and after perfusion of the microbore column with displacer, regenerant, and carrier solutions. If the breakthrough volume following perfusion is less than the initial value, make appropriate changes in the regenerant solution(s) (e.g., pH, salt, organic modifier, etc.) and reexamine the breakthrough volume of the test protein. Clearly, only displacer compounds which can be readily removed from the column during the regeneration step should be examined in the adsorption and displacement studies described below.

6. Measure the adsorption isotherms of potential displacer compounds in the selected carrier mobile-phase using microbore frontal chromatography. Construct the adsorption isotherms of the feed components and potential displacer compounds and select a displacer whose adsorption isotherm lies above the feed components of interest.

7. Select a displacer concentration which will result in an operating line that intersects the adsorption isotherms of all compounds of interest as shown in Fig. 2.

While this general protocol should be used when possible for rational methods development, the researcher is often unable or unwilling to carry out the frontal chromatographic experiments. If this is the case, the following " abbreviated methods development" protocol can be employed.

1.2. Abbreviated Methods Development Protocol

1. Prior to any displacement experiments, linear elution chromatography should be employed to select an appropriate stationary-phase material with sufficient selectivity for the feed components of interest.

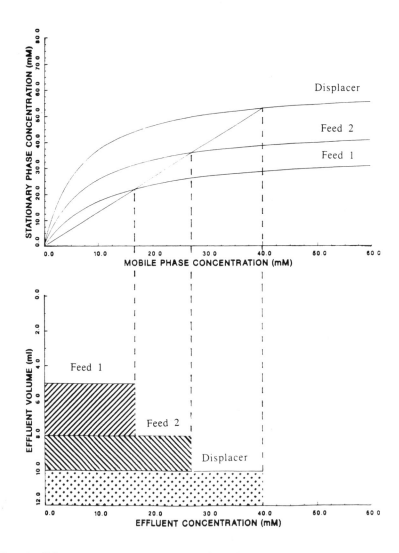

Fig. 2. Schematic representation of isotherms of the feed components and the displacer along with the operating line and the corresponding fully developed displacement train. (Reproduced with permission from ref. *[1]*.)

2. Establish an appropriate mobile-phase condition (e.g., pH, salt, organic solvent, and so forth) that produces at least baseline resolution of the feed components of interest in linear elution chromatography. (*see* Note 2.)

3. Use linear elution chromatography to examine the "regenerability" of the displacer. This can be done by measuring the retention time of a known protein in linear elution chromatography before and after perfusion of the column with displacer, regenerant, and carrier solutions. If the retention time following perfusion is less than the initial value, make appropriate changes in the regenerant solution(s) (e.g., pH, salt, organic solvent, and so on), and reexamine the retention time of the test protein. Again, only diplacer compounds that can be readily removed from the column during the regeneration step should be employed in the displacement experiments.

4. Measure the retention time of potential displacer compounds in the selected mobile phase using linear elution chromatography. Select a displacer whose retention time is greater than the feed components of interest.

Although this abbreviated approach is significantly easier than the more detailed isotherm method described above, it is unable to identify the existence of crossing isotherms or adsorption isotherms not of the Langmuir type. Furthermore, the appropriate displacer concentration can not be established for the subsequent displacement experiments because of the lack of adsorption data. Thus, this abbreviated approach necessitates a "black box" approach to the displacement experiments, which in the long run may actually result in a longer methods development phase. Although this chapter suggests using microbore frontal chromatography to measure single component adsorption isotherms, the reader is referred to alternative methods that have been recently presented for determining single (15) and multicomponent adsorption isotherms (16).

1.3. Displacement Chromatography of Proteins

Once the stationary phase, carrier, regenerant, and displacer have been selected for a given bioseparation, the remaining operating parameters are the feed mass, flow rate, column length, temperature, and displacer concentration. Although the particle diameter is a very important parameter in the optimization and scale-up of the displacement process, the present treatment assumes for simplicity that the particle diameter is fixed by the selection of the stationary phase material.

The relationship among feed mass, displacer concentration, and column length under conditions of ideal chromatography (equilibrium adsorption, plug flow, and constant separation factors) can be obtained using the mathematical transformation developed by Helfferich and Klein *(17)*. The effects of mass transport and axial dispersion on the displacement process have been studied by Phillips et al. *(10)*. Although these mathematical formulations are useful in guiding the optimization and scale-up of the displacement process, the present treatment will focus on general rules of thumb and practical considerations for methods development in displacement chromatography.

The objective of these displacement experiments is to "fine-tune" the operating conditions to achieve the desired objectives for the given separation. In this treatment, we will assume that the separation objectives are the attainment of full development of the displacement train (i.e., adjacent square waves moving at the same velocity) and sharp boundaries separating the displacement zones. In fact, when the separation objectives are maximization of product throughput or minimization of production costs, it is not always desirable to attain full development or establish sharp boundaries.

For the displacement purification of proteins on an analytical-scale column (250×4.6 mm) packed with 5–10 µm porous stationary-phase material, typical ranges of operating conditions are: total feed mass, 5–40 mg; flow rate during loading of the feed, 0.5–1.2 mL/min; flow rate during introduction of the displacer, 0.1–0.5 mL/min; flow rate during equilibration with the carrier and regenerant, 1–2 mL/min; displacer concentration, 20–80 mg/mL. Initial displacement chromatographic conditions should be selected from these typical ranges of conditions, and the displacement experiment carried out using the adsorption conditions established during the methods development phase as described above. Fractions of the column effluent are collected during the displacement run and subsequently analyzed by analytical HPLC to obtain a detailed record of the separation. In order to illustrate how displacement chromatography is actually carried out, specific experimental details for protein displacement on cation exchange materials are presented below.

2. Materials

2.1. Chromatographic Columns

Use analytical columns (4.6 mm id) packed with high-performance cation-exchange chromatographic materials (5–10 µm particle diameter, 300–1000 Å pore diameter). Either weak or strong cation-exchange adsorbents can be used.

2.2. Reagents

Use analytical grade reagents and distilled water. Sodium monophosphate, sodium acetate, and ammonium sulfate can be purchased from Fisher Scientific (Fair Lawn, NJ, USA). Model proteins, such as cytochrome C, lysozyme, α-chymotrypsinogen, and ribonuclease, can be obtained from Sigma (St. Louis, MO, USA). Water-soluble coagulant, Nalcolyte 7105 (now called 8105), is a product of Nalco Chemical Company (Chicago, IL, USA).

2.3. Stock Solutions

1. Carrier; 0.1 M ammonium sulfate in 25 mM phosphate buffer, pH 7.5.
2. Model protein feed solution; 4.0 mg each of ribonuclease, α-chymotrypsinogen, cytochrome C, and lysozyme in 2 mL of the carrier.
3. Displacer; 30 mg/mL Nalcolyte 7105 in the carrier (*see* Note 3).
4. Regenerant; 0.8M ammonium sulfate in 50 mM phosphate buffer, pH 3.0.

2.4. Apparatus

Assemble a chromatographic system that includes an HPLC pump, an injection valve, a UV-VIS detector, a recorder or integrator, column oven or water bath, and a fraction collector. Configure the chromatographic system to enable the sequential perfusion of the column with the carrier, feed, displacer, and regenerant solutions (6). The feed solution is typically introduced into the column via a 2-mL loop in the injection valve.

3. Method

1. Equilibrate the column with the carrier by pumping 10 column vol of the carrier containing 0.1 M ammonium sulfate in 25 mM phosphate buffer, pH 7.5 through the column at a flow rate of 1 mL/min.
2. Prepare the system for introduction of the feed and displacer by priming the pumping system prior to the injection valve with the displacer. This can be done by either disconnecting the inlet tubing to the injector (from the pump) or by the use of a valve immediately prior to the injector. Calculate the dead volume in the system prior to the injector, and make sure that sufficient fluid is pumped to prime the system. During this step, place the feed mixture into the 2-mL sample loop of the injector.
3. Introduce the feed mixture by switching the injection valve position to enable introduction of the feed into the column at a flow rate of 0.5 mL/min. Carefully monitor the absorbance at 280 nm of the column effluent. If the detector indicates the elution of some species

during the introduction of the feed, activate the fraction collector to begin collecting fractions. (Note: At this point it is probably sufficient to collect 300-µL samples.)

4. In order to assure a sharp step change from feed to displacer solution at the column inlet, switch the injector position such that the flow of the displacer bypasses the 2-mL loop and enters directly into the column. Introduce the displacer into the column at a flow rate of 0.1 mL/min. Again, carefully monitor the absorbance at 280 nm of the column effluent. When the detector indicates the emergence of feed components during the displacement process, activate the fraction collector to begin collecting 100-µL fractions.

5. Regenerate the column. Upon the breakthrough of the displacer (*see* Note 4) regenerate the cation-exchange column by passing 50 column vol of a regenerant containing 0.8M ammonium sulfate in 50 mM phosphate buffer, pH 3.0. During this step, collect large volume fractions (10 mL) of the column effluent for future analysis if needed.

6. Reequilibrate with the carrier. Repeat step 1.

7. Analyze displacement fractions by HPLC. Measure the composition of the fractions collected during the displacement chromatographic run using analytical HPLC. Dilute the fractions 50–500-fold with the eluent, and inject 20-µL samples onto an analytical cation-exchange column. The mobile phase for this analytical HPLC system should contain 0.15M ammonium sulfate in 25mM phosphate buffer, pH 7.5. Use a volumetric flow rate of 1.0 mL/min. The effluent from the column should be monitored at 280 nm. Carry out quantitative analysis of the composition of the fractions, and use the data to construct displacement chromatograms. (*See* Note 5.)

8. Examine the displacement chromatogram, and evaluate the efficacy of the separation. If the separation is acceptable, pool the fractions of the desired compositions and purity and repeat steps 1–8. If the separation is not acceptable, the operating conditions must be modified to improve the separation as described in Notes 6–14.

4. Notes

1. For complex feed mixtures, some initial separation may be required to produce approx 10 mg of each feed component for use in the adsorption isotherm experiments described in Step 2 in Section 1.1. Alternatively, one can employ the abbreviated methods development protocol described in Section 1.2.

2. Since the feed components should adsorb strongly enough in the carrier for the displacer front to overtake them, the actual carrier conditions employed in the displacement experiment should have a

lower salt or organic content then the mobile phase established for linear elution chromatography in Step 2 in Section 1.2.

3. Alternative cationic displacers, which adsorb more strongly than the feed protein, can also be employed for displacement experiments described in Section 3. For example, lysozyme can be used as the displacer for a feed mixture containing ribonuclease, α-chymotrypsinogen, and cytochrome C.

4. Since Nalcoylte 8105 and many potential cationic displacers cannot be readily detected with UV-VIS detectors, either collodial titration methods (Nalco Chemical Company, Chicago, IL) or monitoring the column effluent with refractive index can be used to establish the breakthrough volume of the displacer. In addition, adsorption isotherm data for the displacer can be used to estimate the breakthrough volume by means of a simple mass balance on the analytical column.

5. The techniques described in Note 4 can also be employed to determine the displacer content in various fractions.

6. The larger the feed mass, the greater the minimum column length required to achieve full development of the displacement train. Thus, if adjacent square wave zones are not achieved for a given feed mass and column length, either the feed mass can be decreased or the column length increased to achieve the isotachic displacement condition.

7. The slower the linear velocity of the fluid phase, the sharper the boundaries separating the displacement zones. If the objective is to maximize the product throughput, elevated velocities should generally be employed. If the goal is to maximize product yield and purity, slower velocities should be used.

8. Higher column operating temperatures produce sharper boundaries separating the displacement zones.

9. Smaller particle diameter chromatographic supports produce sharper boundaries between the displacement zones. If the boundaries between the displacement zones are too disperse, a reduction in the particle diameter or velocity and/or an increase in temperature can be carried out to decrease the dispersion.

10. The higher the displacer concentration, the more concentrated the components in the displacement zones and the narrower the width of these zones. Thus, if narrow zones result in decreased purity, the displacer concentration can be decreased to widen the zones. Conversely, if more concentration of the feed components is desired, the displacer concentration can be increased. In addition, elevated displacer concentrations will result in an increased linear velocity of the displacer front producing shorter cycle times.

11. Typical problems that can arise in displacement chromatography and

Table 1
Typical Problems and Potential Solutions
for Displacement Chromatography

Problem	Potential solution
1. Diffuse boundries separating the displacement zones	Decrease particle diameter, decrease flow rate, increase temperature
2. Incomplete development of the displacement train	Decrease total feed mass, increase column length
3. Complete mixing of displacement zones	Examine for crossing adsorption isotherms, establish noncrossing adsorption conditions
4. Gap occurs between the feed and displacer zones	Increase the concentration of the displacer or establish alternative adsorptive conditions
5. Precipitation of feed components leads to excessive pressure drop in the column	Decrease displacer concentration change carrier conditions to increase solubility of feed components

suggestions for solving these problems are presented in Table 1. Although Problems 1 and 5 are self explanatory, Problems 2–4 require some additional comments as described in Notes 12–14.

12. Displacement development proceeds from the rear to the front of the feed zones as the displacer shock proceeds down the column. If the column length is insufficient for complete development of the displacement train, the displacement chromatogram will be characterized by region(s) of pure material(s) at the rear of the feed zones and mixing of the earlier eluting feed regions. This condition should be addressed as described for Problem 2 in Table 1.

13. When the adsorption isotherms of the feed components of interest cross, a displacement azeotrope can arise that produces complete mixing of the two feed components in the displacement train. If two of the feed components emerge from the column in a concentrated displacement square wave, but are completely mixed, it is likely that the components have crossing isotherms under these carrier conditions. This condition should be addressed as described for Problem 3 in Table 1.

14. When the operating line for a given displacement does not intersect the feed components of interest, these feed components will move through the column at a faster velocity than the displacer front. If a gap occurs between the elution of the feed components and break-

through of the displacer front, the suggestions given for Problem 4 should be considered.

References

1. Horvath, Cs. (1985) Displacement chromatography: Yesterday, today and tomorrow, in *Science of Chromatography* (Brunner, F., ed.) Elsevier, Amsterdam, pp. 179–203.
2. Knox, J. H. and Pyper, H. M. (1986) Framework for maximizing throughput in preparative liquid chromatography. *J. Chromatogr.* **363**, 1–30.
3. Frenz, J. and Horvath, Cs. (1989) High performance displacement chromatography; in *High Performance Liquid Chromatography, Advances and Perspectives*, Vol. 5 (Horvath, Cs., ed.) Academic, Orlando, FL, pp. 211–314.
4. Cramer, S. M. and Subramanian, G. (1990) *Sep. & Purif. Meth.* **19**, 31–91.
5. Subramanian, G., Phillips, M. W., and Cramer, S. M. (1988) Displacement chromatography of biomolecules *J. Chromatogr.* **439**, 341.
6. Cramer, S. M. and Horvath, Cs. (1988) Displacement chromatography in peptide purification. *Prep. Chromatogr.* **1**, 29–49.
7. Frenz, J., Van Der Schrieck, P., and Horvath, Cs. (1985) Investigation of operating parameters in high-performance displacement chromatography *J. Chromatogr.* **330**, 1–17.
8. Horvath, Cs, Nahum, A., and Frenz, J. (1981) High performance displacement chromatography, *J. Chromatogr.* **330**, 1–17.
9. Tiseliius, A. (1943) *Ark. Kem. Mineral Geol.* **16A**, 1.
10. Phillips, M. W., Subramanian, G., and Cramer, S. M. (1988) Theoretical optimization of operating parameters in non-ideal displacement chromatography. *J. Chromatogr.* **454**, 1–21.
11. Subramanian, G. and Cramer, S. M. (1989) Displacement chromatography of protein under elevated flow rate and crossing isotherm conditions. *Biotechnol. Prog.* **5, No. 3**, 92–97.
12. Subramanian, G., Phillips, M. W., Jayaraman, G., and Cramer, S. M. (1989) Displacement chromatography of biomolecules with large particle diameter systems. *J. Chromatogr.* **484**, 225–235.
13. Frenz, J. and Horvath, Cs. (1985) High performance displacement chromatography: Calculation and experimental verification of zone development. *AIChE J.* **31**, 400–409.
14. Jacobson, J., Frenz, J., and Horvath, Cs. (1984) Measurement of adsorption isotherms by liquid chromatography. *J. Chromatogr.* **316**, 53–68.
15. Golshan-Shirazi, S. and Guiochon, G. (1988) Analytical solution for the ideal model of chromatography in the case of a Langmuir isotherm. *Anal. Chem.* **60**, 2364–2374.
16. Chen, T.-W., Pinto, N. G., and van Brocklin, L. (1989) Rapid method for determining multicomponent Langmuir parameters for displacement chromatography *J. Chromatogr.* **484**, 167–186.
17. Helfferich, F. G. and Kline, G. (1970) *Mutlicomponent Chromatography—Theory of Interference.* Marcel Dekker, New York.

CHAPTER 18

Purification of DNA Binding Proteins by Affinity and Ion Exchange Chromatography

David Hornby, Kevin Ford, and Paul Shore

1. Introduction

It is the exquisite interplay of proteins and nucleic acids within the cell which gives rise to the controlled expression and replication of the genetic material. Our present understanding of these processes is in part the result of the successful purification and characterization of the participating macromolecules. The achievements of the early molecular biologists in obtaining active, homogeneous preparations of low-abundance gene regulatory proteins are outstanding. The strategies employed for the purification of such proteins are, however, no different in principle to those procedures used to purify high-abundance proteins. Moreover, one of the goals of molecular biologists in purifying gene regulatory proteins is to clone the corresponding gene. When this has been achieved, the gene can often be overexpressed, and the purification of the gene product becomes trivial by comparison.

The two physical properties of DNA binding proteins that have been most frequently exploited in their purification are their basic nature and their high affinity for DNA. The incorporation of the cation-exchange resin phosphocellulose into purification schedules was an early success in improving the resolution of these pro-

From: *Methods in Molecular Biology, Vol. 11: Practical Protein Chromatography*
Edited by: A. Kenney and S. Fowell Copyright © 1992 The Humana Press Inc., Totowa, NJ

teins. Although there are some DNA binding proteins whose basic character swamps all other physical properties (such as histones), in general, DNA binding proteins can be resolved on conventional anion (e.g., diethylaminoethyl-based) and cation (e.g., carboxymethyl-based) exchange resins.

Affinity chromatography has revolutionized the purification of both high- and in particular, low-abundance proteins. There are three types of affinity chromatography that have been used to purify DNA binding proteins. The coupling of a specific oligonucleotide duplex to a solid support is the most specific of the three methods and has been used to isolate sequence-specific DNA binding proteins. The other two approaches are less specific and are perhaps better described as pseudo-affinity methods. In one method, a random mixture of DNA fragments is coupled to a solid support, usually cellulose. As an alternative to DNA, heparin, which mimics the phosphodiester backbone, has been coupled to agarose for similar purposes. Both types of pseudo-affinity chromatography have been used in conjunction with other less specific separation steps to purify DNA binding proteins to homogeneity. The main advantage of the pseudo-affinity approach is that it can be applied to all DNA binding proteins. However, there are instances where an extract contains a mixture of very similar sequence-specific DNA binding proteins that cannot be resolved by the pseudo-affinity approach. In this case, heparin-agarose and DNA cellulose are often employed in conjunction with an oligonucleotide affinity step. The use of biotinylated DNA has also been proposed *(1)* for the selective purification of a biotinylated DNA-protein complex using a streptavidin agarose column. A recent modification of the affinity column method is the use of magnetic beads coupled to biotinylated DNA duplexes via a noncovalent streptavidin bridge *(2)*. These affinity beads can be used in a batch approach to obtain homogeneous DNA binding proteins from crude cell extracts.

Finally, in any protein purification program, a convenient and rapid assay should ideally be available. The most widely used methods are filter binding, gel retardation (often called band shift or gel shift), and DNase footprinting. In addition, enzyme activities can be monitored, or immunoreactive material can be detected where appropriate. The generally applicable technique of gel retardation has been described in detail elsewhere in this series (*see* vol. 4) and will therefore not be discussed here, but is used to monitor activity during fractionation procedures.

2. Materials

2.1. Source of the Protein

This is perhaps the most important aspect of any protein purification strategy. The current understanding of the high resolution structure of proteins, such as the serine proteases, is a consequence of their high abundance in slaughter house tissue, which greatly facilitates their purification and crystallization. DNA binding proteins are generally present in very low abundance in both prokaryotes and eukaryotes, since they are often required to interact with just one DNA sequence element in the entire genome of the organism. Thus, DNA binding proteins, such as the lactose repressor of *Escherichia coli*, are present at levels of approx 10 molecules/cell. By comparison, high-abundance proteins may often be present at levels approaching 10^5 molecules/cell. These differences highlight the relative enormity of the task of purification. Clearly then, a judicious choice of cell or tissue type can significantly improve the yield of a DNA binding protein. In the case of the lactose repressor a mutant strain of *E. coli* that contained 2000-fold higher levels of the repressor, was used as a source for the purification. However, such mutants are not always available, and even a twofold improvement in the number of copies of a particular protein can facilitate its purification. The level of a particular protein in a given cell can only be determined by assay. Moreover, it should be remembered that "dormant" species of some DNA binding proteins have been detected after they were initially overlooked *(3)*. We shall describe the purification of two proteins that exemplify the strategies used in the purification of bacterial and eukaryotic DNA binding proteins. The first, *Eco*P1 *(4)*, is a sequence-specific DNA methyltransferase encoded by bacteriophage P1 and has been expressed at a high level in *E. coli*, whereas the second, Sp1 *(5)*, is a human transcription factor that is typical of many low-abundance eukaryotic DNA binding proteins.

2.2. Purification of EcoP1

1. *E. coli* strain expressing the *Eco*P1 *mod* gene.
2. Extraction buffer, comprising: 50 m*M* Tris/Cl, pH 7.4, 10 m*M* magnesium chloride (*see* Note 1).
3. Assay buffer, comprising: 1µg/mL^{-1} DNA (e.g., pBR322 or phage λ), 100 m*M* HEPES, pH 8.0, 14 m*M* 2-mercaptoethanol, 6.4 m*M* magnesium chloride, 0.25 m*M* EDTA, and 100 n*M* (^{3}H-methyl)-*S*-adenosylmethionine (80 Ci/mmol^1).

4. MonoQ or Q-Sepharose (Pharmacia LKB Biotechnology, Milton Keynes, UK) in 50 mM potassium phosphate, pH 7, 7 mM 2-mercaptoethanol, 1 mM EDTA.
5. Affigel-Heparin (Bio-Rad, Herts, UK) in 50 mM Tris HCl pH 8, 100 mM sodium chloride (*see* Note 2). This buffer is also used for the affinity step.
6. Sepharose CL4B (Pharmacia LKB Biotechnology).
7. Cyanogen Bromide (Sigma, Dorset, UK).
8. Sequence-specific oligonucleotide(s).
9. Annealing buffer: 67 mM Tris HCl, pH 7.6, 13 mM magnesium chloride, 9.71 mM dithiothreitol, 1.3 mM spermidine, and 1.3 mM EDTA.
10. Coupling buffers: *see* Section 3.1.3.
11. Reagents for SDS-PAGE (*see* Chapter 20).

2.3. Purification of Sp1

1. 10 g HeLa cells.
2. Extraction buffers:
   A: 10 mM HEPES, pH 7.9, 1.5 mM magnesium chloride, 10 mM potassium chloride, 0.5 mM dithiothreitol.
   B: 300 mM HEPES, pH 7.9, 1.4M potassium chloride, 30 mM magnesium chloride.
   C: 20 mM HEPES, pH 7.9, 25% (vol/vol) glycerol, 0.42M sodium chloride, 1.5 mM magnesium chloride, 0.2 mM EDTA, 0.5 mM phenylmethylsulfonylchloride, and 0.5 mM dithiothreitol.
   D: 50 mM Tris Cl, pH 7.5, 0.42M potassium chloride, 20% (vol/vol) glycerol, 10% (wt/vol) sucrose, 5 mM magnesium chloride, 0.1 mM, EDTA, 1 mM phenylmethylsulfonylchloride, 1 mM sodium metabisulfite and 1 mM dithiothreitol.
   Z: 25 mM HEPES, pH 7.6, 12.5 mM magnesium chloride, 20% glycerol 0.1% (vol/vol) Nonidet P-40, 10 μM zinc sulphate, 1 mM dithiothreitol, and 0.1M potassium chloride.
3. Wheat Germ Agglutinin (WGA) elution buffer:Z buffer containing 0.3M N-acetyl glucosamine (NAG).
4. Z':Z buffer containing 1M potassium chloride.
5. WGA–Sepharose (Pharmacia).
6. Oligonucleotide-affinity column.
7. Gel retardation assay mixture for Sp1 comprises 25 mM Tris HCl, pH 7.9, 6.25 mM magnesium chloride, 0.5 mM EDTA, 0.5 mM dithiothreitol, 50 mM potassium chloride, 10% glycerol, 0.1% Nonidet P40, 0.3 μg poly(dI) • poly (dC), 10 mM end-labeled DNA and 1–10 μL of protein extract (*see* Note 5).

3. Methods

3.1. Purification of EcoP1

Most bacterial cells can be lysed rapidly by sonication, and this is therefore the method of choice for initial preparation of a cell extract. (*see* Note 3 and Chapter 19 for alternative methods of cell breakage.)

1. Dilute the cell paste 2:1 (vol/vol) with extraction buffer, and sonicate for a total period of 5 min at 1-min intervals (*see* Note 4). This approach ensures that the temperature of the cell suspension remains at or below 10°C.
2. Centrifuge the suspension at 48,000g (Sorvall SS-34 rotor at 20,000 rpm) for 20 min.
3. Add solid ammonium sulfate (*see* Notes 5 and 6) to the supernatant to give a final saturation of 70%, and recover the pellet after centrifugation (15 min, SS34 rotor, 4°C).

3.1.1. Ion Exchange Chromatography

The general principles of ion-exchange chromatography have been covered elsewhere (Chapter 16) and are directly applicable to the fractionation of all DNA binding proteins. In view of the low abundance of these proteins, the use of HPLC/FPLC is to be recommended (*see* Note 7). Most cation exchangers, such as Mono S (Pharmacia), MA7S (Bio-Rad, Herts, UK), and Hydropore-SCX (Anachem), are suitable for the purification of basic DNA binding proteins. Of course, there is a broad spectrum of surface charge among DNA binding proteins, and often the use of anion exchangers can be very successful, e.g., Mono Q (Pharmacia), MA7Q (Bio-Rad), Hydropore-AX (Anachem, Bedfordshire, UK). The high resolution of HPLC/FPLC can produce a considerable purification, and often a subsequent affinity step is sufficient to produce a homogeneous preparation.

1. Dialyze the *Eco*P1-containing extract against 1000 vol of column over 1–2 h prior to chromatography. Sample vol of up to 1 mL (at a protein concentration of approx 50 mg/mL) may be loaded onto a 1 mL mono Q column. Larger preparations should be loaded onto a Q-Sepharose column of an appropriately scaled-up volume. Run the columns at a flow rate of 1mL/min, and elute with a 0–1M linear gradient of NaCl.
2. Assay the fractions for DNA binding or DNA methyltransferase activity (*4*). *Eco*P1 elutes at a KCl concentration of 300 mM. The purity of the enzyme can be monitored at this or later stages in the conven-

tional way by SDS-PAGE. The samples may be concentrated for such analysis by precipitation on ice with 10% trichloroacetic acid.

3.1.2. Pseudo-Affinity Chromatography

DNA coupled to cellulose has been used successfully to fractionate DNA binding proteins. However, in view of its long-term instability and susceptibility to nuclease degradation, it has largely been replaced by heparin-Sepharose. This is more stable, produces similar levels of purification, and has much better physical properties including faster flow rates under the same pressure constraints. Moreover, heparin-Sepharose is easy to synthesize and is available commercially.

Pool the active fractions from the Mono Q column, and dialyze against 1000 vol of column buffer for 1–2 h prior to application on to a 5 mL heparin column. The elution protocol is identical to that used for the Mono Q step, and the EcoP1 enzyme elutes at a concentration of 300 mM potassium chloride. At this stage, the protein is approx 80% pure as judged by SDS-PAGE.

3.1.3. Oligonucleotide-Affinity Chromatography

There are three general criteria that will determine the relative success of an affinity step of this kind

1. The oligonucleotide sequence and length.
2. The space between oligonucleotide and the support.
3. The chemical nature of the support matrix.

The choice of sequence is usually governed by information gleaned from biochemical experiments, such as DNase footprinting or genetic experiments. In the case of eukaryotic transcription factors, their extremely high affinity for specific target sequences (binding constants range from 10^9 to 10^{14} M^{-1}) has enabled footprinting experiments to be undertaken in crude preparations. In the case of bacterial repressors, fractionation of the bacterial extract on an ion-exchange column is often a prerequisite for successful footprinting. The minimum length of sequence required to adsorb the protein of interest selectively can only be determined by trial and error. However, in general, the footprinting data should give a good indication of the length of sequence occupied by the protein. In addition to the sequence-specific bases, flanking sequences may also play a role in orienting the protein on the DNA. Therefore, a natural recognition site of approx 25–30 bases is usually a sensible starting point for the affinity ligand. It has also been reported that multiple copies of the recognition sequence can improve the binding of transcription factors to oligonucleotide columns *(5)*. If such an oligonucleotide is to be employed, care should be taken to ensure that the repeated elements are spaced sufficiently far apart so as not to reduce binding affinities.

One of the most neglected aspects of affinity ligand design is the systematic evaluation of the effect of distance between the ligand and the matrix. Optimization of ligand spacing is known to produce dramatic improvements in purification. Optimization of ligand spacing in the case oligonucleotide can be achieved simply by altering the length of the DNA sequence that is to be coupled to the column. There is no need to incorporate inert spacer groups in between the matrix and the ligand. However, the failure of an oligonucleotide column may be a simple consequence of inaccessibility of the sequence to the protein.

The most commonly used matrix for affinity chromatography is the commercially available Sepharose C4B (Pharmacia), although some groups have used the 2B and 6B derivatives, which contain fewer or more crosslinks respectively (*see* Note 8). This matrix has excellent flow characteristics, is largely inert, and can be easily derivatized following "activation" with cyanogen bromide. Coupling of the ligand to such activated "Sepharose" is generally achieved by a facile incubation of the ligand with the resin at a suitable temperature, pH, and salt concentration. Subsequent washing removes excess ligand and serves to equilibrate the column. The coupling reaction involves the following stages:

1. DNA synthesis and subsequent annealing of the complementary strands.
2. Cyanogen bromide (CNBr) activation of the matrix, Sepharose CL-4B (2B or 6B).
3. Coupling of the ligand to the support.
4. Washing and removal of excess CNBr.
5. Estimation of the coupling efficiency.

The procedure for the synthesis of a 10-mL oligonucleotide affinity column is as follows:

1. Synthesize and gel purify two complementary strands of DNA that have four-base pair (or greater) overhangs flanking the DNA binding site *(5)*.
2. Anneal the oligonucleotides (200 µg of each) in annealing buffer by incubating at 88°C for 2 min, then 65°C for 10 min, and finally room temperature for 5 min in a total vol of 75 µL.

3. End-label the DNA with polynucleotide kinase in the presence of 5 µCi (γ ^{32}-P) ATP (in a final vol of 100 µL containing 100 U T4 polynucleotide kinase and 20 mM ATP) at 37°C for 2 h.
4. Stop the reaction by addition of 100 µL 5M ammonium acetate, pH 5.5, and precipitate the DNA with 5 vol of absolute ethanol cooled on dry ice. Wash the DNA pellet with 100 uL 70% ethanol and lyophilize.
5. Redissolve the DNA in 100 uL of sterile, distilled water. In our experience, it is not necessary to covalently link the oligomers together using DNA ligase.
6. Wash 10 g (or 15 mL settled vol) Sepharose CL-4B with water using a sintered glass funnel. A course grade is preferable, since the slurry can be quickly washed without recourse to the use of a vacuum pump, thereby avoiding the potential destruction of some of the Sepharose beads.
7. Carefully remove the Sepharose from the filter, avoiding too much mechanical damage and resuspend in 20 mL water. It is always advisable to recover the Sepharose from the filter using an excess of washing buffer and subsequently allowing the resin to settle, rather than scraping it from the filter with a spatula.
8. Place the slurry together with a magnetic stirring bar in a beaker jacketed with a larger beaker containing water maintained at 15°C with regular additions of ice. Over a 1-min period add 2 mL of cyanogen bromide (1.1 g freshly dissolved in N,N-dimethylformamide) dropwise to the gently stirred resin *in a fume cupboard.*
9. Over a 10-min period, add dropwise 1.8 mL of 5M NaOH.
10. Finally, stop the reaction by the addition of 100 mL of ice-cold, sterile, distilled water, and wash successively with 300 mL of ice-cold water followed by 100 mL of ice-cold potassium phosphate, pH 8, as before on a sintered funnel.
11. All glassware should be soaked in a large beaker containing 1 L of 200 mM glycine before washing.
12. The DNA is coupled to the activated resin as follows. Transfer the resin in a total vol of 14 mL of 100 mM potassium phosphate, pH 8, to a polypropylene tube, and add the DNA in 100 µL of water. Stir on a rotary shaker for around 16 h at room temperature.
13. Wash the resin with 200 mL of water followed by 100 mL of 1M ethanolamine HCl, pH 8, and resuspend in 14 mL of 1M ethanolamine HCL pH 8. Stir the resin at room temperature for 4–6 h in order to block the unreacted Sepharose beads completely.
14. Wash the affinity resin successively with 100 mL of 10 mM potassium phosphate, pH 8; 100 mL of 1M potassium chloride, 100 mL of

water, and finally with 100 mL of storage buffer containing 10 mM Tris HCl, pH 7.6, 0.3M sodium chloride, 1 mM EDTA, and 0.02% (wt/vol) sodium azide.

15. The efficiency of coupling can be approximately estimated by determining the radioactivity recovered from the wash after step 12.

*Eco*P1 can be purified to apparent homogeneity using an oligonucleotide affinity column containing multiple copies of the enzyme's recognition sequence, AGACC. The heparin fractions after dialysis against column buffer are applied to 1 mL of affinity column, which is then washed with 10 column vol of buffer. Elution is effected by simply washing the column with 2 column vol of 1M potassium chloride. The pure *Eco*P1 is dialyzed against column buffer containing 50% glycerol and stored at –20°C.

3.2. Purification of Sp1

3.2.1. Preparation of Nuclei

A major difference between bacteria and eukaryotes is the localization of the genetic material in a nucleus in the latter. This partitioning of the genome has clear implications for the subcellular distribution of DNA binding proteins. Often the starting material for a preparation is cell nuclei, rather than the whole cell. On the other hand, a number of transcription factors are reported to be present at significant levels in the cytosol. This will clearly determine whether the nuclei or the whole cell extract is used as the starting material for a particular purification, and this can only be determined empirically.

The method used to obtain a nuclear extract is that of Dignam et al. *(6)*.

1. Wash the Cells (10 mL packed cell vol in this example, but the method can be scaled up or down as appropriate) in 50 mL of Buffer A, and leave at 4°C for 10 min. Harvest by centrifugation at 2000 rpm for 10 min in a Sorvall HG4L rotor (or similar), and resuspend the pellet in 20 mL of Buffer A.

2. Homogenize gently with 10 strokes of a glass Dounce homogenizer. Centrifuge as above for 10 min to pellet the nuclei, and remove the supernatant carefully. The cytoplasmic, S100 fraction *(6)* often contains a significant proportion of Sp1, and is obtained by mixing the supernatant with 0.11 vol of Buffer B, centrifuging at 100,000g (Beckman Type 42 rotor, or equivalent), and dialyzing against 20 vol of Buffer D.

3. Prepare the nuclear fraction by centrifugation of the crude nuclear pellet for 20 min at 25,000g (Sorvall SS-34), resuspending the pellet in 30 mL of Buffer C (3 mL/10^9 cells), and lysing with 10 strokes of a glass Dounce homogenizer as before.

4. Stir the suspension by inversion for 30 min at 4°C before centrifuging for 30 min at 25,000g (Sorvall SS-34) to obtain the nuclear extract supernatant. Finally dialyze against several changes of Buffer D over a 2–3 h period. Freeze the nuclear extract (at a protein concentration of 5–10 mg/mL^{-1}) is then frozen in aliquots at –80°C.

Using this method, both nuclear and cytoplasmic fractions are obtained. It should be noted that, in our hands, both extracts often contain Sp1 as determined by gel retardation assay, and purification from both fractions has been achieved.

3.2.2. Wheat Germ Agglutinin Chromatography

It has been observed that a number of eukaryotic transcription factors are glycosylated (7). Therefore, a convenient method for their rapid purification has emerged from Tjian's laboratory (8) and is described here. (See also Chapter 6.)

1. Following dialysis against buffer D, mix the nuclear extract with a 1 mL slurry of WGA-Sepharose in a 10-mL plastic column using a Pasteur pipet. Wash the column is then washed with 10 mL of Z buffer, and retain the wash for assay.

2. Add 10 mL of WGA elution buffer, and mix with the resin using a Pasteur pipet. Let the slurry settle over a 30-min period. Run the column dry, and collect the effluent. This material can now be applied directly onto a DNA affinity column. (In our hands, the WGA resin is not suitable for subsequent use.)

3.2.3. Oligonucleotide-Affinity Chromatography

The effectiveness of the Sp1 affinity column is improved by the addition of competitor DNA. It has been suggested (5) that the fraction to be loaded onto the affinity column should include a 20-fold excess of competitor DNA in order to improve sequence-specific resolution. The nature of the competitor, calf thymus, salmon sperm DNA or, poly(dI).poly(dC) may influence the success of the purification. It is of course possible to couple a small affinity column of this kind to an FPLC system in order to undertake a systematic investigation of such parameters in optimizing the elution conditions. Some authors report that several cycles of application and elution of the protein extract can improve the level of purification given by both affinity and pseudo-affinity approaches. Again, this can only be determined empirically.

For the purification of Sp1 from 10 g of HeLa cells, use the following protocol:

1. Add 60 μg poly(dI). poly(dC) (Pharmacia, Milton Keynes, UK) to the WGA-purified material, and incubate at 4°C for 15 min.
2. Add the extract to, and mix with, 1 mL of affinity resin. Stir the resin into the WGA fraction in a 10-mL plastic column, and allow the resin to settle over 30–60 min.
3. Wash the column thoroughly with 20 mL of Z buffer.
4. Elute the Sp1 in a total vol of 1 mL by the addition (and mixing) of 1.2 mL of Z' buffer.
5. Dialyze the eluted protein against 100 vol of Z buffer and store at –70°C.

Both *Eco*P1 and Sp1 prepared by these methods are electrophoretically pure, and are active in sequence-specific DNA methylation and sequence-specific transcriptional activation, respectively.

4. Notes

1. When attempting the purification of a new DNA binding protein, the extraction buffer should contain the following cocktail of stabilizing reagents in addition to any specific requirements: EDTA [1 mM], dithiothreitol[1 mM] (or dithioerythritol[1 mM] or β-mercaptoethanol [7 μM]), Phenylmethylsulphonyl fluoride (PMSF;) [1 mM]. NB PMSF is both difficult to dissolve and has a very short half-life in aqueous solution (approx 30 min). It is therefore advisable to dissolve the PMSF in 1 mL of acetone (or isopropanol) at a concentration of 100 mM, add it to 9 mL of hot buffer, and subsequently add this to 91 mL of ice-cold extraction buffer and use immediately. Further additions of PMSF are usually unnecessary since the inactivation of serine proteases by this reagent is irreversible. PMSF is very toxic and may be replaced by benzamidine at a final concentration of 1 mM. However, the reversibility of benzamidine inhibition means that it must be added at all stages throughout the purification schedule.
2. "Affigel Heparin" as supplied by Bio-Rad Laboratories (Herts, UK) is probably the most stable and durable of the commercial resins, since the ligand is coupled via an amide linkage to the crosslinked agarose matrix. The advantage of this coupling is that it affords greater resistance to heparin displacement by primary amino groups, which are found in many commonly used buffers and in the protein mixtures applied to the column.
3. As an alternative to sonication, the cells can be dried down overnight in a large desiccator containing sulfuric acid (100 mL is generally sufficient). The dried cells can be lysed by manual grinding with a

mortar and pestle. Other methods, such as shaking with glass beads, shearing with either the Hughes press or the French pressure cell, and enzymic lysis, are also used routinely, but do not lyse all microorganisms. (*See also* Chapter 19.)

4. The syrupy/yellow appearance of the supernatant is a good indication of successful lysis. Yields of protein can be significantly improved by the second round of sonication of the cell pellet. The supernatants are then pooled before the next step.

5. The supernatant contains most of the soluble proteins of the bacterium, most of which are unwanted. A convenient method of both purifying and concentrating the protein at this stage is salt fractionation with ammonium sulfate. Ideally, salt is added until the protein just remains soluble, and the precipitate is removed by centrifugation at 48,000g for 20 min. Next, sufficient ammonium sulfate is added to precipitate the protein, and this is then harvested and the supernatant discarded. The appropriate saturation of ammonium sulfate required can only be determined on an empirical basis. In practice, ammonium sulfate is often used primarily to concentrate the protein extract rather than produce a significant fractionation. Furthermore, it should be noted that precipitation by ammonium sulfate varies with respect to ionic strength, temperature, pH, and protein concentration. It is important therefore to standardize these parameters in each preparation. Care should be taken to ensure that the local pH changes that occur during the addition of large pieces of ammonium sulfate are avoided. It is advisable to grind the ammonium sulfate to a fine powder and to add small amounts sequentially in order to overcome denaturation of the protein mixture.

6. Ammonium sulfate fractionation can either be performed directly or after removal of cellular DNA by selective precipitation with polyethyleneimine (PEI) (1% final concentration) or steptomycin sulfate (0.1–5%). Some DNA binding proteins coprecipitate with the DNA, and this property has been exploited to purify these proteins selectively. More often, however, the supernatant is adjusted to 200 mM with respect to sodium (or potassium) chloride, and the DNA precipitate produced by PEI or streptomycin sulfate is discarded following a 10-min centrifugation at 48,000g (Sorvall SS-34 rotor).

7. The optimal conditions of pH and ionic strength that produce the best chromatographic resolution for a particular protein can only be determined empirically. The distinct advantage of automated FPLC or HPLC over conventional, soft-gel chromatography is the ease of modifying elution conditions and the rapidity of the separation. In general, an elution buffer at a pH of between 7–8 and a linear salt

gradient of between 0.1–1M (either sodium or potassium chloride) is a good starting point. There is no reason why other salts should not be employed in the elution procedure; indeed, this may be necessary if sodium or potassium ions are inhibitory. In order to stabilize the eluted fractions (which should be approx 1 mL), the fraction collector can be jacketed with ice and the collection tubes can be filled with an appropriate volume of glycerol to give a final concentration of 50%. Note: The sample should be applied to the column in the equilibrating buffer, which itself should not contain glycerol (or anything of a similar viscosity) or any reagents that absorb at 280 nm.

8. The disadvantage of this particular matrix is that the oligonucleotide is susceptible to slow displacement during protein chromatography. As a consequence, it is generally advisable to replace the column after four or five runs. If, on the other hand, the oligonucleotide column is used at a late stage in a protein purification schedule, then the column may last for 20–30 runs. The use of other matrices for affinity chromatography has been explored with a range of nucleotide ligands but in view of the convenience of Sepharose, it will probably remain the most widely used support.

References

1. Chodosh, L. A., Carthew, R. W., and Sharp, PA. (1986) A single polypeptide possesses the binding transcription activities of the Adenovirus major late transcription factor. *Mol. Cell. Biol.* **6,** 4723–4729.
2. Gabrielson, O., Hornes, E., Korsnes, L., Ruet, A., and Oyen, T. (1989) Magnetic DNA affinity purification of yeast transcription factors. *Nucl. Acids Res.* **17,** 6253–6258.
3. Baeurle, P. A. and Baltimore, D. (1988) Activation of a DNA binding activity in an apparently cytoplasmic precursor of the NF-kappa B transcription factor. *Cell* **53,** 211–217.
4. Hornby, D. P., Muller, M. and Bickle, T A. (1987) High level expression of the *Eco*P1 modification methylase gene and characterisation of the gene product. *Gene* **54,** 239–245.
5. Kadonaga, J. T. and Tjian, R. (1986) Affinity purification of sequence-specific DNA binding proteins. *Proc. Natl. Acad. Sci USA* **83,** 5889–5893.
6. Dignam, J. D., Lebovitz, R. M., and Roeder, R. G. (1983) Accurate transcription initiation by RNA poymerase II in a stable extract from isolated mammalian nuclei. *Nucl. Acids Res.* **11,** 1475–1489
7. Jackson, S. P. and Tjian, R. (1988) O-glycosylation of eukaryotic transcription factors: Implications for the mechanism of transcriptional regulation. *Cell* **55,** 125–133.
8. Jackson, S. P. and Tjian, R. (1989) Purification and analysis of RNA polymerase II transcription factors by using wheat germ agglutinin affinity chromatography. *Proc. Natl. Acad. Sci. USA* **86,** 1781–1785.

CHAPTER 19

Making Bacterial Extracts Suitable for Chromatography

Roger F. Sherwood

1. Introduction

Despite the exotic appearance and odor of bacterial extracts, it should be remembered that they are predominantly water with only a few percent w/v solids. The twin objectives of preparation prior to chromatography are, therefore, clarification and concentration. It is not the purpose of this chapter to describe cell disruption techniques, but it is worth noting that the method used, whether physical, mechanical, or chemical, can affect the properties of the extract and subsequent clarification, for instance, in relation to viscosity on membrane filtration. The concentration stage can incorporate a primary purification step, for example, in the form of a protein precipitation/resuspension or a batch binding of protein, normally to an ion-exchange matrix. The more recent development of fast-flow chromatographic matrices does, however, allow larger process volumes to be applied directly to columns, for example, 200-L bed vol columns can operate at loading flow rates of up to 1000 L/h. The general techniques used for clarification and concentration are summarized in Table 1, and practical examples are described as they relate to pilot and large-scale processing rather than laboratory techniques.

From: *Methods in Molecular Biology, Vol. 11: Practical Protein Chromatography*
Edited by: A. Kenney and S. Fowell Copyright © 1992 The Humana Press Inc., Totowa, NJ

Table 1
General Techniques
Used for Clarification and Concentration

Physical	Chemical
Clarification	
Centrifugation	Two phase
Filtration	pH
	Polymers
Concentration	
Filtration	Precipitation
	Batch binding

1.1. Clarification (Cell Debris Removal)

This is a key stage in any process, since significant losses of product can occur through ionic and hydrophobic interaction of proteins with membrane fragments and other particulate matter. It is also a stage where product aggregation is most likely to take place, and coprecipitation of product with other proteins and proteolytic attack of product by the cells endogenous proteases can occur. Although there are many methods for reducing these effects, choice is often limited by the acceptability of the reagents used. This is particularly so for proteins destined for clinical use and made under Good Manufacturing Practice. Process addition of enzymes, inhibitors, and other chemicals often requires "proof positive" of no effect on the nature of the product and complete removal during the subsequent purification process.

1.1.1. Nature of the Extract

In very general terms, bacterial extracts will contain the following: Protein, 40–70%; Nucleic Acids, 10–30%; Polysaccharide, 2–10%; and Lipid, 10–15%.

High viscosity is most often caused by nucleic acid, particularly if heating has occurred during cell disruption. Viscosity can be reduced by addition of deoxyribonuclease (e.g., Sigma Deoxyribonuclease I from bovine pancreas) at a concentration of 1 mg/L. More traditional methods involve the precipitation of nucleic acids with positively charged compounds, such as polyethyleneimine (e.g., BDH Polymin P). A 50% aqueous stock solution adjusted to pH 6–8 with HCl and

Table 2

Effect of Detergents on metHGH Assay in Lysed Cell Extracts of *E. Coli*[a]

Soluble met HGH assay, µg/mL				
Control	Triton X-100	Nonidet P40	Tween 80	Brij 58
105	220	163	117	140

[a]Detergents were added to cell culture samples at a final concentration of 0.05% v/v. The sample was then adjusted to pH 11.5 with 4N NaOH and stirred for 15 min before neutralization with 20% phosphoric acid solution. Cell debris was removed by centrifugation for 10 min at 5000 g. met HGH was assayed using a rapid turbidometric method based on precipitation of metHGH as antibody complex in the presence of sheep antiHGH serum. Optical density at 600 nm was measured after incubation for 30 min at room temperature and compared with a standard curve in the range of 0.1–16 µg/mL clinical-grade HGH calibrated against NIBSC international standard (80/514) rated at 1.6 mg/vial.

Note: Increasing Triton X-100 concentration to 0.5% gave no significant improvement in metHGH assay.

added to a final extract concentration of 2–3% in the presence of 0.02M NaCl will precipitate 85–95% of the nucleic acid without loss of protein or enzyme activity *(1)*.

Interaction with membrane fragments can be reduced by working at alkaline pH, though this carries the risk of deamidation of asparagine and glutamine residues in the protein, and by the use of detergents. Methionyl human growth hormone (metHGH) produced as a recombinant DNA product in *Escherichia coli* shows strong association with membranes in cell extracts. This can be overcome if cells are lysed by alkali treatment (above pH 11.0) and the yield enhanced further, by reducing product precipitation when extract is neutralized, with detergents (Table 2).

Proteolytic attack by endogenous cell proteases can result in aberrant forms of product, often difficult to separate from the native protein at later stages of purification. A good example is provided again by metHGH where proteolytic clip at residues Thr-142:Tyr-143 in the amino acid sequence causes a conformational change, although the disulfide bridge is maintained *(2)*. The survival of product in *E. coli* extracts is extremely variable; interferon has a $t_{1/2}$ around 12 h, whereas the $t_{1/2}$ for insulin is only 20 min under the same conditions. *E coli* has at least eight proteases, five confined to the cytoplasm and two to the periplasm *(3)*. For major cloning hosts, such as *E. coli* and *Bacillus subtilis*, mutation/deletion of the protease genes has been used to reduce the

risk of product proteolysis. Alternatively, introduction of the proteolytic inhibition (pin) gene cloned from phage T_4 into *E. coli* has been
employed *(4)*, but with some loss in growth performance. Such genetic approaches are host limited, and the many proteins still extracted
from "natural" microbial isolates often require some form of protec-
tion during cell disruption. Since most of the proteases encountered
are neutral or alkaline in activity, and are either of the serine protease
class or require divalent metal ions for full activity, the chelating agent
EDTA at 0.5 mM concentration can be added together with
phenylmethylsulfonyl fluoride (PMSF) at 0.1 mM, acting via a fluo-
rine atom complex with the serine hydroxyl. Proteolysis can also be
minimized by introducing a protein precipitation stage early in the
process (*see later*).

Polysaccharide and lipid contents of bacterial extracts are normally
low and are not highlighted in most protein extractions. There are
exceptions: Certain pseudomonads produce large cell capsules when
grown on specific substrates for enzyme induction, for example, *Pseudomonas cepacia* for salicylate monooxygenase (used as basis of an aspirin
diagnostic kit) fermented with a salicylate feed *(5)*. The presence of
capsular material makes removal of cell debris difficulty, only partially
resolved using a proprietary cellulose "flocculent" (Whatman CDR),
requiring high gravity forces and extended centrifugation times. Similarly, the processing of extracellular streptavidin from *Streptomyces
avidinii* is affected by the presence of high levels of polysaccharide
(predominantly glycan) secreted into the culture supernatant. Although 90% can be removed during supernatant concentration, significant quantities come through an ion-exchange step in the
streptavidin fraction (Table 3) and can severely reduce the efficiency
of an affinity purification step based on iminobiotin-agarose. In such
cases, modification of the fermentation conditions has proven to be
the most successful approach. Growing *S. avidinii* on a defined rather
than complex medium reduces polysaccharide level from 5–6 g/L to
1–2 g/L and prevents the acute membrane pore blocking encountered during the filtration.

1.2. Centrifugation

Most proteins isolated from bacterial extracts are in soluble form,
but a significant number of proteins expressed at high level in *E. coli*
from cloned, recombinant DNA may be expressed in the form of

Table 3
Purification of Streptavidin from Culture Supernatant
of *Streptomyces avidinii* Grown on Defined Medium

	Volume, L	Streptavidin, g	Polysaccharide, g
Culture supernatant	5	0.42	8.95
Crossflow filtration			
Filtrate	4.7	0.02	8.13
Concentrate	0.3	0.40	0.79
DE52 cellulose pool	0.57	0.36	0.72
Iminobiotin-agarose			
Wash	1.23	–	0.68
Eluate	0.24	0.14	0.05

inclusion bodies *(6)*. These are microcrystalline or amorphous masses containing product in a biologically inactive form and can be removed from cell debris by differential centrifugation; for example, sedimentation of prochymosin inclusions at $12,000g$ for 5 min at 4°C leaves much of the cell debris in the supernatant fraction *(7)*. The size (about 1 µm) and density of protein inclusion bodies found from prochymosin and γ-interferon and their sedimentation properties have been well reviewed by Taylor et al. *(8)*.

For soluble proteins, centrifugation still remains the most frequently used method for removal of cell debris. Theoretical aspects of sedimentation rate have been described by Atkinson et al. *(9)*, and for practical purposes, centrifuges used for pilot/large-scale separation of cell debris fall into two main types.

1.2.1. Disk Centrifuges

As a first step to extract clarification, disk centrifuges are in routine use based on their ability to handle large process volumes and can contain typically 20–60 kg wet wt solids. The bowl contains a central stack of coned disks and particulate matter is thrown outwards, impinging on the disks and sedimenting on the wall of the bowl. In this manner, flow rate is maintained, and the separation efficiency only marginally reduced as solids build up. The flow rate that can be applied is highly dependent on the nature of the extract, but a Westfalia KA25, six-chamber disk bowl centrifuge capable of 400 L/h, generating 8000*g* and with a bowl capacity of 25 kg solids is typically run at 200L/h. The predominant suppliers of such centrifuges are Westfalia Separation Ltd (Wolverton, UK), Alfa-Laval Separation Engineering Ltd (Brentford, UK), and Bird Machine Co. (S. Walpole, MA, USA) and the latest models offer *in situ* steam sterilization, continuous or intermittent slurry discharge, and RCF up to 14,000*g*.

1.2.2. Hollow Bowl Centrifuges

Hollow bowl centrifuges have a tubular section to allow a sufficient flow path for particulates to sediment on the wall as the extract moves upwards through the bowl. Residence time is critical in determining the efficiency of separation and is controlled by pump flow rate. As separation proceeds, the effective diameter of the bowl is reduced with consequent loss of RCF, and flow rate may have to be adjusted. The attraction of this type of centrifuge is the high gravity force generated; for example, the Sharples AS26 (Pennwalt Ltd., Camberley, UK) can be operated at 16,000*g* at a flow rate of 30–60 L/ h with a bowl capacity of up to 8 kg wet wt solids. Compared with disk centrifuges, the solids are more easily removed, particularly when using inert (e.g., PTFE fabric) bowl liners. Similar centrifuges are supplied by Carl Padberg GmbH, Lahr, FRG.

1.3. Membrane Filtration

The removal of cell debris by membrane-based separations has been an attractive proposition for many years based on process economics (lower capital outlay and energy consumption) and safety (reduction/elimination of aerosols) if applied to pathogenic or genetically engineered bacteria. Since cell debris is not retained internally, the apparatus can remain small, or be readily scaled upwards. Typical flow rates used are 50 L/h for a 1-m^2 and 1000 L/h for a 8-m^2 membrane. The main limitation has been fouling of the filtration pores, particularly in the case of bacterial extracts where viscosity and a wide range of particle size (unless a flocculent of filter aid is first employed) can result in a rapid deterioration of performance. The problem can be alleviated by using high flow rates, creating turbulence, in a tangential (cross) flow apparatus where the liquid (retentate) flow is at right angles to the direction of filtration. The basic principles of tangential flow filtration for whole cell and debris separation have been described by Kroner et al. *(10)*, who also made the important point that higher than anticipated protein retention on filters, as well as

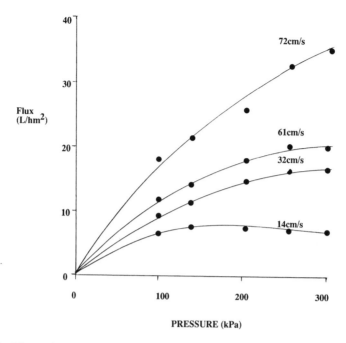

Fig. 1. The relationship between applied pressure (feed velocity) and flux during tangential flow microfiltration of cell extracts from *Pseudomonas fluorescens*. *P. fluorescens* cells (15% wet wt/vol) in 150 m*M* potassium phosphate buffer, pH 7.5 were treated with lysozyme (1 mg/g cells) for 15 min at room temperature in the presence of 10 m*M* EDTA and 1 mg/L-deoxyribonuclease. The extract was clarified using the apparatus and method described in Table 5, and measurements were started after an elapse time of 30 min to eliminate flux variations caused by the high initial rate of membrane fouling.

reduced flux rate, is caused by the formation of secondary layers of particulate matter on the filter surface. The dependence of flux on the pressure applied to cell extract for various feed velocities is illustrated in Fig. 1 and described in more detail by Le and Atkinson *(11)*.

Tangential flow filtration has been used experimentally in our process plant to clarify extracts of *Pseudomonas* and *Erwinia* (Table 4). It is interesting to note that the percentage enzyme recovery using the Asypor membrane was ordered with protein mol wt; asparaginase (132,000 kDa, tetramer) 59%; carboxypeptidase G_2 (83,000 kDa, dimer) 66%; and aryl acyl amidase (52,000 kDa, monomer) 94%. Nonspecific enzyme interaction with cell debris during filtration can be

Table 4

Efficiency of Tangential Flow Filtration Compared with Centrifugation
for the Separation of Three Bacterial Enzymes from Cell Debris

Organism	Enzyme	Breakage method	Membrane type	Filtrate[a]		
				Enzyme yield, %	Specific activity, U/mg	Turbidity, (OD_{600})
Erwinia chrysanthemi	Asparaginase	M[b]	Aspor[e]	60	22	0.35
			Centrifuge	100	16	1.22
		A[c]	Aspor[e]	57	31	0.05
			Centrifuge	100	34	0.15
Pseudomonas fluorescens	Aryl Acyl Amidase	M[b]	Aspor[e]	87	0.97	0.09
			Centrifuge	100	0.55	0.70
		E[d]	Aspor[e]	98	0.50	0.50
			Centrifuge	100	0.49	0.67
Pseudomonas sp. RS-16	Carboxypeptidase G₂	M[b]	Aspor[e]	66	0.89	ND
			Centrifuge	100	0.39	ND
		M[b]	Diapore (0.45)[f]	52	0.39	1.8
			Centrifuge	100	0.27	1.8
		M[b]	Diapore (0.60)[f]	78	0.23	6.6
			Centrifuge	100	0.21	4.4

[a]Filtration was at room temperature in an Amicon TC5E assembly containing five membranes (150 mm diameter) with a total filtration area of approx 690 cm². Cell suspension was circulated with an air-driven gear pump and air pressure adjusted to give a starting liquid pressure of 20 psi. Results after Quirk and Woodrow (12).
[b]Mechanical (Manton Gaulin or Dynomill).
[c]Alkali.
[d]Enzymic (lysozyme).
[e]Asymmetrical membrane tight side (0.45 μm), open side (1.5–2.0 μm) used with open side towards retentate. Supplier: Domnick-Hunter Filters, Birtley, UK.
[f]Pore size 0.45 or 0.6 μm. Supplier: Amicon Ltd., Stonehouse, UK.

reduced by using higher ionic strength buffers, as measured by percentage enzyme transmission through the membrane (Table 5). The effects of extract pH, temperature, and ionic strength on enzyme recovery have been described by Le et al. *(13, 14)*.

1.4. Aqueous Two-Phase Separations

The application of aqueous two-phase separation to the extraction and purification of enzymes has been established for some years (for a review of theory and practice in biotechnology see Hustedt et al. *[15]*). Despite numerous examples based on model systems, the technique has not been applied as extensively as one would expect for cell debris separation from proteins, perhaps based on the fact that, in practice, there can still be a largely empirical element, for example, the anomalous behavior of aryl acyl amidase in polyethylene glycol (PEG)/dextran systems *(16)*. The general rules that have emerged are as follows:

1. The affinity of a protein for a phase increases as either the mol wt or concentration of the polymer in that phase decreases. Thus, the partition coefficient *(K)*, which is the concentration of the enzyme in the top phase divided by the concentration of the enzyme in the bottom phase, will increase if the mol wt or concentration of PEG in the upper phase decreases, and to a lesser extent decrease with decreasing mol wt or concentration of dextran in the lower phase.
2. Addition of neutral salts to a PEG/dextran phase system will decrease *(K)* of negatively charged proteins and increase *(K)* of positively charged proteins. The size of the effect increases in the order SO_4^{2-}, Cl^-, NH_4^+, Na^+, K^+. Altering pH of a salt-containing system, therefore, alters *(K)* as the charge on the protein alters.
3. For ionic species with multiple dissociations, such as phosphate, the situation is more complex with the $H_2PO_4^-$ and HPO_4^{2-} exhibiting different *(K)* values. This generates an interfacial potential, resulting in the bottom phase becoming negatively charged at or above pH 7.0, such that the *(K)* for proteins with a pI below 7.0 will increase under conditions where phosphate at neutral or alkaline pH is added to the two-phase system.

Partitioning of proteins can be improved by the use of affinity ligands, normally attached to PEG (upper phase) in PEG-Dextran systems. The selection of ligands and chemistry of attachment have been reviewed by Harris and Yalpani *(17)*.

The empirical nature of two-phase systems can be demonstrated by studies that we performed with carboxypeptidase G_2 in PEG-Dext-

Table 5
Effect of Ionic Strength on Interaction of Asparaginase and Aryl Acyl Amidase
with Cell Debris During Tangential Flow Microfiltration[a]

Enzyme	Buffer strength, mM	Time, min	Enzyme transmission, %
Asparaginase[b]	7.5	5	98
		10	90
		20	62
		40	28
		60	20
	200	5	90
		10	85
		20	80
		40	75
		60	65
Aryl acyl amidase[c]	5	5	72
		10	62
		20	45
		30	34
	150	5	97
		10	95
		20	92
		30	90

[a]Filtration was performed using a housing fitted with four cells with rectangular channels 55 mm wide × 1 mm. The cells each held a membrane of surface area 32 cm^2 and were operated in parallel using a gear pump to provide a common feed via a flow manifold. Results shown were obtained using an Asypor 0.45-μm membrane with the open side toward the retentate.

[b]Asparaginase extract was prepared as a 10% wet wt/vol cell suspension, lysed at pH 11.5 (NaOH) in potassium phosphate buffer (7.5 or 200 mM), and readjusted to pH 7.0. Temperature was 25°C, and pressure was maintained 103 kPA.

[c]Aryl acyl amidase exract was prepared as a 10% wet wt/vol cell suspension, disrupted in a Manton Gaulin homogenizer at 550 kg/cm^2 in potassium phosphate buffer (5 or 150 mM) at pH 7.0. Temperature was 25°C, and pressure was maintained at 207 kPa.

ran, PEG-ammonium sulfate, and PEG-potassium phosphate systems (Table 6). In all cases, (K) was low and the majority of the enzymes were partitioned to the lower phase with the cell debris, but it was found that treatment with PEG alone (6–12% w/v PEG 6000) precipi-

Table 6
Behavior of Cell Extracts Containing Carboxypeptidase G_2
in Aqueous Two-Phase Systems and Precipitation with PEG 6000[a]

System	Mol wt PEG	Partition coeffificient, K
PEG (10%)—Dextran T500 (7.5%)	2000	No separation
	3500	0.190
	4000	0.110
	6000	0.040
PEG (10%)—Ammonium sulfate	2000	0.020
(1M)	3500	0.020
	4000	0.016
	6000	0.010
PEG (10%)—Potassium phosphate	2000	No separation
(0.7M)	3500	0.023
	4000	0.013
	6000	0.006
PEG (10%)	6000	N/A
		Cell debris precipitated
		Enzyme in supernatant

[a]Cell extract was prepared by mechanical disruption of *Pseudomonas sp.* RS-16 cells in 10 mM Tris-acetate buffer pH 7.3. Partition experiments were performed at room temperature. Results after ref. *18*.

tated cell debris, leaving carboxypeptidase G_2 in the clear, supernatant fraction. Increasing PEG concentration to 20% w/v precipitated the enzyme with a twofold purification (increase in specific activity), or conversely, adjustment of PEG concentration to 5% w/v and addition of 1.5M potassium phosphate (pH 7.8) brought the majority of the enzyme, but very little other protein, into the bottom phase yielding a 15-fold purification.

2. Protein Concentration

Clarified extracts containing soluble protein may well require concentration before application to a chromatographic column. As stated in the introduction, the modern use of fast-flow matrices has given a greater latitude on volumes that can be applied, while maintaining short process times, but an initial concentration step can also provide a useful purification stage, lower the protein burden on the column (i.e., reduce the size of the column required — matrices are expensive materials), and remove proteases at an early stage.

2.1. Filtration

Concentration of protein solutions by ultrafiltration at pilot/large scale normally employs a crossflow rather than dead-end filtration apparatus fitted with membrane or hollow-fiber cartridges of between 5000–20,000 mol wt cutoff. The variety of membranes and cartridges available is extensive and has been reviewed by McGregor *(19)*. In addition to concentration, this step can be used as a diafiltration to exchange buffers prior to column chromatography. Typical procedures and apparatus can best be described with reference to concentration steps used during the experimental purification of four proteins from microbial extracts: asparaginase from *Erwinia chrysanthemi*, streptavidin from *Streptomvces avidinii*, and metHGH and carboxypeptidase G$_2$ as recombinant products from *Escherichia coli* (Table 7).

2.2. Precipitation

Primary purification of proteins by precipitation can be applied either in the form of precipitating unwanted proteins, leaving the required product in solution, or precipitating the product, which also represents an effective concentration step. Selective precipitation depends on the particular properties of the protein and the careful control of pH, temperature, and the composition of any salt solution or nonaqueous solvent used. The most important characteristic of the protein that can be exploited is its isoelectric point, dependent on the number of free carboxyl groups of aspartyl and glutamyl residues, and basic groups of histidyl, arginyl, and lysyl residues. As pH of the solvent phase is varied, the protein is least soluble when it carries a zero net charge. Ions can also affect solubility. High concentrations of salts lead to dehydration of the protein molecule and an effect traditionally known as "salting-out" of proteins. Reduction in "water activity" (dielectric constant of the medium) by the addition of nonaqueous solvents, such as ethanol, acetone, and propan-2-ol, has been used historically, but the denaturation of many enzymes in solvent mixtures, the general need to keep temperature below 0°C and the special requirements of handling flammable solvents on a large scale have limited their use for most processes. Other organic precipitants that can be used are the water-soluble polymers, notably polyethylene glycol, which have the advantage of being nontoxic and nondenaturing to proteins. The use of PEG-6000 for clarification and subsequent precipitation of carboxypeptidase G$_2$ was described earlier.

Table 7
Concentration Methods Used During the Experimental Purification
of Four Microbial Proteins

Protein	System	Membrane/cartridge, buffer	Operation, at 8°C
Streptavidin	Pellicon (Millipore Harrow, UK)	PTGC (Polysulfone) 10,000 mol wt cutoff 4 m² (Culture supernatant)	400L–40 L at flow rate, 800 L/h (100 kPa) in retentate, and 42 L/h filtrate 70–590 mg/L in retentate, and 4 mg/L in filtrate
metHGH	Pellicon (Millipore Harrow, UK)	PTGC (polysulfone) 10,000 mol wt cutoff 8 m² (50 mM Tris-HCl, pH 8.0)	150L–55 L at flow rate, 1800 L/h (130 kPa) in retentate, and 180 L/h filtrate 20–53 mg/mL retentate, and 0.02 mg/mL in filtrate
Asparaginase	DC50 (Amicon Stonehouse, IK)	HP 10–20, ID 0.5 mm, 10,000 mol wt cutoff 4.4 m² (10 mM sodium phosphate, pH 6.0, plus 1 mM EDTA)	100L–12 L at flow rate 600 L/h in retentate 10–140 mg/mL in retentate, and 0.2 mg/mL in filtrate
Carboxypeptidase G₂	DC10 Amicon (Stonehouse, UK)	HP 10–20, ID 0.5 mm, 10,000 mol wt cutoff 0.88 m² (25 mm Tris-acetate, pH 9.0, plus 0.2mM ZnCl₂)	10–2.4 L at flow rate 75 L/h in retentate 5–17 mg/mL in retentate, and 0.03 mg/mL in filtrate

2.2.1. Precipitation by Heat

Heat treatment of protein solutions, in which the product obviously has to exhibit heat tolerance, has been proven to be useful by precipitating large quantities of denatured proteins, including proteases that were otherwise active in degrading the product. A good example is provided in the purification of protein A as a recombinant product in *E. coli (20)*, where 50% of the unwanted protein was removed by holding clarified cell extract (in $0.1 M$ potassium phosphate buffer pH 8.0) at 80°C for 10 min (Table 8).

2.2.2. Precipitation by pH

As with heat treatment, this takes the form of precipitating unwanted proteins by adjustment of bacterial cell extracts, which contain predominantly negatively charged (acidic) proteins, to acid pH. As a general observation, 60% of bacterial proteins have isoelectric points between pH 5–6 and 80% between pH 4.5–6.7 *(21)*. Adjustment of extracts of *Bacillus stearothermophilus* (in 50 mM potassium phosphate buffer) to pH 5.1 gave an effective removal (70%) of unwanted protein (Table 8) prior to ion-exchange chromatography for the purification of glycerokinase *(22)*. Precipitation at pH 4.8 is also used during the initial purification of asparaginase from cell lysates yielding a fivefold increase in specific activity (*see* Table 9).

2.2.3. Precipitation with Ammonium Sulfate

Ammonium sulfate is the most commonly used salt for protein precipitation, fulfilling the necessary criteria of high solubility, low cost, and having little or no effect on the activity of most proteins. It should be remembered, however, that ammonium sulfate is not a buffer and, being a weak acid combined with the possible slow loss of ammonia, may exhibit changes in pH and is very temperature sensitive. It does provide the possibility of selective precipitation (precipitation of discrete fractions as concentration is increased) by following Eqs. (1) and (2), which define the amount required to be added to a liter of solution at 20°C to take the molar concentration from C_1 to C_2 and the percentage saturation from $S_1\%$ to $S_2\%$.

$$g = \frac{553\,(C_2 - C_1)}{4.05 - 0.3\,C_2} \tag{1}$$

Table 8

Precipitation Techniques Used for the Purification/Concentration of Proteins from Clarified Cell Extracts

Protein product	Protein Precipitation method		Product conc (n) g/L	Product conc (n) g/L	recovery %
Protein A	Heat (80°C)	Before	62	8.6	100
		After	34	7.1	82
Glycerokinase	pH (5.1)	Before	19	0.3	100
		After	6	0.3	100
Streptavidin	$(NH_4)_2SO_4$[a]	culture supnt.	NR[b]	0.11	100
		80% ppt.	NR[b]	0.10	92
metHGH	$(NH_4)_2SO_4$[a]	20% supnt.	3.2	0.2	100
		20–50% ppt.	0.9	0.2	100
Carboxypeptidase G_2	$(NH_4)_2SO_4$[a]	supnt. (nil salt)	9.6	0.05	100
		35–55% supnt	2.7	0.05	100

[a]Streptavidin was precipitated directly from culture supernatant using $(NH_4)_2SO_4$ at 80% saturation with constant stirring at 4°C overnight and separated by centrifugation at 10,000g for 20 min. Pellet dissolved in water at 40 mL/L original supernatant volume. MetHGH was precipitated from a clarified pH 12.0 cell lysate by a two-step addition of $(NH_4)_2SO_4$ first to 20% saturation and then to 50% saturation. At each step, the precipitate was "aged" for 60 min at 4°C before collection by centrifugation at 10,000g for 30 min. The 50% precipitate was resuspended in 50 mM Tris-HCl buffer, pH 8.5. Carboxypeptidase G2 was isolated from a clarified cell extract prepared by mechanical disruption of cells (Manton-Gaulin homogenizer, type 15M-8BA, two passes at 50 MPa in 20 mM Tris-acetate buffer, pH 7.0, containing 0.2 mM ZnCl₂ and debris removal by centrifugation in Sharples AS-6 tubular bowl centrifuge. Proteins were selectively precipitated at 4°C by a two-step addition of $(NH_4)_2SO_4$, first to 35% saturation and then to 55% saturation, with each step held for 30 min before centrifugation. Carboxypeptidase G2 was normally recovered from the 35–55% supernatant fraction, following removal of precipitate by centrifugation at 10,000 g for 30 min, or could be precipitated by addition of further $(NH_4)_2SO_4$ to 85% saturation.

[b]NR = not recorded

$$g = \frac{553 \, (S_2 - S_1)}{100 - 0.3 \, S_2} \qquad (2)$$

Precipitate can be removed by centrifugation at between 8–10,000g and many proteins are very stable in precipitates providing a convenient long-term storage form. Following resuspension in minimum volume buffer, ammonium sulfate can be readily removed by gel filtration (desalting) or diafiltration, or in some instances, applied directly to hydrophobic chromatography matrices where loading in high salt is required. Examples of ammonium sulfate application are numerous; streptavidin precipitated directly from culture supernatant of *Streptomyces avidinii (23)*, and metHGH and carboxypeptidase G_2 during purification from extracts of *E. coli* and *Pseudomonas sp.*, respectively (Table 8).

2.3. Batch Adsorption of Proteins

Batch adsorption to resins provides a useful concentration method for dilute protein solutions at the early stages of large-scale protein purification. Most commonly, an ion-exchange resin is added, mixed, and then either recovered by centrifugation prior to batch elution with a minimum "desorption" buffer volume or packed directly into a column for a chromatographic elution with buffer. The recovery of asparaginase from a pH 11.5 cell lysate of *Erwinia chrysanthemi*, subsequently acidified to pH 4.8 and clarified by centrifugation (Westfalia KA25, 300 L/h flow rate and 6000g), provides an example of a batch adsorption/concentration process (Table 9). Adsorption/elution of asparaginase using CM-cellulose represents a 75-fold concentration.

3. Conclusion

The early stages of protein purification following release of product from microbial cells are often critical in determining the efficiency of subsequent chromatographic steps. Incomplete clarification can lead to rapid decline in column performance, and protein aggregation, precipitation, and proteolysis can result in both low recoveries and multiple product forms, which may be difficult to separate. Concentration of clarified extracts by a precipitation procedure can be useful in removing significant quantities of contaminating proteins, provided the method does not in itself lead to denaturation of the product. At pilot and large-scale process time, "scalability" and economics are key

Table 9
Concentration of Asparaginase
from Clarified Cell Extract by Batch Adsorption onto CM-Cellulose

Purification stage	Total activity, mega U	Specific activity, U/mg	Yield, %
Alkaline extract at pH 11.5	222	19.8	100
Clarified extract[a] following acid precipitation at pH 4.8	197	104	89
CM-cellulose[a] eluate	146	166	66

[a]The clarified extract at pH 4.8 was diluted to a conductivity value of < 2.6 mS. The solution (3000 L) was held at 8°C, and 15 kg (wet wt) CM-cellulose (previously equilibrated in 40 mM sodium acetate buffer, pH 4.8) added and stirred continuously for 90 min. The CM-cellulose, with bound enzyme, was recovered by centrifugation through a basket centrifuge (Padberg GZ651, Carl Padberg, Lahr, FRG) fitted with a porous cloth liner at 1000 L/h flow rate and 500g. Asparaginase was eluted by suspension of the resin in 20 L of 40 mM NaOH containing 1 mM EDTA initially adjusted to pH 10.3 with CO_2 gas and maintained at pH 10.3 during elution by automatic addition of 2M NaOH. The resin was collected as before, subjected to an identical second elution/recovery step, and the eluates combined and adjusted to pH 6.0 using 20% orthophosphoric acid.

factors when designing protocols, but the overriding requirement is a thorough understanding of the properties and tolerances of the desired protein product.

References

1. Atkinson, A. and Jack, G. W. (1973) Precipitation of nucleic acids with polyethyleneimine and the chromatography of nucleic acids and proteins on immobilised polyethyleneimine. *Biochem. Biophys. Acta* **308,** 41–52.
2. Chang, Y. Y.-H., Pai, R.-C.,Bennett, W. F., and Bochner, B. R. (1989) Periplasmic secretion of human growth hormone by *Escherichia coli. Biochem. Soc. Trans.* **17,** 335–337.
3. Swamy, K. H. S. and Goldberg, A. L. (1982) Subcellular distribution of various proteases in *E. coli. J. Bacteriol.* **149,** 1027-1033.
4. Simon, L. D., Randolph, B., Irwin, N., and Binkowski, G. (1983) Stabilisation of proteins by a bacteriophage T4 gene cloned in *Escherichia coli. Proc. Natl. Acad. Sci. USA* **80,** 2059–2062.

5. Hammond, P. M., Ramsay, J. R., Price, C. P., Campbell, R. S., and Chubb, S. A.P. (1987) A simple colorimetric assay to determine salicylate ingestion utilising salicylate monooxygenase. *Annals NY Acad. Sci.* **501,** 288–291.

6. Kane, J. F. and Hartley, D. L. (1988) Formation of recombinant protein inclusion bodies in *Escherichia coli. Trends in Biotechnology* **6,** 95–101.

7. Marston, F. A. O., Lowe, P. A., Doel, M. T., Schoemaker, J. M., White, S., and Angal, S. (1984) Purification of calf prochymosin (prorennin) synthesised in *Escherichia coli. Bio/Technology* **2,** 800–804.

8. Taylor, G., Hoare, M., Gray, D. R., and Marston, F. A. O. (1986) Size and density of protein inclusion bodies. *Bio/Technology.* **4,** 553–557.

9. Atkinson, T., Scawen, M. D., and Hammond, P. M. (1987) Large-scale industrial techniques of enzyme recovery, in *Biotechnology: A Comprehensive Treatise. vol. 7a: Enzymes* (Rehm, H. J. and Reed, G., eds.), VCH Verlagsgesellschaft, Weinheim, FRG, pp. 279–323.

10. Kroner, K. H., Schutte, H., Hustedt, H., and Kula, M. R. (1984) Cross-flow filtration in the downstream processing of enzymes. *Process Biochem.* **19,** 67–74.

11. Le, M. S. and Atkinson, T. (1985) Cross flow microfiltration for recovery of intracellular products. *Process Biochem.* **20,** 26–31.

12. Quirk, A. V. and Woodrow, J. R. (1984) Investigation of the parameters affecting the separation of bacterial enzymes from cell debris by tangential flow filtration. *Enyme Microb. Technol.* **6,** 201–206.

13. Le, M. S., Spark, L. B., and Ward, P. S. (1984) The separation of aryl acyl amidase by crossflow microfiltration and the significance of enzyme/cell debris interaction. *J. Membrane Sci.* **21,** 219–232.

14. Le, M. S., Spark, L. B., Ward, P. S., and Ladwa, N. (1984) Microbial asparaginase recovery by membrane processes. *J. Membrane Sci.* **21,** 307–319.

15. Hustedt, H., Kroner, K. H., and Kula, M. R. (1986) Applications of phase partitioning in biotechnology, in *Partioning in Aqueous Two-Phase Systems: Theory, Methods, Uses and Applications in Biotechnology* (Walter, H., Brooks, D. E., and Fisher, D., eds.), Academic, New York, pp. 529–587.

16. Woodrow, J. R. and Quirk, A. V. (1986) Anomalous partitioning of aryl acyl amidase in aqueous two-phase systems. *Enzyme Microb. Technol.* **8,** 183–187.

17. Harris, J. M. and Yalpani, M. (1986) Polymer-ligands used in affinity partitioning and their synthesis, in *Partitioning in Two-Phase Aqueous Systems: Theory, Methods, Uses and Applications in Biotechnology* (Walter, H., Brooks, D. E., and Fisher, D., eds.), Academic, New York, pp. 589–626.

18. Woodrow, J. R., Guiver, J. D., Quirk, A. V., Hughes, P., and Sherwood, R. F. (1983) Behavior of carboxypeptidase G2 in polyethylene glycol systems. *Proceedings of the Third International Conference on Partitioning in Two Polymer Systems.* University of British Columbia, Vancouver, Canada.

19. McGregor, W. C. (1986) Selection and use of ultrafiltration membranes, in *Membrane Separations in Biotechnology* (McGregor, W. C. ed.), Marcel Dekker, New York and Basel, pp. 1–36.

20. Hammond, P. M., Philip, K. A., Hinton, R. J., and Jack, G. W. (1990) Recombinant protein-A from *Escherichia coli. Annals. NY Acad. Sci.* **613**, 863–867.
21. Atkinson, T. (1973) The purification of microbial enzymes. *Process Biochem.* **8**, 9–13.
22. Scawen, M. D., Hammond, P. M., Comer, M. J., and Atkinson, T. (1983) The application of triazine dye affinity chromatography to the large-scale purification of glycerokinase from *Bacillus stearothermophilus. Anal. Biochem.* **132**, 413–417.
23. Suter, M., Cazin, J., Butler, J. E., and Mock, D. M. (1988) Isolation and characterisation of highly purified streptavidin obtained in a two-step purification procedure from *Streptomyces avidinii* grown in a synthetic medium. *J. Immunol. Methods.* **113**, 83–91.

CHAPTER 20

Determination of Purity and Yield

Sudesh B. Mohan

1. Introduction

The dual concepts of protein purity and yield are so basic in protein chemistry that it is easy to forget that both of them are almost impossible to define in absolute terms. A protein is totally pure only when it is known to contain only a single species uncontaminated with salts or adventitious water. Samples of such a "pure" protein will yield a single band after electrophoresis on a one- or two-dimensional SDS-PAGE gel, will elute from a gel filtration, HPLC, or ion exchange column as a single symmetrical absorbance peak, will yield a single set of mass spectrometric, NMR, or UV absorbance spectral signals, and where appropriate, will be free of contaminating enzyme activities. However, many tests are used to assess the purity of a product. Ultimately, they all share one property: Rather than prove that the preparation is absolutely pure, each additional test simply provides another example of the failure to detect any contaminating species that might be present. Since absolute purity can never be established, a simple criterion of purity is used routinely, namely, the inability to detect more than a single band of protein after SDS-PAGE. Since one-dimensional SDS-PAGE is technically far easier than two-dimensional SDS-PAGE, this technique will be described in detail.

From: *Methods in Molecular Biology, Vol. 11: Practical Protein Chromatography*
Edited by: A. Kenney and S. Fowell Copyright © 1992 The Humana Press Inc., Totowa, NJ

Table 1
Types of Methods Used to Determine the Concentration
of Specific Protein in a Sample

Property used	Typical examples and references
Enzyme activity	Acetyl Co-A carboxylase *(1)*
	Nitrite reductase *(2)*
	Dextransucrase *(3)*
Absorption of visible light	Cytochromes *(4)*
	Flavoproteins *(5)*
Binding to substrate	DNA binding proteins *(6,7)*
Immuno-complex	Antigen assay using antibody *(8,9)*
Activity stain after electrophoresis	Hydrogenase activity *(9)*
	L-phenylalanine ammonia lyase *(10)*

The purity of a sample is defined as the quantity of the protein of interest expressed as a percentage of the total concentration of protein. The various methods available to quantify the required protein are summarized in Table 1. Essentially, there are almost as many variations on these themes as there are proteins, but clearly the major types of specific assay are listed in this table. The following example illustrates the difficulty in determining the second parameter required to assess purity, namely, the total protein concentration of a sample.

Imagine that you have bottles of "pure," salt- and water-free samples of three proteins. In one bottle is the most commonly used protein standard, bovine serum albumin (BSA). In the second bottle is a pure sample of a small protein rich in basic arginine residues and small amino acids, such as glycine, serine, and alanine. The third bottle contains a high-mol-wt acidic protein rich in aromatic amino acids. It is a trivial matter to prepare with great accuracy 1.00 mg/mL solutions of each of these three proteins, but uninitiated research workers would be surprised by the results if three commonly used assays were each used in turn to determine the concentrations of proteins two and three relative to the BSA standard. It is highly unlikely that the Folin method, the Bradford dye-binding assay, or the Biuret method would indicate a concentration of 1.00 mg/mL for either of the test samples. The reason is that each method detects a different characteristic of a protein solution: the concentration of amide linkages, the ability to bind a particular dye, or the concentration of aromatic amino acid residues. Once again, the pragmatic solution is to adopt one simple method that is convenient to the task in hand: For this reason, the Folin and dye-binding methods will be described in this chapter.

The concept of protein yield is again extremely easy to define in theory. One simply needs to know how much of the protein you are interested in is present in a purified sample compared with the quantity of that protein in the starting material. Complications arise if an active enzyme or a protein in its native conformation is required, but the assay used is unable to discriminate between active and inactive forms or between native and denatured or fragmented forms. In general, the greater the specificity of the method used to determine its concentration in a sample, the more precise the information concerning purity and yield will be.

The purification of proteins involves a series of steps that typically include the concentration of the initial extracts, for example, by ammonium sufate precipitation, followed by chromatographic techniques, such as gel filtration, ion exchange chromatography, and affinity chromatography. Throughout the purification, it is essential to determine the amount of the total protein present and, if possible, to estimate the specific activity of the protein. Such determinations allow data for the overall purification and yield of the protein of interest to be calculated. A typical example is shown in Table 2, which lists the various parameters measured after each step used to purify nitrite reductase from *Escherichia coli (2)*. The purity achieved after each step can also be monitored visually by electrophoretic techniques, which involve the separation of the proteins on the basis of their size as in SDS polyacrylamide gels, or charge as in isoelectric focusing, followed by their staining for proteins or for specific activity.

Electrophoretic methods, therefore, serve to assess the actual number of protein species removed during the various stages of purification. The assay of specific activity would depend on the protein of interest, and such an assay can be modified to detect activity on gels as, for example, in the case of the glucosyltransferase enzyme dextransucrase *(3)*. The methods generally used to determine protein concentration and for their electrophoretic separation are described in the following sections.

Table 2
Purification of Nitrite Reductase from *E. coli* Strain K12 OR75 *Ch*15

Step	Volume, mL	Total protein, mg	Total activity, μkat	Yield, %	Specifific activity, kat/kg⁻¹	Purification, fold
Crude extract	298	3220	42.0	100	0.0087	1
$(NH_4)_2SO_4$ fractionation	54	1250	42.8	102	0.00356	4.1
DEAE- cellulose	373	112	23.5	56	0.214	24.7
$(NH_4)_2SO_4$ fractionation	22.6	32	19.4	46	0.617	71.0
DEAE-sephadex	128	8.3	8.6	22	1.04	121.0

2. Materials

2.1. Lowry Assay Reagents

1. Solution A. 2% (w/v) Sodium carbonate in 0.1*M* sodium hydroxide.
2. Solution B. 1% (w/v/) Copper sulfate.
3. Solution C. 2% (w/v) Sodium potassium tartrate.
4. Solution D. Copper reagent:
   0.5 mL of Solution B.
   0.5 mLof Solution C.
   50 mL of Solution A.
5. Solution E. Folin-Ciocalteu regent: Dilute with water until 1*M* acid. Determine the molarity of the reagant by titrating a known volume with a standard sodium hydroxide solution, and dilute as necessary.

Store solutions A, B, and C at 4°C. Prepare solutions D and E just before use.

2.2. Dye Binding (Bradford) Reagent

Dissolve 100 mg of Coomassie brilliant blue G-250 in 50 mL of 95% ethanol. Add 100 mL of 85% (w/v) phosphoric acid, and dilute to 1 L with distilled water.

2.3. SDS-Page

1. The slab gel apparatus is available commercially from many different suppliers, for example, Pharmacia LKB AB, Uppsala, Sweden or Bio-Rad Laboratories, Inc., Richmond, CA, USA. It consists of two glass plates 100–200 mm long and 100–150 mm wide, a set of three spacers

(1–3 mm thick) to fit the sides and the base of the slab, combs for forming wells of the same thickness as the spacers, and an electrophoresis chamber to hold electrode buffers and the cast gel.

2. A DC powerpack again available from the companies mentioned above.
3. Acrylamide-bisacrylamide solution. Dissolve 30 g acrylamide and 0.8 g of bisacrylamide in 100 mL water, filter, and store at 4°C. **Caution: Acrylamide is toxic.** Wear gloves while handling and, if possible, weigh in a fume cupboard.
4. Stacking gel buffer stock: $0.5M$ Tris-HCl (pH 6.8); store at 4°C.
5. Resolving gel buffer: $3.0M$ Tris-HCl (pH 8.8); store at 4°C.
6. Reservoir buffer: $0.25M$ Tris, $1.92M$ glycine (pH 8.3) containing 1% SDS. Store at 4°C, and dilute 10-fold before use.
7. Sample buffer: $0.5M$ Tris-HCl (pH 6.8), containing 2% SDS 5% 2-mercaptoethanol, 10% glycerol, and 0.001% bromphenol blue. Prepare a stock solution without 2-mercaptoethanol to an appropriate volume just before use.
8. 10% (w/v) SDS solution. Store at room temperature.
9. 1.5% (w/v) Ammonium persulfate. Prepare fresh.
10. N,N,N',N', -tetramethyl-ethylenediamine (TEMED).
11. Butan-2-ol solution saturated with water: Mix butan-2-ol with water for a few seconds in a bottle; allow the mixture to separate and use the top phase.

2.4. Coomassie Brilliant Blue Staining of Gels

2.4.1. Apparatus

A shaker and a gel drier, both of which can be bought commercially from Pharmacia LKB AB. The latter is required only if the gels are to be dried.

2.4.2. Reagents

1. Staining solution: 0.1% Coomassie brilliant blue R 250 (Pharmacia) or Page blue (Merck) in water:methanol:acetic acid in the ratio of 5:5:2.
2. Destain: as above, but without the dye.

2.5. Silver Staining of Gels

2.5.1. Apparatus

A shaker and drier as described in Section 2.4.1.

2.5.2. Reagents (see Note 11)

1. Fixative: 50% ethanol, 10% acetic acid.
2. Dithiothreitol solution: 5 mg/L. Prepare fresh.
3. Ammoniacal silver nitrate: prepare immediately before use by mixing in the following order (*see* Note 12):
   a. 1.2 mL Ammonia solution to 21 mL of 0.36% sodium hydroxide
   b. Silver nitrate solution (400 mg in 4 mL water, added dropwise)
   c. Make up to 100 mL with high-purity water.
4. Developer: prepare immediately before use by adding 2.5 mL of 1% citric acid and 200 µl formaldehyde to 500 mL water.
5. Preservative: 50 mL methanol plus 50 mL of 5% acetic acid.

2.6. Gradient Gel Electrophoresis

In addition to a gradient former, which can be bought commercially from, for example, Bio-Rad, the apparatus and reagents required for this technique are the same as for the 10% gels described above. The gradient former consists of two reservoirs connected at the base and has a stop-cock to prevent the mixing in the two chambers. One of the reservoirs has an outlet for pouring the gel.

3. Methods

3.1. Estimation of Protein Concentration

3.1.1. Estimation by UV Absorption

This method estimates the concentration using the absorption of tyrosine and tryptophan residues present in the protein. It involves the measurements of the absorption of protein solutions at 260 nm and 280 nm. The protein concentration is calculated from the equation of Layne *(11)*, which is based on the data of Warburg and Christian *(12)*:

$$\text{Protein concentration} = 1.55\,A_{280\,nm} - 0.76\,A_{260\,nm} \qquad (1)$$

This method is useful when salt is present, for example, in ammonium sulfate precipitated fractions, and when only small quantities of protein are available.

3.1.2. Estimation by Lowry Assay

The method of Lowry et al. *(13)* is the most widely used procedure for the quantitative determination of protein. It is based on the phenol reagent of Folin and Ciocalteau *(14)*, the active constituent of which is phosphomolybdic-tungstic mixed acid. Proteins reduce the mixed acid to produce reduced tungstate and/or molybdate species, which have a characteristic blue color. In addition, a copper reagent is used in the assay that is chelated by the peptide structure, facilitating the electron transfer to the mixed acid chromogen. The final color is the result of a rapid reaction with the aromatic residues tyrosine and tryptophan, plus a slower reaction with the copper chelates.

1. To solution (1mL) containing 20–200 µg protein, add 5 mL of solution D, mix, and stand at room temperature for 10 min.
2. Add 0.5 mL Folin reagent, mix rapidly, and incubate for 30 min at room temperature.
3. Remove any insoluble material by centrifugation, and measure the absorption at 750 nm against a reagent blank. Calculate the protein in the sample by reference to a standard curve prepared simultanously by incubating known concentrations of BSA.

Many substances interfere with the Lowry assay. Refer to Notes 1–5 for ways of reducing some of the interference.

3.1.3. Estimation by Dye-Binding Assay

This method was described by Bradford *(15)* and depends on the observation that the absorbance maximum for an acidic solution of Coomassie brilliant blue G-250 shifts from 465 nm to 595 nm when binding to protein occurs. Interactions are chiefly with arginine rather than primary amino groups; other basic and aromatic residues give a slight response. To 0.1 mL of sample, add 1.0 mL of reagent, mix, and read absorbance at 595 nm after 5–60 min. Use BSA as a standard protein to establish a calibration curve. Additional information about this method is given in Notes 6–10.

3.2. Determination of Purity Using Electrophoretic Methods

3.2.1. Polyacrylamide Gel Electrophoresis

Polyacrylamide gels are produced by polymerization of acrylamide with N,N'-methylenebisacrylamide. The gel density is determined by the total concentration of both components, the gel crosslinking being dependent on the percentage of bisacrylamide. The most frequently used concentrations are 5–15% gels with 3–5% crosslinking. The polymerization reaction is promoted by a suitable redox system: Ammonium persulfate or riboflavin in combination with N,N,N',N'-

tetramethylene diamine (TEMED) is used generally. The technique was developed originally by Davies (16) and Ornstein (17), and involved electrophoresis with gels in glass tubes that contained a large pore gel at the top and a small pore gel at the bottom. The samples to be tested were layered onto the upper surface of the gel in a dense solution (0.5 M sucrose) and subjected to electrophoresis in Tris-glycine buffer, pH 8.3. The large pore gel, at pH 6.7, served as a stacker: Proteins migrated through this gel rapidly, but were retarded at the interface of the two gels. The effect of this was to concentrate the samples applied to it. The small pore gel, at pH 8.3, then served as a sieve, and separated the proteins on the basis of their molecular size and charge. Shapiro et al. (18) and Weber and Osborn (19) introduced sodium dodecylsulfate (SDS) polyacrylamide electrophoresis, in which proteins are dissociated into their constituent polypeptides by breaking the hydrogen bonds. The binding of SDS to the polypeptide introduces one negative charge per bound molecule of SDS. At neutral pH, the net charge of the protein-SDS complex is almost entirely dependent on the charge of the SDS molecules. It has been found that the charge per unit mass is approximately constant (1.4 g SDS/1 g protein) and that the mobility of the complex is, therefore, dependent on the mol wt of the protein. The mol wt of the polypeptides can be determined by comparing their electrophoretic mobilities with the mobilities of known marker proteins. A linear relationship is obtained if the logarithm of the mol wt is plotted against the relative mobility. Furthermore, addition of 2-mercaptoethanol to the dissociation buffer breaks any sulfydryl bonds that hold some polypeptides together.

The three systems described above can therefore provide useful data on the molecular properties of the proteins of interest. An example shows the patterns obtained for the purified enzyme, acetyl coenzyme A carboxylase in a 5% gel (Fig.1A). This enzyme was stabilized with 0.1% BSA, which was present throughout the purification protocol: Two intense bands corresponding to the enzyme and BSA were detected on the gel. The subunit structure was determined by eluting the enzyme from native gels in dissociating buffer, in the absence or presence of 2-mercaptoethanol (lanes 1 and 2, respectively, Fig. 1B; [1]). Three subunits were obtained with SDS alone, but an additional polypeptide was detected in sample dissociated with 2-mecaptoethanol, indicating the presence of a sulfydryl bond. The mol wt of the various polypeptides were obtained by electrophoresis of a number of marker proteins simultaneously (Fig. 2).

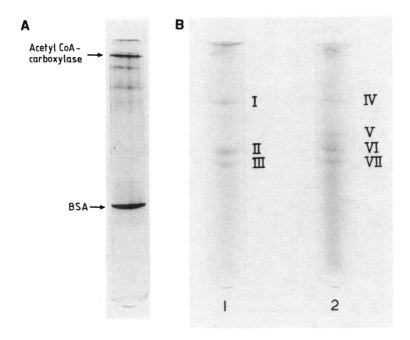

Fig. 1. Polyacrylamide gel electrophoresis of acetyl coenzyme A carboxylase. **(A)** Patterns obtained in 5% native gel. **(B)** patterns obtained in SDS gel in the absence and presence of 2-mercaptoethanol (lanes 1 and 2, respectively). Mol wt of polypeptides I-VII were determined from the standard curve shown in Fig. 2.

Raymond *(20)* introduced a slab gel system, in which large and small pore gels are poured between two glass plates and different samples are compared in the same gel. The method for running a 10% resolving gel is described below and is essentially that of Laemmli *(21)*. It uses SDS and 2-mercaptoethanol in the dissociating buffer. Nonreducing gels (in the absence of 2-mercaptoethanol) can be run by omitting this reagent from the sample buffer. It can be adapted for native gels by omitting SDS from all the buffers and 2-mercaptoethanol from the sample buffer.

1. Assemble the slab apparatus. Clean and dry the plates thoroughly. Wipe them with paper tissue soaked in acetone to remove any traces of grease. Clean and dry the spacers. Apply grease (petroleum jelly) thinly to both sides of one edge of the spacers, and carefully align them on the edges of one glass plate. Place the second glass plate on the spacers, and secure the slab with clamps. Stand the slab in a vertical position to pour the gel.

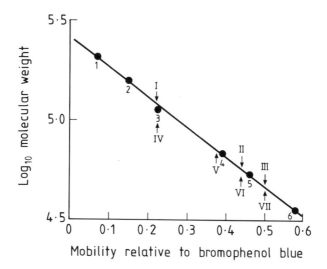

Fig. 2. Estimation of the mol wt of the polypeptides derived from acetyl-CoA carboxylase. Relative mobilities of marker proteins (labeled 1–6) were used to obtain the standard curve as described by Shapiro et al. and Weber and Osborn *(18, 19)*. 1, 2, 3, and 5 were standard markers (Merck, Dorset, UK) of mol wt 212, 159, 106, and 55 kDa, respectively. 4 and 6 were BSA (68 kDa) and lactate dehydrogenase (36 kDa), respectively. I–VII denote polypeptides shown in Fig. 1B.

2. Pour the resolving gel. The volume of gel required will depend on the size of the slab. For 30 mL (sufficient for a gel $100 \times 150 \times 0.1$ cm), mix the following in a 50–100 mL beaker:

Stock acrylamide solution	10.0 mL
Resolving gel buffer	3.75 mL
10% SDS	0.3 mL
Ammonium persulfate	1.5 mL
Water	14.75 mL

Add TEMED (10 µL), mix, and carefully pour the solution into the slab to a level about 2–3 cm below the surface, (air bubbles should come to the surface but if the mixture is too frothy, degas before adding TEMED). Add butan-2-ol gently to cover the surface of the acrylamide. Leave at room temperature for about 10 min for the gel to polymerize.

3. Pour the stacking gel: Decant the butan-2-ol layer, wash the slab with distilled water, and remove as much water as possible by blotting with

3 MM filter paper. Mix the following:

Acrylamide solution	1.25 mL
Stacking gel buffer	2.5 mL
10% SDS	0.1 mL
Ammonium persulfate	0.5 mL
Water	5.65 mL

Add 7.5 μL TEMED; mix and fill the slab with this solution. Insert the well-forming comb taking care to avoid air bubbles. The stacking gel polymerizes quickly (5 min at room temperature). When the gel has set, remove the comb and the spacer at the base of the slab (this spacer is usually slightly longer than the width of the slab and is, therefore, easily removed). Clamp the gel to the electrophoresis apparatus, and fill the reservoirs with the diluted reservoir buffer. Rinse the wells with buffer by squirting it from a syringe connected to a fine hypodermic needle. At this stage, the gel can be left at room temperature until required (overnight if necessary).

4. Add equal volumes (50 μL) of sample and sample buffer to an Eppendorf tube, mix well, and heat at 100°C for 5 min to dissociate the proteins. When using native gels, mix the sample with sample buffer (containing no SDS or 2-mercaptoethanol), but do not heat since this would denature the proteins. Centrifuge the samples to remove any insoluble material, and carefully layer an aliquot in each well. (The amount applied would depend on the protein concentration but as much as 50 μL can be applied without any significant loss of the resolution on the resolving gel.) Start electrophoresis immediately at a constant 50 V overnight or a constant 100 mA for 3–4 h until the bromphenol blue band is about 1 cm from the base of the gel. At the end of the run, discard the buffer, remove the slab, and place it on the bench. Gently ease out the spacers (a flat spatula helps), and with the help of a spatula, lift the top glass plate. Stain the gel either with Coomassie brilliant blue or silver as described below.

3.2.2. Coomassie Blue Gel Staining

1. Place the gel in the staining solution (about 100 mL/gel contained in a suitable sized sandwich box). Shake it gently on a shaker for 1 h or longer.

2. Decant the stain, and destain the gel in several washes of fresh destain until the background is clear. The destained gel can be dried onto filter paper using a gel drier that consists of a flat plate with a rubber flap connected to a vacuum line. The gel is placed on the plate, covered first with filter paper and then with the flap, and connected to

vacuum line. Some commercial apparatus have facilities for heating the plate while the gel dries, thus shortening the drying time.

3.2.3. Silver Staining of PAGE Gels

This technique is more sensitive than Coomassie stain and can detect nanogram quantities of protein. Gels previously stained with Coomassie blue can also be stained with this method as long as they are thoroughly destained. The method of Wray et al. *(22)* modified to include a dithiothreitol incubation is described.

1. Fix the gel in the fixative for at least 1h or longer.
2. Wash the gel in water, preferably overnight.
3. Shake the gel in dithiothreitol solution for 1h.
4. Wash the gel in water for 1 h.
5. Shake the gel in ammoniacal silver nitrate solution for 30 min.
6. Develop the gel in the developer for 10–15 min until the bands appear (*see* Note 13).
7. Stop the development with the preservative. If required, the gel can be dried as described above for the Coomassie stained gels (Section 3.2.2.).

3.2.4. Gradient Gel Electrophoresis

The main problem with discontinuous gel electrophoresis is the diffusion with time of the bands observed in the gels. This is particularly a problem if the gels are run at high voltage or if a large volume of sample is applied because the protein concentration of the samples is low. Gradient gel electrophoresis, in which the gels contain a gradient of acrylamide concentration (the lower concentration being at the top of the gel), solves the diffusion problem. The proteins stop migrating through the gel when they reach the pore size corresponding to their mass. Any trailing protein now concentrates in this area to produce sharp zones on staining.

1. Assemble the slab and rest it vertically against a support on which a magnetic stirrer could be placed to provide enough pressure head to pour the gel.
2. Make sure that the gradient former is clean and has no blockages in the connecting tube. (This can be ensured by applying a vacuum to the outlet tubing.)
3. The concentration range of acrylamide used depends on the samples tested. Thus, for a 6–18% gel, the following solutions would be used.

	Solution A, 6% gel	Solution B, 18% gel
Stock acrylamide solution	3.0 mL	9.0 mL
Resolving gel buffer	1.88 mL	1.88 mL
10% SDS	0.15 mL	0.15 mL
Ammonium persulfate	0.35 mL	0.35 mL
Water	9.63 mL	2.38 mL
Sucrose	–	2.25 g

Pipet the various reagents for the two solutions, and mix thoroughly. Add 5 µL TEMED, mix, and pour the gel immediately.

4. With the outlet tubing of the gradient mixer and the stop-cock between the two chambers closed, add solution B to the chamber with the outlet. Tilt the gradient former slightly, open the stop-cock, and allow the solution to fill the connecting tubing. Close the stop-cock, and add solution A to the second chamber. Place the gradient maker on the magnetic stirrer, and stir solution B with a small magnetic follower. Open the stop-cock and the outlet tubing. Hold the latter in the middle of the slab, and allow the gel to pour by gravity. Cover the gel with water-saturated butan-1-ol, and leave the gel to polymerize at room temperature. The rest of the procedure and staining is also the same as that described for the 10% gel (Note 14).

3.2.5. Electrophoresis in Mini-Gels

Apparatus is available commercially for running small slab gels (gel volume approx 10 mL; Midgit Electrophoresis System, Pharmacia LKB AB). The assembly of the slab, reagents, and the techniques are the same as described for the 10% and gradient gels. The sample volumes used are correspondingly small, as is the time taken to run the gel (about 30 min).

For those who can afford it, the PhastSystem sold by Pharmacia LKB AB offers a system that comes complete with ready-poured native or SDS, homogeneous or gradient mini-gels and the corresponding buffers in agarose strips. The system is programmed to run 0.5–1.0 µl samples in about 30 min, and to stain the gels with either Coomassie blue or silver stain in approx 30 min and 2 h, respectively. The protocols to be followed are provided in the instruction manual that accompanies the instrument.

4. Notes

1. The key to successful protein determination by the Lowry method is clean glassware. The test tubes should preferably be acid washed, or failing this, they should be thoroughly rinsed in water to remove traces of detergent.
2. Many substances interfere with the Lowry assay described. These include lipids, chelating agents such as EDTA and EGTA, detergents such as Triton X-100 and Tween 20, sulfydryl reagents, and aromatic amino acids. Some of the interference can be minimized by dialyzing samples before assay or by using reagent controls containing all of the substances present in the test solutions.
3. This method can be used when protein concentrations of crude samples, such as cell homogenates that contain nucleic acids and plant material, that contains chlorophyll, are measured. Proteins are precipitated by trichloroacetic acid as described below prior to assay; substances that interfere with the assay are hence removed.
   a. To 1 mL of solution containing 0.5–3 mg protein, add 1 mL of 10% trichloroacetic acid. Leave the solution at 4°C for 30 min. Centrifuge the precipitated protein at 4000g for for 10 min, and remove the supernatant.
   b. Dissolve the precipitate in 1 mL 2M potassium hydroxide, and assay for proteins as described above.
4. If the samples to be tested contain lipids or chlorophyll, these can be removed by extracting with diethyl ether containing 20% (v/v) ethanol prior to assay. The number of extractions required will depend on the concentration of these substances, but normally three extractions should prove adequate. This extraction method would also remove Triton X-100.
5. The Lowry assay has also been modified to measure low protein concentrations in the presence of sulfydryl reagents *(23)*. This method is particularly useful when small quantities of valuable material are available.
   Reagents:
   a. Stock alkaline copper reagent: Prepare by mixing equal vol of copper sulfate (1% w/v) and potassium D-tartrate (5.4% w/v). Mix 1 vol of this solution with 10 vol of sodium carbonate (10% w/v in 0.5M sodium hydroxide).
   b. 0.03% Hydrogen peroxide solution.
   c. Folin-Ciocalteau reagent diluted 10-fold with water just before use.
   Method:
   a. To a 0.2 mL sample (containing 5–20 µg protein,) add 0.25 mL of the alkaline copper reagent, mix and add 50 µL hydrogen perox-

ide solution. Mix and leave at room temperature for 30 min.

b. Add 0.5 mL of the diluted Folin reagent, mix, and incubate at room temperature for a further 30 min to complete color formation. Read absorbance at 725 nm against a reagent blank, and calculate the protein concentration with reference to a standard curve obtained with BSA.

6. Assay kits for this method are available commercially, for example from Bio-Rad Laboratories, Richmond, CA and Pierce, IL, USA. These kits describe two standard procedures depending on the protein concentrations of the samples to be tested: macroassay (for 20–150 µg/mL protein) and microassay (for 1–25 µg/mL protein).

7. The Bradford assay is very rapid and is recommended because various reagents that limit the use of the Lowry method do not interfere with this assay (Bio-Rad Laboratories Bulletin, 1979).

8. In the author's experience, however, the color produced with plant or fungal extracts, both of which contain phenolic compounds, was not stable. Matoo et al. *(26)* also describe similar problems.

9. Recently, Loffler and Kunze *(27)* have described a modification of the Bradford assay, which is 25X more sensitive than the original procedure. These authors also reported interference with KCl, NaCl, EDTA, 2-mercaptoethanol, ethylene glycol, sucrose, acetone, and SDS.

10. Another disadvantage of the Bradford assay is that there is quite a large variation in the absorbance produced by different proteins. Stoscheck *(28)* has shown that the addition of NaOH to the reagent reduced this variation: 20 µL of $10 M$ NaOH/mL of reagents obtained from Pierce and Bio-Rad produced maximal response by the proteins tested.

11. The success of the staining depends on the use of good quality water, which should be distilled and deionized.

12. Add silver nitrate dropwise, and mix the ammonia solution thoroughly between each addition to ensure that no silver hydroxide precipitates. Brown color will take longer to disappear as more silver nitrate is added. If the color persists add a few drops of ammonia solution but this should not be required. The solution used should be colorless.

13. If the developer turns orange when it is added to the gel, pour it off and rinse the gel in water before adding more developer. If the gels do not develop, add a further 200 µL of formaldehyde to the solution, but not directly onto the gel.

14. Strictly speaking, a stacking gel is not required for a gradient gel, but some people prefer to have one. If so, the stacking gel components will be the same as for the 10% gel.

15. The methods described in this chapter should provide a general guide to the determination of yield and purity. However, the reader should be aware of the possibility that, unless the right techniques are chosen, detection of some proteins might not be easy. A good example to illustrate this point is the outer membrane protein H.8 of *Neisseria*. This protein does not stain with Coomassie blue, migrates on SDS-PAGE as a cone-shaped band of 18–30 kDa because it copurifies with lipid, and does not absorb at 280 nm because it has a high content of alanine and proline with no aromatic acids (8,23,24). conventional methods would not have detected this protein: H.8 was detected by immunological methods using monoclonal antibodies raised against gonococcal outer membrane components.

References

1. Mohan, S. B. and Kekwick, R. G. O. (1981) Acetyl-Coenzyme A carboxylase from Avocado *(Persia americana)* plastids and spinach *(Spinacia oleracea)* chloroplasts. *Biochem. J.* **187**, 667–676.
2. Jackson, R. H., Cornish-Bowden, A., and Cole, J. A. (1981) Prosthetic groups of the NADH-dependent nitrite reductase from *Escherichia coli* K12. *Biochem. J.* **193**, 861–867.
3. Kenney, A. C. and Cole, J. A. (1983) Identification of a 1,3-glucosyltransferase involved in insoluble glucan synthesis by a serotype c strain of *Streptococcus mutans*. *FEMS Microbiol. Letts.* **16**, 159–162.
4. Kajie, S.-I. and Anraku, Y. (1986) Purification of a hexaheme cytochrome c552 from *Escherichie coli* K12 and its properties as a nitrite reductase. *Eur. J. Biochem.* **154**, 457–463.
5. van Berkel, W. J. H., van Den Berg, W. A. M., and Muller, F. (1988) Large scale preparation and reconstitution of apo-flavoproteins with special reference to butyryl-CoA dehydrogenase from *Megasphaera elsedenii*. *Eur. J. Biochem.* **178**, 197–207.
6. Hagen, D. C. and Magasanik, B. (1973) Isolation of the self-regulated repressor protein of the *hut* operons of *Salmonella typhimurium*. *Proc. Natl. Acad. Sci.* **70**, 808–812.
7. Wilcox, G., Cl emetson, K. J., Santi, D. V., and Englesberg, E. (1971) Purification of the *araC* protein. *Proc. Natl. Acad. Sci. USA* **68**, 2145–2148.
8. Cannon, J. G., Black, W. J., Nachamkin, I., and Stewert, P. W. (1984) Monoclonal antibody that recognises an outer membrane antigen common to the pathogenic *Neisseria* species but not to most nonpathogenic *Neisseria* species. *Infect. Immun.* **43**, 994–999.
9. Sawers, R. G. and Boxer, D. H. (1986) Purification and properties of membrane-bound hydrogenase isoenzyme 1 from anaerobically grown Escherichia coli K12. *Eur. J. Biochem.* **156**, 265–275.
10. de Cunha, A. (1988) Purification, characterisation and induction of L-phenylalanine ammonia-lyase in Phaseolus vulgaris. *Eur. J. Biochem.* **178**, 243–248.

11. Layne, E. (1957) Spectrophotometric and turbidimetric methods for measuring proteins. In *Methods in Enzymology*, vol. III, (Colowick, S. P. and Kaplan N. O., eds.) Academic, New York, pp. 447–451.

12. Warburg, O. and Christian, W. (1941/42) *Biochem. Z.* **310**, 384–421.

13. Lowry, O. H., Rosenbrough, N. Y., Farr, A. L., and Randall, R. Y. (1951) Protein measurement with the Folin phenol reagent. *J. Biol. Chem.* **193**, 265–275.

14. Folin, O. and Ciocalteau, V. (1927) On tyrosine and tryptophan determination in proteins. *J. Biol. Chem.* **73**, 627–649.

15. Bradford, M. M. (1976) A rapid and sensitive method for the quantitation of microgram quantities of proteins utilizing the principle of protein-dye binding. *Anal. Biochem.* **72**, 248–254.

16. Davies, B. J. (1964) Disc electrophoresis — II. Method and application to human serum proteins. *Ann. NY Acad. Sci.* **121**, 404–427.

17. Ornstein, L. (1964) Disc electrophoresis - I Background and theory. *Ann. NY Acad. Sci.* **121**, 321–349.

18. Shapiro, A. L., Vinula, E. and Maizel, J. V. (1967) Molecular weight estimation of polypeptide chains by electrophoresis in SDS polyacrylamide gels. *Biochem. Biophys. Res. Commun.* **28**, 815–820.

19. Weber, K. and Osborn, M. (1969) The reliability of molecular weight determinations by dodecyl sulfate polyacrylamide gel electrophoresis. *J. Biol. Chem.* **244**, 4406–4412.

20. Raymond, S. (1962) Acrylamide gel electrophoresis. *Ann. NY Acad. Sci.* **121**, 350–365.

21. Laemmli, U. K. (1970) Cleavage of structral proteins during the assembly of the head of bacteriophage T4. *Nature* **227**, 680–685.

22. Wray, W. Bouikes, T., Wray, V. P., and Hancock, R. (1981) Silver staining of proteins in polyacrylamide gels. *Analyt. Biochem.* **118**, 197–203.

23. Geiger, P. J. and Bessman, S. P. (1972) Protein determination by Lowry's method in the presence of sulphydryl reagents. *Analyt. Biochem.* **49**, 467–473.

24. Hitchcock, P. J., Hayes, S. F., Mayer, K. W. Shafer, W. M., and Tessier, S. L. (1985) Analysis of gonococcal H.8 antigen: Surface location, inter- and intrastrain electrophoretic heterogeneity, and unusal two-dimentional electrophoretic characteristics. *J Expt. Med.* **162**, 2017–2034.

25. Bhattacharjee, A. K. Moran, E. E., Ray, J. S., and Zollinger, W. D. (1988) Purification and characterization of H.8 antigen from Group B *Neisseria meningitidis*. *Infect. Immun.* **56**, 773–778.

26. Matoo, R. L., Ishaq, M., and Saleemuddin, M. (1987) Protein assay by Coomassie Brilliant Blue G-250-binding method is unsuitable for plant tissues rich in phenols and phenolases. *Anal. Biochem.* **163**, 376–384.

27. Loffler, B.-M. and Kunze, H. (1989) Refinement of the Coomassie Brilliant Blue G assay for quantitative protein determination. *Anal. Biochem.* **177**, 100–102.

28. Stoscheck, C. M. (1990) Increased uniformity in the response of the Coomassie Blue G protein assay to different proteins. *Anal. Biochem.* **184**, 111–116.

Index